D0502290

2000 MC
no

COMPREHENSIVE URBAN PLANNING

COMPREHENSIVE URBAN PLANNING

A Selective Annotated Bibliography with Related Materials

by MELVILLE C. BRANCH, *Professor of Planning,*
University of Southern California, Los Angeles
Member, Los Angeles City Planning Commission

 SAGE PUBLICATIONS / BEVERLY HILLS, CALIFORNIA

Copyright © 1970 by Sage Publications, Inc.

All rights reserved. No part of this book may
be reproduced or utilized in any form or by
any means, electronic or mechanical, including
photocopying, recording, or by any information
storage and retrieval system, without permission
in writing from the Publisher.

For information address:

SAGE PUBLICATIONS, INC.
275 South Beverly Drive
Beverly Hills, California 90212

Printed in the United States of America

Standard Book Number 8039-0041-4

Library of Congress Catalog Card No. 73-92349

Designed by HANFT / LAGARDERE / HOFFMAN

First Printing

for

M.P.B.

1884-1944

*who first encouraged me
toward scholarship*

40094

CONTENTS

ABOUT THE AUTHOR

MELVILLE C. BRANCH is Professor of Planning in the Graduate Program of Urban and Regional Planning at the University of Southern California in Los Angeles. He is also a member of the Los Angeles City Planning Commission. He holds Bachelor and Master of Fine Arts degrees from Princeton University and its Graduate College. His Doctor of Philosophy in Regional Planning from Harvard University is the first advanced graduate degree awarded in the field of planning.

In addition to the above positions, Dr. Branch's activities have included (in the order of most recent occurrence): Lecturer in Engineering (Planning) at the University of California, Los Angeles; Planning Consultant to the Aircraft Division of Douglas Aircraft Company; Corporate Associate for Planning and Member of the Senior Staff (West Coast), Thompson Ramo Wooldridge. Inc. (now TRW Inc.); Associate Professor of Planning, Secretary, and Acting Director of the Graduate Program of Education and Research in Planning, Division of the Social Sciences, University of Chicago; Director of the Bureau of Urban Research at Princeton University; and Research Assistant at the U.S. National Resources Planning Board in the Executive Offices of the President.

Dr. Branch is the author of many professional papers on various aspects of planning, and of several books, including *Planning: Aspects and Applications* (1966); *Selected References for Corporate Planning* (1966); *The Corporate Planning Process* (1962); *and Aerial Photography in Urban Planning and Research* (1948).

INTRODUCTION AND EXPLANATION

City planning has existed for over five thousand years. During this long period, its development has generally followed the progress of man and his institutions. As part of this dependent association, the content of city planning in the United States during the past fifty years has experienced extensive change comparable to that brought about in most professions by the rapid growth of knowledge. In city planning, this substantive change has been compounded by the unprecedented migration of people to cities and consequent aggravation of long-existing municipal difficulties. This period of rapid development has not yet run its contemporary course; a resolution of present trends must be expected before a plateau of comparative equilibrium and consolidation will occur. What are some of these present trends?

City planning is abandoning its spatial overemphasis on location, three-dimensional form, and aesthetics of physical facilities and features. Conceptually, it is becoming much more comprehensive—incorporating economic, political, social, legal, environmental, communicative, and scientific-technological considerations far more than it has in the past. Most recently, urban blight, unemployment, poverty, and social tensions have underscored the spreading conviction that cities cannot be planned successfully from any arbitrarily limited or partial point of view.

Urban studies and urban environment or ecology have emerged as particular substantive emphases. The first covers the great variety of materials past and present which deal generally with cities. In essence, urban studies seek to describe and explain cities as complex and diverse organisms requiring deliberate study and observation. Environmental, ecological, or "comprehensive health" study and planning have come to the fore recently because of the increasing awareness of the harmful effects of the pollution of land, water, and air. There is concern that we may be seriously impairing the future of mankind in ways and to a degree that will become apparent only after it is too late to rectify the ecological consequences. Urban studies can be considered a background or prerequisite to urban planning, environmental study a subsystem or composite of several aspects or elements of urban planning.

Coincident with this expansion of concept and substance has been geographical extension. The functional artificiality of most municipal boundaries has long been recognized. Only recently, however, has the necessity of planning on an areal scale which is consistent with actual urbanization on the land

and with the geographical context of the problems under consideration, reached the point of metropolitan governments or binding coordination between different municipal jurisdictions comprising one metropolitan region. Water supply, flood control, air pollution, and air traffic control are familiar examples.

With this expanded scope and substantive intensification, policies are assuming greater significance in urban planning. At one time, master city plans had to do almost exclusively with physical facilities and features. Nowadays, policy plans are proposed or incorporated as part of the broader context of master planning, in order to direct and coordinate those activities of municipal departments which clearly shape the future of the city, and to suggest consistent lines of action for municipal legislatures. Programmatic involvement was also once regarded as outside the scope of city planning or too short-ranged for its consideration. City planning departments now seek to integrate the separate functional plans and operating programs of municipal departments and agencies most affecting the community's future.

The principal instrument of these efforts—by federal and state governments as well as cities—is the comprehensive master plan, currently the subject of much debate. The traditional intention of formulating a master plan depicting the "end-state" of the entire city as it should be twenty or more years in the future, and of programming in advance all specific actions necessary to achieve this end, is being challenged as intellectually invalid and operationally unrealistic. Concepts of more continuous master planning have been advanced which do not pretend to define the complete and totally desirable end-state of a city many years hence. Also proposed is more direct responsibility and involvement by the municipal legislatures which make city planning decisions.

Increasingly, it is being said that the traditional instruments of master city planning—land-use controls by zoning and subdivision—must be supplemented, not only by much more effective capital project and other program planning, long advocated and rarely achieved, but by new mechanisms. Examples of these new mechanisms would be: tax policies supporting rather than conflicting with urban planning; increased incentives or controls directing private investment; equitable ways of assembling land more easily and quickly; higher standards of urban environmental health and amenity; municipal "land banks" for unforeseen future uses, planned communities, or to forestall undesirable development.

At the same time, greater participation in city planning by the urban public is being discussed. In part reflecting the widening informational and political gap between the average citizen and his government, in part responsive to the plight of the socially and politically disadvantaged, "advocacy planning" is proposed as a means of bringing the people of a community more actively into the urban planning process. Some seek to engage the public at large in the setting of goals and objectives for city planning, or to decentralize the process as much as possible.

Probably most important of all is the thrust toward more adequate theory and greater analytical capability to match the extended content of city planning. There are few generally accepted theoretical formulations of urban form, dynamics, or planning. There is no recognized body of general urban and urban planning theory. Repetitive analysis and widely different conclusions under closely comparable conditions are therefore the order of the day. In many ways, it is also being shown that the range and reliability of the coordinative analysis which is possible today cannot support comprehensive urban planning as now conceived. Meaningful and reliable correlation of the multitude of diverse variables to be taken into account in present-day urban planning cannot yet be accomplished. Thus, systems analysis, mathematical models, electronic data processing and critical-path scheduling are centers of attention today. Underlying current interest in quantification is the continuing effort to extend the application of scientific method to the range and diversity of urban analysis, procedure, decision-making, and implementation which are necessarily involved in city planning viewed comprehensively. Vastly augmented research directed toward greater urban analytical capability will be needed for many years.

The extraordinary analytical difficulties of expanded urban planning also focus attention on the selection, education, and in-service training of planning personnel. Even partial or gradual improvement of cities requires professional competence of the highest order: attitudinally, intellectually, substantively, technically, strategically, managerially, and administratively. Graduate education in planning must be strengthened in depth as well as broadened in scope; academic programs must be expanded at least temporarily to meet a demand which has progressively exceeded supply for over fifteen years. At the same time, sufficient knowledge of city planning must be made available within the many closely related fields whose contributions are essential to its success. More and more it is claimed that comprehensive city planning will be accomplished by interdisciplinary teams of professional city planners and representatives drawn from participating fields such as civil, traffic or systems engineering, operations research and environmental sciences, urban sociology, land economics, municipal planning law, administration-management, statistics and applied mathematics, and, of course, the long allied fields of architecture and landscape architecture. This need for a variety of specialists is reflected in this bibliography by the many books with more than one author. Finally, if the body politic is to participate more actively in city planning, as advocated by some, public education in city planning must be many times expanded.

It is this type of comprehensive urban planning, still in the process of resolution, which is the subject of these fifteen hundred selected references. Use of the adjective "comprehensive" does not imply that *all* elements and aspects of cities and their planning must be analyzed and correlated. This is impossible now, and will not be possible for a long time to come. Rather, "comprehensive" signifies that all primary components are borne in mind, and as many as possible or practicable are integrated into the urban planning process. The fact that it is an intellectually incomplete process is characteristic not only of planning and other fields of broad coverage, but, in a strict sense, of every field of knowledge.

There are several purposes and intended audiences for these references and related material. Professional practitioners of city planning should find them helpful and time saving when they are drawn into aspects of the field new to them, or when they simply wish to extend their awareness and knowledge. Persons in the many professional and academic fields that are directly involved in comprehensive urban planning or closely related to it will find these references useful when they want to learn more about city planning as a distinct effort; these fields now comprise a lengthy list ranging from political philosophy to applied mathematics. Certainly, researchers, historians, students of urban planning, and ordinary citizens seriously interested in the field should find this document a ready reference to comprehensive urban planning as it is broadly conceived. It may lighten the task of teachers in preparing reading lists and otherwise providing references for instructional purposes; reference reading can be designated simply by a list of numbers referring to this bibliography. Also, in the opinion of the compiler, these selected references constitute an excellent list of initial acquisitions for a city planning library starting *de novo*. If necessary, they would suffice to support a graduate educational program; better yet, of course, they could form an excellent nucleus and general content guide for a much larger library on city planning. They can also be used to check an existing library with respect to its coverage of comprehensive urban planning, serve as a separate selected catalog to the field, or provide a basic classification system for some special purpose.

Finally, these selected references, together with their tables of contents and other annotations, describe the intellectual and analytical content of the field of comprehensive urban planning as it is evolving today and will exist for some time to come. Perusing the different titles and indications of subject matter serves to define comprehensive urban planning and provides a useful conceptual referent to the scope and content of the field.

ORGANIZATION, SELECTION, AND CITATION OF REFERENCES

Three categories of references are incorporated:

1. *Those dealing directly with comprehensive urban planning as described generally in previous paragraphs. Because this more inclusive form of planning is new and still in process of formulation, the more advanced subject matter is usually found in articles or monographs rather than books. Some of these materials may include the words "comprehensive planning" in the title or abstract; others may employ synonymous terms such as master, development, general, overall, or system planning. The relevance of some references is revealed in the contents rather than the title.*

2. *References not specifically on comprehensive urban planning as such, but a short step removed and closely relevant. This considerable grouping of materials includes various forms of functional and subsystem planning—dealing, for example, with land use, transportation, utilities, housing, legislation, or project programming. These materials have to do with elements or subsystems which comprehensive urban planning seeks to integrate into one process of overall consideration, analysis, and coordinated action.*

3. *Environmental and other background materials generally underlying comprehensive urban planning and a higher order of municipal management. They include references on urban planning of the past, urbanization, outstanding city problems of today, economic geography, geology, social-psychological knowledge, technological trends, cultural factors, human intercommunication, or public administration. Since the content of comprehensive urban planning is open-ended in that it can be extended progressively almost without end, comparatively few of these underlying references can be included. This type of material, which in the aggregate could comprise a very large library, is best sought out as needed. The aim here is to choose several introductory or broadly explanatory references, which do not require special prerequisite knowledge, for those subjects most important as background for comprehensive urban planning. In addition, whenever possible, bibliographies are cited as leads to further information. When these are not available, as many substantive references are listed as can be included within the overall limitations of length for this document.*

Library of Congress catalog card (LC) numbers are included, whenever available, within brackets after the reference citation of books [66-12345]. Although this number has no significance in the widely used card-catalog *classification* system of the Library of Congress, it is the only available numerical designation for almost all of the books among these selected references. It is included as an identification with potential use for processing and retrieval in electronic or manual reference systems which provide more precise access to material than by author and title alone. Up until December 1, 1968, the first two digits normally give the year of publication, the remaining numbers the sequential order of catalog recording. British national bibliographic numbers are given for a few references published in England; these are prefaced by the letters BNB.

A Standard Book Numbering (SBN) system, adopted in Britain in 1966, is now being adopted by publishers throughout the United States. Since it is not compatible with Library of Congress catalog card numbers, the SBN is being added to LC cards and may eventually replace them. Specific information concerning the Standard Book Numbering System is available from R. R. Bowker Co. (1180 Avenue of the Americas, N.Y. 10036) or Standard Book Numbering Agency Ltd. (13 Bedford Square, London, W.C. 1).

Unfortunately, Library of Congress catalog card numbers are rarely assigned to documents issued by municipal, state, and federal governments in time for inclusion in their first (and usually last) printing. In part, this is because governmental publications are considered most appropriately in the public domain, with no restrictions on quotation or reproduction; therefore, Library of Congress catalog card numbers are not assigned in advance as they frequently are for commercial publishers when copyright is requested. State and municipal governments tend to discount the vital function of libraries as depositories and central references for documentation of lasting value. Since many important documents relating to comprehensive urban planning are issued today—and will be issued increasingly in the future—by regional, state, and local governments, Library of Congress catalog card numbers or Standard Book Numbers should be available for inclusion in the initial printing. Otherwise, any electronic reference and data-processing system for city planning which utilizes general reference or standard numbers as descriptors, may be seriously incomplete until the assignment of these numbers to documents is received . . . perhaps long after publication.

With the enormous increase in published materials and growing specialization, efficient reference to existing literature is essential. State and local governments can do much to help themselves in this respect by more careful titling and description of documents. It is surprising how often state and local governmental publications are undated, have no clear indication of source, are paginated in some irregular way, are described differently on cover and title page, or are so lengthily titled, and in such a fashion, that uniform citation by different catalogers or computer programmers is unlikely. Either governmental and public librarians should be consulted before documents are issued, or greater attention and care should be given to the incorporation of necessary reference information and proper titling.

These selected references are derived from diverse sources. Many were noted by the author-compiler in connection with his various activities over a period of years. Some were obtained from other bibliographies, some from colleagues. Still others were noted in miscellaneous sources: periodicals and books, publication announcements, newspapers and newsletters, student papers and graduate work, occasionally personal correspondence. The wide range of sources should not be surprising. With the concept of comprehensive urban planning more widely accepted, and with mounting municipal difficulties forcing attention on cities, relevant material can now be found far and wide. Fields of knowledge once disassociated from city planning are now pertinent—such as applied mathematics and general systems analysis, or biochemistry and metropolitan air pollution. At the same time, awareness of various "urban crises" has reached the point where responsible people can no longer disregard them to the extent they could in the past. Genuine alarm, aroused interest, sensitive response, or simply following the leader, concentrate attention on cities and planning in order to alleviate their difficulties.

It will be noted that many of the selected references in this bibliography are drawn from the literature of business, management-administration, the management sciences, operations research, the military services, and various physical and social sciences. Not only has business long engaged in its applications of planning, but in recent years "corporate planning" has come to the fore as the business world's equivalent of comprehensive urban planning. The military services have contributed concepts of line and staff, strategy and tactics, and techniques of graphical display and communication for portraying a complex situation in real life and real time, command control, or other decision-making. Businesses, military operations, and cities share many more planning problems of information collection, subsystems and systems analysis, optimum allocation of available resources, and projection-forecasting than is commonly supposed. Furthermore, as urban planning becomes more effective in shaping cities, such long-time areas of emphasis by business as organization management, decision-making, strategy, public relations, and programming will become of primary importance. Hence, certain work in management science is relevant as background information, analytical technique, or indirectly suggestive for city planning. The day has never existed, of course, when the knowledge required for meaningful urban planning could be supplied from the literature generated within the profession itself.

Not all materials with appropriately worded titles are included in this document, not only because of limited space in which to reference so vast a body of applicable knowledge and activity, but because some articles written in recent years do little more than advocate comprehensive urban planning, deal very generally or even superficially with the concept, or reiterate previous work. Where a choice existed between titles clearly referring to comprehensive city planning but mainly advocative or repetitious in content, and material less directly identified with comprehensive planning but with more substantial content concerning technique or practice, the latter was selected. In substantive areas closely related to comprehensive city planning, available literature is often so extensive that more than a few selections would lengthen this publication several times.

The study of cities, consideration of their environment in the broadest sense of the word, and urban-metropolitan planning, constitute an enormous field indeed. These references are selected to scan that field and provide points of entry from numerous viewpoints and particular interests. Various objectives entered into their selection. Significant content and the inclusion of material representing all major elements and aspects of comprehensive urban planning were the main criteria of reference selection. Additional objectives were to cover as many as feasible of the areas of knowledge which relate significantly to urban planning theory and practice, to include leading contributors to the field, indicate the range of sources of city planning information, represent different approaches and conclusions on important topics, and minimize materials of short-lived relevance or usefulness. Not all references were selected because they are outstandingly good; sometimes, they reflect the problems or limitations of the existing state of the art or planning situation.

Materials written exclusively in a foreign language are·excluded, and only a few are included which must be purchased outside the United States. The focus is almost entirely on experience, problems, and the future of cities in the United States. Almost all references are believed to be reasonably available from publisher, library, or other source in this country.

To accommodate differences in reader interest and background, there are references which are short and relatively simple, and others comprising longer and more rigorous reading. At one end of the spectrum of selection are materials—usually articles—which can be absorbed quickly by people who are only mildly interested in city planning or by practitioners and students moderately concerned with the subject. At the other extreme are reports, studies, and books requiring careful assimilation by experienced or strongly motivated readers. Both "academic" and "practical" works are included. Most references have been published in recent years, but works which apparently are the best available, and some which are unique or landmark contributions, are included regardless of age.

To describe the material's source and main characteristics for readers, readily available bibliographical detail is given in each reference, and in such form that it will be clear to readers abroad as well as to those in the United States. Source of publication is given so the reader can write for the book or periodical without further ado. If the address is not included in the List of Publishers and Sources with Addresses beginning on page 325, it will normally be found in the citation itself. For the benefit of foreign readers who are not familiar with the location of cities in the United States, and to differentiate between the same place names located in more than one country or state, places of publication are fully identified—except in a few cases such as New York City which are known far and wide. "Paperback" editions are usually identified as such at the end of the citation; the word "paper" signifies that the document has paper rather than hard covers, but is not a paperback in size or other usual characteristics. When confronted with a choice between references considered comparable except for cost, the less expensive work or edition has usually been selected. Specific prices of documents, however, are not given, since they are so often subject to discount or change.

Three kinds of references are included under each subject heading: first, articles and papers in periodicals; second, books, reports, and pamphlets; and last, bibliographies. It is expected that readers with a beginning or casual interest in city planning will turn to the periodical literature; practitioners and serious students will probably favor books and other longer documents; those deeply involved with a specialization or engaged in research will find the bibliographies valuable. When a good bibliography is contained in a book or article cited, this fact is noted at the end of the annotation; the number of pages devoted to the bibliography is given as a clue to its probable completeness. The best of these reference compilations are also listed separately under the appropriate section for Bibliographies; usually, this is the same heading as for the parent document.

With few exceptions, the annotations after almost every reference consist of its table of contents. Although the type of review or critical comment which ordinarily comprises annotation is occasionally helpful, it is far more often characterized by digression, misleading overreaction, or insufficient description of the actual contents of the document. Book reviews are often prejudicial, rather than objectively useful. Therefore, rather than search out the few sound reviews or present the compiler's judgment of quality, the author's table of contents, textual headings, or abstract has been chosen as the best single indicator of what the document is about—what it professes to do, rather than another man's opinion of what it achieves.

Chapter or section numbers and quotation marks are omitted from such reporting of headings from tables of contents. The language is taken directly from the source, except that grammatical articles which are informationally useless or repetitious are eliminated, and capitalization has been made editorially consistent throughout this book. A definite sequence of punctuation is used in reporting tables of contents. Semicolons separate the major subdivisions of the book. Colons are employed after a section or chapter heading to indicate what is contained within it. Dashes are used where further punctuation is needed subordinate to the above. The authors of different chapters in books which are compilations are separately identified and included in the author index.

When short descriptive comments are included other than headings from table of contents or text, their enclosure within quotation marks indicates a source other than the compiler: preface, another bibliography, publisher's description, book, or other reportive opinion. The quotation marks do not, however, also designate exact quotation; the material descriptive of the selected reference is sometimes condensed or otherwise altered without change of content and with appropriate editorial indications. No further editorial modification is attempted to make the various descriptive comments incorporated throughout this document consistent.

Selected references are grouped under the following main headings:

Background	Institutionalization
Process	Management, Decision-Making
Theory	Effectuation
Information-Communication	System Elements
Research-Analysis	Subsystems
Methodology	Particular Forms of Urban Planning

References under the first eight of these headings have to do more directly with comprehensive urban planning per se. When references of an overall, inclusive, or systems nature are not available because the comprehensive approach is still new and integration of the many diverse elements involved enormously difficult, examples of present levels of attainment are given. References on Effectuation are comparatively weak because the implementation of urban planning has been unimpressive, usually difficult to document or measure separately from the activities of other agencies and forces, and rarely reporting failures as well as successes.

Throughout the entire bibliography, no attempt is made to duplicate items under two or more headings to which they may apply; nor are different subjects comprising chapters in a book listed separately and individually, except in a few instances especially worthwhile and when the reference is also available as a reprint. However, the separate chapters and their authors listed for books are included in the author and title indexes, and incorporated into the subject index.

Municipal and Metropolitan Agencies in the United States and Canada, 1968

Colleges and Universities in the United States and Canada Offering Programs in Urban-Regional Planning, 1968

While the selected references comprising the substantive base of comprehensive urban planning are the primary purpose of this book, municipal and metropolitan planning agencies constitute a "bibliography" of a different sort. They are important sources of unpublished information and shared experience. Therefore, this brief section summarizes a few important facts and figures concerning local governmental agencies as the best sources of specific information. For essentially the same "bibliographical" reasons, the section on urban and regional planning education lists the programs which provide not only professional training in city planning, but also much of the basic and applied research presently undertaken within the profession.

Each of the three indexes has a purpose. Since it is impossible to divide the subject matter of comprehensive urban planning into categories equally descriptive for all potential readers regardless of their particular background, the subject index provides another means besides the table of contents for readers to find the specific references which interest them. The subject index can also be more detailed than is desirable for a table of contents.

Besides enabling the reader to find what work of a particular person is included in these references, the author index comprises a list of the outstanding writers in comprehensive urban planning, closely related fields, and indirectly or generally supportive areas. It is also likely that these people can be looked to for the most important contributions in the immediate future, until they are replaced gradually by the next generation of significant producers. It is interesting to note what a large percentage of the significant materials concerning city planning are written by a rather small group of individuals.

The List of Publishers and Sources with Addresses is included to make it easier for the reader to order a book. Rather than track down the current address himself—an especially bothersome task for a foreign reader—it is provided for him here. The List of Periodicals, besides also giving specific addresses, represents those journals most likely to contain material relevant to comprehensive urban planning and therefore worth regular review by those sufficiently concerned.

Acknowledgments

The author-compiler first undertook this type of selected bibliography for the initial class in corporate business planning included in the Engineering and Management Course at the University of California, Los Angeles. It was believed that a careful selection of several hundred references relating to the then comparatively new topic of comprehensive corporate planning would save time and effort for businessmen taking the course who might wish to pursue the subject further upon their return home. The demand for the document which developed through this limited distribution and by word of mouth indicated clearly that a selected bibliography on corporate planning would have wider usefulness. Accordingly, the book entitled *Selected References for Corporate Planning* was published by the American Management Association in 1966. This present bibliography is intended to fill the similar need which exists for comprehensive urban planning. Because the literature and task of city planning are much larger than those of corporate planning, three times as many references are included to attain comparable introductory and illustrative coverage.

Librarians at the various libraries visited to review material have been unfailingly gracious and helpful. Happily, these traits appear to be characteristic of this profession. If any single library can be singled out, it is the Municipal Reference Library of the City of Los Angeles. The assistance is appreciated of Ruth E. Palmer as Head Librarian, and Sarah D. Wolf as Librarian in both the Municipal Reference Library and the recently established "branch library" in the Los Angeles City Planning Department.

Over a period of almost two years, professional compatriots too numerous to list were generous in supplying or calling attention to particular documents. Some opened their personal bookshelves to my perusal. In a sense, the Graduate Program of Urban and Regional Planning at the University of Southern California contributed indirectly to this document, as the locus and occasion for university courses given by the compiler which required considerable investigation of reference materials. Similarly, the concept of comprehensive urban planning represented by this volume has been shaped by experiences as an initial faculty member of the Program of Education and Research in Planning at the University of Chicago, as Corporate Associate for Planning and Member of the Senior Staff (West Coast) of Thompson Ramo Wooldridge, Inc. for seven years, followed by eight years as a Member of the Los Angeles Board of Planning Commissioners.

SELECTED
REFERENCES

COMPILER'S NOTE

Full citations of shortened references listed under "Bibliographies" are to be found under "Books, Reports, Pamphlets" or "Articles, Periodicals" immediately preceding.

1. BACKGROUND

Articles, Periodicals

1. Anthony, Harry Antoniades, "Le Corbusier: His Ideas For Cities," *Journal of the American Institute of Planners*, Vol. XXXII, No. 5, September 1966, pp. 279-288.

Building on Stilts; Contemporary City; Radiant City; Modular; Voisin Plan; Other City Plans; Housing Designs; Le Corbusier on America; Writings; Critique; Concluding Comments.

2. Baker, John A., " 'The Rural Alternative' to Overcrowded Cities," *University*, A Princeton Quarterly, No. 37, Princeton, N.J. (Stanhope Hall, Princeton University, 08540), Summer 1968, pp. 9-14.

"We can . . . implement the rural alternative and provide a new creative opportunity for life in higher quality communities of tomorrow so that large numbers of Americans, *many of them ghetto residents,* will not have to be swallowed up in megalopolis."

3. Boesen, Victor, "And the Hills Came Tumbling Down," *Frontier*, Los Angeles (1434 Westwood Blvd., Cal. 90024), April 1966, pp. 1-6.

Gravity Cannot Be Mocked; Post-World War II Gold Rush; Builders Ignore Repeated Warnings; Faulty Code and Faulty Enforcement; Rising Profits, Falling Houses; "You Got to Expect Problems"; Examples of Laxness by the City; Wonder-Stricken After the Fact; Study Filed and Forgotten.

4. Bose, Nirmal Kumar, "Calcutta: A Premature Metropolis," *Scientific American*, Vol. 213, No. 3, September 1965, pp. 90-102; also in: Scientific American (Periodical), *Cities*, New York (Knopf), 1965, 211 pp. Paperback. [65-28177]

"Calcutta has become a metropolis without benefit of the industrial revolution that gave rise to cities in advanced nations."

5. *Daedalus,* The Conscience of the City, Vol. 97, No. 4, Fall 1968, pp. 1091-1433.

Traditional City in Transition: Post-City Age (Melvin M. Webber), The City as an Element in the International System (Kenneth E. Boulding), Remaking the Urban Scene—New Youth in an Old Environment (John R. Seeley); Processes and Goals for Change: Eight Goals for an Urbanizing America (Lyle C. Fitch), Urban Process (Edmund N. Bacon), On Coping with Complexity—Planning and Politics (Donald N. Michael), Political Leadership and Social Change in American Cities (Alexander L. George), Managerial Innovation and the Cities (Franklin A. Lindsay), Why Government Cannot Solve the Urban Problem (Edward C. Banfield); New Functions in Urban Communities: Worker and Work in the Metropolis (Peter F. Drucker), Service Society (Adam Yarmolinsky), Where Learning Happens (Stephen Carr; Kevin Lynch), Metropolis as a Transaction-Maximizing System (Richard L. Meier); Ways Out of the Ghetto: Like It Is in the Alley (Robert Coles), Alternative Futures for the American Ghetto (Anthony Downs), Liberal Approach–Its Weaknesses and Strengths (Michael Young), The Negro American and His City–Person in Place in Culture (Max Lerner); Urban Policy: Reforming Reform (Martin Meyerson).

6. Davis, Kingsley, "The Urbanization of the Human Population," *Scientific American*, Vol. 213, No. 3, September 1965, pp. 41-53; also in: Scientific American (Periodical), *Cities*, New York (Knopf), 1965, 211 pp. Paperback. [65-28177]

"More than half the world's people will probably be living in cities of 100,000 or more by 1990."

7. *Fortune*, A Special Issue On Business and the Urban Crisis, Vol. LXXVII, No. 1, January 1968, Entire Issue, pp. 132-234.

Deeper Shame of the Cities (Max Ways); The City: More Dollars and More Diplomas (Edmund K. Faltermayer); New Negro Mood (Roger Beardwood); How John Johnson Made It (A. James Reichley); Systems Engineering Invades the City (Lawrence Lessing); New Business for Business: Reclaiming Human Resources (Gilbert Burck); Upward Bound: Case Against the Unions (Thomas O'Hanlon); Putting Out the Fires Next Time (Harold B. Meyers); St. Louis Economic Blues (William S. Rukeyser).

8. Green, Carleton (Editorial Director), "The City in Transition," *Journal of the Stanford Research Institute*, Vol. 4, Fourth Quarter 1960, pp. 114-152.

Problems Facing the City in Transition (Frank W. Barsalou); Urban Mobility of People, Information, and Things (Carleton Green); The City: A Creator of Wealth (Robert A. Sigafoos); Changing Land Use in the Central City (Ernest M. Fisher); Metropolitan Government for Metropolitan Needs (J. Knight Allen).

9. Hancock, John L., "Planners in the Changing American City, 1900-1940," *Journal of the American Institute of Planners*, Vol. XXXIII, No. 5, September 1967, pp. 290-304.

Modern Challenge; American City Surface, 1800-1940; Pressure for Reorganization and Reform; The Profession Emerges, 1907-19; Flux and Flow, 1920-40; Retrospect–The Past As Index of Planning's Place in Future Policy.

10. Ingraham, Joseph C., "Car Production Outpaces Births," *New York Times*, International Edition, Wednesday, 31 July 1963, p. 9.

"There are now 22 autos for every mile of U.S. road." The Road Gap; 10-Year Expectancy; Age of Cars Watched; Postwar Babies; Obituary of an Auto; Scrap Business Off.

11. Nolen, John, "Twenty Years of City Planning Progress in the United States, 1907-1927," *ASPO Newsletter*, American Society of Planning

Officials, Chicago, Ill. (1313 East 60th St., 60637), Part 1, Vol. 32, No. 6, June-July 1966, pp. 69-70; Part 2, Vol. 32, No. 7, August 1966, pp. 90-93.

Conditions Twenty Years Ago; Record of Progress in City Planning, 1907-1927; City Planning Commissions, City Planning Publications, Progress in City Planning; New Towns, Garden Cities, Suburbs, or Satellite Towns; Landmarks; Conclusion; Educational Institutions with City Planning Lectures and Courses; Cities That Have Comprehensive Plans (Planning Reports Before 1910).

12. *Saturday Review,* Making Cities More Livable, Vol. XLIX, No. 2, 8 January 1966, pp. 37-99.

Presented with the Committee for Economic Development. The City As a Work of Art (Paul D. Spreiregen); Surviving the Age of the City (Ralph Lazarus); Tales of Three Cities: Philadelphia—The Economics of Social Problems (Howard C. Petersen),—Behind the Renaissance (Morris Duane),—A Look at Techniques (William H. Wilcox), San Francisco—Getting Together (S. Clark Beise),—Telling the Bay Area Story (Stanley McCaffrey),—Traveling in High Style (B. R. Stokes), Milwaukee—A Fair Deal (Elmer L. Winter),—Education for Industry (Harold S. Vincent); The Future of the American City (John V. Lindsay).

13. *Scientific American,* Cities, Vol. 213, No. 3, September 1965, pp. 40-219; also in: Scientific American (Periodical), *Cities,* New York (Knopf), 1965, 211 pp. Paperback. [65-28177]

Urbanization of the Human Population (Kingsley Davis); Origin and Evolution of Cities (Gideon Sjoberg); Modern Metropolis (Hans Blumenfeld); Calcutta: Premature Metropolis (Nirmal Kumar Bose); Stockholm: Planned City (Göran Sidenbladh); Ciudad Guayana: New City (Lloyd Rodwin); New York: Metropolitan Region (Benjamin Chinitz); Uses of Land in Cities (Charles Abrams); Transportation in Cities (John W. Dyckman); Metabolism of Cities (Abel Wolman); Renewal of Cities (Nathan Glazer); The City As Environment (Kevin Lynch).

14. Sjoberg, Gideon, "The Origin and Evolution of Cities, "*Scientific American,* Vol. 213, No. 3, September 1965, pp. 55-63; also in: Scientific American (Periodical), *Cities,* New York (Knopf), 1965, 211 pp. Paperback. [65-28177]

"The first cities arose some 5,500 years ago; large-scale urbanization began about 100 years ago. The intervening steps in the evolution of cities were nonetheless a prerequisite for modern urban societies."

15. Ulman, Neil, "The Metrophobes—Many Executives Are Hesitant to Take Jobs In the Biggest Cities," *The Wall Street Journal,* Pacific Coast Edition, Vol. LXXX, No. 58, Monday, 24 March 1969, pp. 1 and 10.

Cost of Living, Urban Woes Cited; Some Men Decline Raises to Avoid New York; Tired Blood at Headquarters.

16. Zile, Zigurd L., "Programs and Problems of City Planning in the Soviet Union," *Washington University Law Quarterly,* No. 1. February 1963, pp. 19-59.

Introduction; Post-Revolutionary Period of Disorganization and the First Steps Toward Recovery; Establishment of the Principles of Urban Administration in the RSFSR; Early Building Regulations in the RSFSR; Preparations for Systematic City Planning in the RSFSR in the Face of Rapid Urban Growth; Landmark Events of the 1930s; Impact of the Second World War and Postwar Reconstruction; Recent Standards and Practices; Administration of City Planning; Conclusion.

Books, Reports, Pamphlets

17. Adrian, Charles R.; Press, Charles, *Governing Urban America*, New York (McGraw-Hill), Third Edition, 1968, 530 pp. [68-11599]

Trend Toward an Urban Nation; Suburbia; Local Government Ideology; Municipal Elections; Intergroup Activity and Political Power; Urban Political Organization; Law of Municipalities; Forms of Government; Executive Officers; City Council; Government in Metropolitan Areas; Intergovernmental Relations; Administration; Personnel Systems; Revenues and Expenditures; Public Safety, Civil Rights and Liberties; Public Education; Public Utilities and Transportation; Land and Housing Policies; Public Health and Welfare.

18. Beckinsale, R. P.; Houston, J. M., *Urbanization and Its Problems, Essays in Honour of E. W. Gilbert,* New York (Barnes & Noble), 1968, 443 pp.

Urbanization in England to A.D. 1420; Spa Towns of Britain (J. A. Patmore); Boroughs in England and Wales of the 1830s (T. W. Freeman); Railways and the Morphology of British Towns (J. H. Appleton); Expanded Town in England and Wales (D. I. Scargill); Mining Settlements in Western Europe—Landscape and Community (R. T. Jackson); Urban Development in Belgium Since 1830 (E. C. Vollans); Recent Changes in the Morphology of West German Townscapes (M. Blacksell); Preface to Chinese Cities (Y. F. Tuan); Growth of Cities in Pre-Soviet Russia (D. J. M. Hooson); City and Region in the Soviet Union (C. D. Harris); Urbanization in India (G. C. K. Peach); Development of Urbanization in Southern Africa (N. C. Pollock); Colonial District Town in Uganda (D. N. McMaster); Foundation of Colonial Towns in Hispanic America (J. M. Houston); Towns of Lower Canada in the 1830s.

19. Benevolo, Leonardo, *The Origins of Modern Town Planning*, Cambridge, Mass. (M.I.T.), 1967, 170 pp. [67-17494]

Growth of the Industrial Town; Great Expectations (1815-48); Nineteenth Century Utopias; Beginnings of Town Planning Legislation in England and France; 1848 and Its Consequences.

20. Berry, Brian J. L.; Meltzer, Jack, (Editors), *Goals for Urban America,* Englewood Cliffs, N. J. (Prentice-Hall), 1967, 152 pp. [67-28390]

City—Work of Art and of Technology (August Heckscher); Urbanization in the Developing World (David Owen); The New Urbanism (Senator Joseph S. Clark); States and Cities—Unfinished Agenda (Terry Sanford); National Urban Policy Appropriate to the American Pattern (Martin Meyerson); Race in the City (Nathan Glazer); Desegregation—What Impact on the Urban Scene? (Whitney M. Young,

Jr.); Policies to Combat Negro Poverty (Senator Robert F. Kennedy); Social and Physical Planning for the Urban Slum (Jack Meltzer, Joyce Whitley).

21. Blake, Peter, *God's Own Junkyard*—The Planned Deterioration of America's Landscape, New York (Holt), 1964, 142 pp. [63-22178]

Textual and photographic indictment of American urban development and physical scene, written not in "anger" but in "fury."

Preface; God's Own Junkyard; Lords of the Soil; Improvements and Comforts of Social Life; Thin Alabaster Cities; Townscape; Landscape; Roadscape; Carscape; Skyscape; To Determine That the Community Should Be Beautiful.

22. Buder, Stanley, *Pullman—An Experiment in Industrial Order and Community Planning, 1880-1930*, New York (Oxford), 1967, 263 pp. [67-25456]

The Pullman System; Model Town, 1880-93; Community and Company; Impact of Strike and Boycott on the Community; From Model Town to Urban Community, 1894-1930.

23. Burke, Gerald L., *The Making of Dutch Towns—A Study in Urban Development from the Tenth to the Seventeenth Centuries*, London, England (Cleaver-Hume), 1956, 176 pp.

Background to Development until 1400; Mediaeval Town Foundation and Development: Dike and Dike-and-Dam Towns, Seaports, Bastides, Water Towns and Geestgrond Towns; Background to Development, Fourteenth to Seventeenth Centuries; Renaissance and Dutch Town Planning; Town Extensions; Amsterdam; Comparisons and Conclusions; Appendix; General Bibliography, pp. 169-170.

24. Collins, George R.; Collins, Christiane Crasemann, *Camillo Sitte and the Birth of Modern City Planning*, Columbia University Studies in Art History and Archaeology, New York (Random House), 1965, 232 pp. [64-15893]

Camillo Sitte's Background, Life, and Interests; State of City Planning in Germany and Austria; Important Contemporaries of Sitte: Baumeister, Stübben, Buls, and others; Transformation of Vienna; Some other Sources of Sitte's Ideas; Sitte's Point of View and Analytical Procedures; Successive Editions and Translations of the Book; Sitte's Impact on German City Planning; Influence of Sitte and German Planning Abroad; Conclusion.

25. Creese, Walter L., *The Search for Environment—The Garden City, Before and After*, New Haven, Conn. (Yale University), 1966, 360 pp. [66-12492]

Where Does the City Stand?; Bradford-Halifax School of Model Village Builders; Bad Example of Leeds; Oasis of Bedford Park; Neat and Clean at Port Sunlight and Bournville; Morris and Howard—Boston and Chicago; Parker and Unwin; New Earswick; First Garden City of Letchworth; Hampstead Garden Suburb; Wythenshawe; Architecture of Parkin and Unwin; Image Overseas; New Towns.

26. Crump, Spencer, *Ride the Big Red Cars*, Los Angeles, Cal. (Crest Publications), November 1962, 239 pp. [65-17582]

"How Trolleys Helped Build Southern California."

27. Duhl, Leonard J., M.D. (Editor), *The Urban Condition—People and Policy in the Metropolis,* New York (Basic Books), 1963, 410 pp. [63-12844]

"This . . . collection of 29 papers . . . sweeps across a big field, attempting to show how the problems of a metropolis—mental health, slums, poverty, transportation, crime, delinquency, racial tensions—are interrelated and require a massive effort by many disciplines if effective solutions are to be found. City planning, like other fields, is admonished for its inability to see how its solution is completely dependent upon solutions in spheres entirely foreign to its own."

Man and His Environment; Renewal and Relocation; Urbs and Suburbs; Social Action—and Reaction; Strategy of Intervention; Ecology of the Social Environment.

28. Eldredge, H. Wentworth (Editor), *Taming Megalopolis,* Vol. 1, What Is and What Could Be, 584 pp., Vol. 2, How to Manage an Urbanized World, pp. 579-1166, New York (Doubleday), 1967. Paperback. [67-12878]

What Is: America's Super-Cities (Christopher Tunnard), Most Dynamic Sector (Alan K. Campbell), Culture, Territoriality, and the Elastic Mile (Melvin M. Webber; Carolyn C. Webber), The Metabolism of Cities—Water and Air (Abel Wolman), Nature of Cities Reconsidered (Edward L. Ullman), People—Urbanization and City Growth (H. Wentworth Eldredge), Cities in Developing and Industrial Societies (Gideon Sjoberg), Urban Economics (Wilbur R. Thompson), The Contributions of Political Science to the Study of Urbanism (Robert C. Wood); What Could Be: The Declaration of Delos, Beauty for America (Lyndon B. Johnson), The Planner as Value Technician—Two Classes Utopian Constructs and Their Impacts on Planning (Thomas A. Reiner), Societal Goals and Planned Societies (John W. Dyckman), The Townscape—A Report to the President (Edmund N. Bacon), Visual Development of Ciudad Guayana (Wilhelm von Molka), Open Spaces in Urban Growth (S. B. Zisman), Administering the Spread City (Alan K. Campbell, Seymour Sacks), Conurbation Holland (Jac. P. Thijsse), Building Blocks for the New York Tri-State Metropolitan Region (Arthur T. Roe), Outlook for Traffic in Towns (Colin Buchanan), Terminal Planning (Roger H. Gilman), Urban Transportation Planning—An Overview (Ralph A. Gakenheimer), Economic Aspects of Business Blight and Traffic Congestion (William L. Baldwin), Why the Skyscraper? (Jean Gottmann), Economic Feasibility of the Chestnut Street Mall, Philadelphia (Arthur C. Kaufmann and Assoc.), Cluster Development (William H. Whyte), Hong Kong Housing (Michael Hugo-Brunt), Report of the Housing and Urban Renewal Task Force to Mayor John V. Lindsay (Charles Abrams, Chairman), Toward Equality of Urban Opportunity (Bernard J. Freiden), Place of Man in Nature and Place of Nature in the Environment of Man (Ian McHarg), Recapturing the Waterfront (Christopher Tunnard), Maintenance in Urban Parks and Open Spaces (Jane Jacobs), The Arts and the National Government—Report to the President (August Heckscher); How to Manage an Urbanized World: Cities and City Planners (William Alonso), Advocacy and Pluralism in Planning (Paul Davidoff), Urban Development Models—New Tools for Planners (Britton Harris), Review of Analytical Techniques for the CPR (Wilbur A. Steger), Roles of Intelligence Systems in Urban-Systems Planning (Melvin M. Webber), Metro Toronto—A

Decade Later (Frank Smallwood), Effectiveness of Metropolitan Planning (William W. Nash, Jr.), Uses and Limitations of Metropolitan Planning in Massachusetts (Edward C. Banfield), Metropolitan Planning—Comments on Papers by Nash and Banfield (Harvey S. Perloff), Existing Techniques for Shaping Urban Growth (F. Stuart Chapin, Jr.), Requiem for Zoning (John W. Reps), Zoning—What's the Good of It? (Dennis O'Harrow), Reconstruction of Coventry (R. W. G. Bryant), Dynamics of Urban Renewal (Reginald R. Isaacs), San Francisco Community Renewal Program (Louis K. Loewenstein, Cyril C. Herrmann), Britain's Place in Town Planning (Sir Frederic J. Osborn), Lessons Learned from the British New Towns Program (H. Wentworth Eldredge), Continental Europe Offers New Town Builders Experience (Frederick Gutheim), City of Columbia, Maryland (James W. Rouse), Land Development Program for California (Charles Abrams), Aerospace Project Approach Applied to Building New Cities (John H. Rubel), New Directions in Social Planning (Harvey S. Perloff), Planning—and City Planning—for Mental Health (Herbert J. Gans), Urban Poverty (Janet S. Reiner, Thomas A. Reiner), Social Aspects of Town Development (Maurice R. Broady), Indispensable One Hundredth of One Per Cent (John W. Bodine), Circle of Urban Participation (David R. Godschalk), Role of Universities in City Renewal (K. C. Parsons), Latin American Shantytown (Sam Schulman), Measuring Housing Needs in Developing Countries (Lloyd Rodwin), Approach to Mass Housing in India—with Special Reference to Calcutta (Alfred P. Van Huyck), Regional Planning in an Urbanizing World—Problems and Potentials (Charles Abrams), Research for National Urban Development Planning (Lawrence D. Mann), Urban Planning in the Soviet Union and Eastern Europe (Jack C. Fisher), State Development Planning in a Federal Economic System (John W. Dyckman), Regional Development—Problems and Lines of Advance in Europe (Myles Wright), Dutch National Land Use Planning (Cornelius de Cler), Total Environmental Planning (H. Wentworth Eldredge).

29. Eyring, Henry (Editor), *Civil Defense*, Publication No. 82, American Association for the Advancement of Science, Washington, D.C. (1515 Massachusetts Ave., NW, 20005), 1966, 136 pp. [66-25382]

Basic Case for Civil Defense (Fred A. Payne); Civil Defense as Insurance and as Military Strategy (Wolfgang K. H. Panofsky); Effect of Civil Defense on Strategic Planning (Owen Chamberlin); Possible Effectiveness of Civil Defense (Eugene P. Wigner); Medical Aspects of Civil Defense (Victor W. Sidel); Agricultural Problems in Civil Defense (John H. Rust); Feasibility of Biological Recovery from Nuclear Attack (Barry Commoner); Panel Discussions.

30. Fabos, Julius Gy.; Milde, Gordon T.; Weinmayr, V. Michael, *Frederick Law Olmsted, Sr.—Founder of Landscape Architecture in America*, Amherst, Mass. (University of Massachusetts), 1968, 114 pp. [68-19670]

Introduction; Urban Parks; State and National Parks; Community Design; Regional Planning; Campus Design; Estate Design; Urban Design; America's Designer; Appendices; Bibliography, pp. 113-114.

31. Fogelson, Robert M., *The Fragmented Metropolis, Los Angeles 1850-1930*, Cambridge, Mass. (Harvard University), 1967, 362 pp. [67-20876]

Los Angeles—1850-1930: From Pueblo to Town, Private Enterprise, Public Authority, and Urban Expansion, Rivalry between Los Angeles and San Diego, Great Migration, Transportation, Water, and Real Estate, Commercial and Industrial Progress; Fragmented Metropolis: Urban Landscape, Failure of the Electric Railways, Quest for Community, Politics of Progressivism, Municipal Ownership Movement, City and Regional Planning; Conclusion—"The Simple Life"; Bibliography, pp. 279-295.

32. Foley, Donald L., *Controlling London's Growth*—Planning the Great Wen, Berkeley, Calif. (University of California), 1963, 313 pp. [63-19958]

". . . a sophisticated account of how a metropolitan plan—the Greater London Plan, finished in 1944—was embodied into government policy and action. . . . Describes how, in response to unforeseen social and economic changes, development policies were altered. . . . One of the rare documentations of a continuing planning program on a metropolitan scale. . . ."

33. Frazer, Douglas, *Village Planning in the Primitive World*, New York (Braziller), 1968, 128 pp. [68-24701]

Introduction; Mbuti Pygmies; Bushmen; Cheyenne Indians; Haida of the Pacific Northwest; Mailu, New Guinea; Trobriand Islanders, New Guinea; South Nias Islanders, Indonesia; Yoruba, Nigeria; Conclusions.

34. Geen, Elizabeth; Lowe, Jeanne R.; Walker, Kenneth (Editors), *Man and the Modern City*, Pittsburgh, Penna. (University of Pittsburgh), 1963, 134 pp. Paperback. [62-19452]

Preface (Otto F. Kraushaar); Introduction; Limitations of Utilitarianism as a Basis for Determining Urban Joy (John Ely Burchard); The City Image (Edmund N. Bacon); The Promise of the City (Lawrence K. Frank); The Anti-City (William H. Whyte); Living with the Coming Urban Technology (Richard L. Meier); What Are the Economic, Social, and Political Choices? (Joseph H. Clark); Challenge of the City to the Liberal Arts College (Richard C. Lee); Continuing Education for Urbanism (Eugene I. Johnson); American Suburb: Boy's Town in a Man's World (Robert C. Wood); Suburbia Reconsidered: Diversity and the Creative Life (Dorothy Lee).

35. Glaab, Charles N., *The American City—A Documentary History*, Dorsey Series in American History, Homewood, Ill. (Dorsey), 1963, 478 pp. Paperback. [63-19883]

Collection of historical readings. The City in Early American History: Meaning of American Urban History, Atlantic Ports in the Eighteenth Century, Municipal Services in Philadelphia, Cities with Plans, Cities in the West, Views of the City; Growth of Cities in the Early Nineteenth Century: Transportation and the Growth of Cities 1800-1850, City Visited 1824-1846, Health of the City, Rise of Manufacturing Towns, Town Promotion in the Wilderness, Urbanization and the Future of the West, Plea for Southern Cities, Urban Trend Discerned; Growth of Cities in the Late Nineteenth Century: Urbanization in the Nineteenth Century, Natural Advantages vs. Enterprise, City in the West, American and British Cities Compared, Urban Rivalry—St. Louis and Chicago, Suburbs (Chicago, 1873), Manufacturing

City, Parks and Planning; City Examined, 1850-1910; Slum, Sins of the City, Uto-
pian City, Dangers of City, Two Literati Assess Our Largest Cities, Youth of the
City, City Defended; The City in an Era of Reform, 1890-1914; Boss, Machine, and
Political Corruption, Municipal Reform Goals, City Examined in Depth; Era of the
Super-City: Spatial Expansion of the Metropolis; The New Deal—Changing Atti-
tudes Towards Cities, Metropolis 1960; Suggested Additional Readings.

36. Glaab, Charles N.; Brown, A. Theodore, *A History of Urban America*,
New York (Macmillan), 1967, 328 pp. Paperback. [67-15198]

Colonial Matrix; Cities in the New Nation; City in American Thought, 1790-1850;
Urban Milieu; Completion of the Urban Network, 1860-1910; Transformation and
Complexity—Changing City, 1860-1910; Web of Government; Bosses and Re-
formers; Urban Community Examined; Emergence of Metropolis; Further Reading,
pp. 309-318.

37. Goodman, Percival and Paul, *Communitas, Means of Livelihood and
Ways of Life*, New York (Vintage), 1960, 248 pp. Paperback.
[60-6381]

Introduction; Manual of Modern Plans: Green Belt, Industrial Plans, Integrated
Plans; Three Community Paradigms: Introduction, City of Efficient Consumption,
New Community—Elimination of the Difference between Production and Consump-
tion, Planned Security with Minimum Regulation, Conclusion.

38. Gordon, Mitchell, *Sick Cities—Psychology and Pathology of American
Urban Life*, Baltimore, Md. (Penguin), 1963, 444 pp. Paperback.
[63-14600]

By Way of Introduction—A Few Stepping-Stones to a Preface; Traffic Jam—The
Concrete Spread; Beware the Air; Water—Filthier and Farther; No Place for Fun;
Help, Police!; Fire!; School Bells—and Burdens; Libraries and a Couple of Nuisances
—Noise and Birds; City Dump; Public Purse; Too Many Governments; City Limits;
Urban Blight and Civic Foresight; Conclusions.

39. Gottehrer, Barry; Staff of the New York Herald Tribune, *New York
City in Crisis: A Study in Depth of Urban Sickness*, New York (Pocket
Books), 1965, 212 pp. Paperback. [65-8773]

"This eye-opening indictment of New York's bureaucracy, mismanagement, and
malappropriation of city resources is a rewritten, reorganized, and up-to-date version
of a full-scale investigation carried out and published early in 1965 by the New
York Herald-Tribune. . . . It is not an 'in depth' scientific investigation and analysis
of urban ills [but] journalism of a high order. . . . As the book moves into the actual
working of city government, highway programs, urban renewal, relocation of industrial sites, etc., it makes a valuable contribution toward exposing New York's myriad
problems. Many fine examples of Parkinsonian bureaucracy underscore the gross
inefficiency common to many municipal governments."

40. Gottmann, Jean, *Megalopolis—The Urbanized Northeastern Seaboard
of the United States*, Cambridge, Mass. (M.I.T.) 1961, 810 pp. Paperback. [61-17298]

"A geographer's views and conclusions from a four-year study of the urbanized Northeastern seaboard."

Introduction: Main Street of the Nation; Dynamics of Urbanization: Prometheus Unbound, Earthly Bounds, Continent's Economic Hinge, How the Cities Grew and the Suburbs Scattered; Revolution in Land Use; Symbiosis of Urban and Rural, Megalopolitan Agriculture, Woodlands—Their Uses and Wildlife, Urban Uses of the Land; Earning a Living Intensely: Manufacturing in Megalopolis, Commercial Organization, White-Collar Revolution, Transportation and Traffic; Neighbors in Megalopolis: Living and Working Together, Sharing a Partitioned Land, Conclusion —Novus Ordo Seclorum.

41.   Greer, Scott, *The Emerging City*, Myth and Reality, New York (Free Press), 1962, 232 pp. Paperback. [62-11851]

City in Crisis; Order and Change in Metropolitan Society; Citizen in the Urban Worlds; Community of Limited Liability; Urban Policy; Problems of the Metropolis; Changing Image of the City.

42.   Greer, Scott; McElrath, Dennis L.; Minar, David W.; Orleans, Peter (Editors), *The New Urbanization*, New York (St. Martin's), 1968, 384 pp. [67-12239]

". . . organizational transformation of the total society that results in cities . . . internal differentiation of urban subareas in large-scale societies . . . mechanism of integration . . . how we do, and do not, generate policies within the contemporary American metropolis . . . processes of innovation in cities. . . ."

Urbanization as Process; Shape and Texture of the City; Urban Polity as Process; Struggle to Govern the Metropolis.

43.   Gutkind, E. A., International History of City Development, New York (Free Press), London, England (Collier-Macmillan),

Vol.   I, *Urban Development in Central Europe*, 1964, 518 pp.

Vol.   II, *Urban Development in the Alpine and Scandinavian Countries*, 1965, 518 pp.

Vol. III, *Urban Development in Southern Europe: Spain and Portugal*, 1967, 544 pp.

Vol. IV, *Urban Development in Southern Europe: Italy and Greece*, 1969, 552 pp.

Vol.   V, *Urban Development in Western Europe: France and Belgium*, 1969, c. 564 pp.   [64-13231]

44.   Hadden, Jeffrey K.; Masotti, Louis H.; Larson, Calvin J. (Editors), *Metropolis in Crisis—Social and Political Perspectives*, Itasca, Ill. (Peacock), 1967, 521 pp. [67-28510]

Challenge of Metropolis—Introduction; From Mudhuts to Megalopolis—Historical Development of the Urban Area: Emergence of Cities, Process of Urbanization and Its Consequences; Edeological Perspectives—Anti-Urban Bias; Dimensions of the Crisis: Race, Housing, Poverty, Education, Crime, Transportation, Air Pollution; Adaptation, Innovation, and Reform—Organizing for Urban Change: Distribution and Exercise of Power and Influence, Reform and Reorganization of Government, Intergovernmental Relations—Role of the States and Nation.

45. Hall, Peter, *The World Cities,* New York (McGraw-Hill), 1966, 256 pp. [64-66181]

Metropolitan Explosion; London; Paris; Randstad Holland; Rhine-Ruhr; Moscow; New York; Tokyo; Future Metropolis; Bibliography, pp. 246-254.

46. Handlin, Oscar; Burchard, John (Editors), *The Historian and the City,* Cambridge, Mass. (M.I.T.), 1963, 299 pp. Paperback. [63-18004]

The Modern City as a Field of Historical Study; The City in Technological Innovation and Economic Development: Crossroads Within the Wall (Robert S. Lopez), Economic Significance of Cities (Shigeto Tsuru), City Economics—Then and Now (Alexander Gerschenkrow), Innovation and the Industrialization of Philadelphia 1800-1850 (Sam B. Warner, Jr.), Economics of Urbanization (Aaron Fleisher), Organization of Technological Innovation in Urban Environments (Richard L. Meier); The City in the History of Ideas: Two Stages in the Critique of the American City (Morton White), Idea of the City in European Thought—Voltaire to Spengler (Carl E. Schorske), Boosters, Intellectuals, and the American City (Frank Freidel), The City as the Idea of Social Order (Sylvia L. Thrupp); History and the Contemporary Urban World: Death of the City—Frightened Look at Postcivilization (Kenneth E. Boulding), Implications of Modern City Growth (Denis W. Brogan); City as an Artifact: Urban Forms (Sir John Summerson), Proprietary Philadelphia as Artifact (Anthony N. B. Garvan), Form of the Modern Metropolis (Walter L. Creese), Visible Character of the City (Henry Million); Planners and Interpreters of the City: Customary and Characteristic—Note on the Pursuit of City Planning History (Christopher Tunnard), Urbanization and Social Change—on Broadening the Scope and Relevance of Urban History (Eric E. Lampard), Building Blocks of Urban History (Frederick Gutheim; Atlee E. Shidler); Conclusion: Some Afterthoughts (John Burchard), Selection of Works Relating to the History of Cities (Philip Dawson; Sam B. Warner, Jr.).

47. Hardoy, Jorge, *Urban Planning in Pre-Columbian America,* New York (Braziller), 1968, 128 pp. [68-24700]

Preface; Criteria for Defining a Planned City in Pre-Columbian America; General Characteristics of Urbanization in Pre-Columbian America; The Golden Age of Urbanization in Central Mexico; The Mayas; Post-Classic Urbanization in Middle America; Early Attempts at Urban Planning in South America; The Golden Age of Urbanization in South America; Conclusions.

48. Hauser, Philip M.; Schnore, Leo F. (Editors), *The Study of Urbanization,* New York (Wiley), 1965, 554 pp. [65-24223]

Urbanization—Overview (Philip M. Hauser); Study of Urbanization in the Social Sciences: The Historian and the American City—Bibliographic Survey (Charles N. Glaab, Survey of Urban Geography (Harold M. Mayer); American Political Science and the Study of Urbanization (Wallace S. Sayre; Nelson W. Polsby); Theory and Research in Urban Sociology (Gideon Sjoberg); Economic Aspects of Urban Research (Raymond Vernon; Edgar M. Hoover); Comparative Urban Research: Cities in Developing and in Industrial Societies—Cross-cultural Analysis (Gideon Sjoberg), Political-Economic Aspects of Urbanization in South and Southeast Asia (Nathan Keyfitz), Urban Geography and "Non-Western Areas" (Norton S. Ginsburg), On the Spatial Structure of Cities in the Two Americas (Leo F.

Schnore); Selected Research Problems: Research Frontiers in Urban Geography (Brian J. L. Berry), Urban Economic Growth and Development in a National System of Cities (Wilbur R. Thompson), Folk-Urban Ideal Types—Further Observations on the Folk-Urban Continuum and Urbanization with Special Reference to Mexico City (Oscar Lewis), Observations on the Urban-Folk and Urban-Rural Dichotomies as Forms of Western Ethnocentrism (Philip M. Hauser), Historical Aspects of Urbanization (Eric E. Lampard).

49. Helmer, Olaf, "The Use of the Delphi Technique in Problems of Educational Innovations," Publication P-3499, Santa Monica, Cal. (RAND Corporation), December 1966, 22 pp. Paper. Mimeo.

"The so-called Delphi Technique is a method for the systematic solicitation and collation of expert opinions. It is applicable whenever policies and plans have to be based on informal judgment, and thus to some extent to virtually any decision-making process."
Delphi Technique; Range of Applications of the Delphi Technique to Educational Planning; Description of a Delphi Pilot Experiment Related to Educational Innovations.

50. Herber, Lewis, *Crisis in Our Cities*, Englewood Cliffs, N.J. (Prentice-Hall), 1965, 239 pp. [65-12920]

Urban Pace; Dirty Skies; Silent Disasters; Disease from the Air; Cool, Refreshing—and Filthy; Disease from the Streams; Living on Nervous Energy; Crude Art of Cracking Up; Immobilized Urbanite; Road Ahead; In the Long Run.

51. Hirsch, Werner Z. (Editor), *Urban Life and Form*, New York (Holt, Rinehart, and Winston), 1965, 248 pp. Paperback. [63-10967]

Introduction (Werner Z. Hirsch); Emergence of City Form (Joseph R. Passonneau); A New Vision in Law: The City as an Artifact (William L. Weismantel); The City in History—Some American Perspectives (Richard C. Wade); The Philosopher and the Metropolis in America (Morton White); The Contributions of Political Science to Urban Form (Robert C. Wood); Urban Government Services and Their Financing (Werner Z. Hirsch); Urban Form—The Case of the Metropolitan Community (Leo F. Schnore); Social Problems Associated with Urban Minorities (Lee N. Robins); Foundations of Urban Planning (F. Stuart Chapin, Jr.).

52. Jacobs, Jane, *The Death and Life of Great American Cities*, New York (Random House), 1961, 458 pp. Paperback. [61-6262]

". . . an attack . . . on the principles and aims that have shaped modern, orthodox city planning and rebuilding."
Peculiar Nature of Cities; Conditions for City Diversity; Forces of Decline and Regeneration; Different Tactics.

53. Jarrett, Henry (Editor), *Environmental Quality in a Growing Economy*, Washington, D. C. (Resources for the Future, Inc.), December 1966, 184 pp. [66-28505]

Economics of the Coming Spaceship Earth (Kenneth E. Boulding); Pressures of Growth upon Environment (Harold J. Barnett); Promises and Hazards of Man's Adaptability (René Dubos); Mental Health in an Urban Society (Leonard J. Duhl);

Side Effects of Resource Use (Ralph Turvey); Some Problems of Criteria and Acquiring Information (Roland N. McKean); Research Goals and Progress Toward Them (Allan V. Kneese); Welfare Economics and the Environment (M. Mason Gaffney); Formation and Role of Public Attitudes (Gilbert F. White); Assumptions Behind the Public Attitudes (David Lowenthal); New Tasks for All Levels of Government (Norton E. Long); Some New Machinery to Help Do the Job (Jacob H. Beuscher).

54. Lampl, Paul, *Cities and Planning in the Ancient Near East,* New York (Braziller), 1968, 128 pp. [68-24699]

Introduction; Mesopotamia; Egypt; The Levant; Anatolia and the Syrian Foothills; Armenia and Persia; Urartu, Elam, Iran.

55. League of Women Voters, *Crisis: The Condition of the American City,* Washington, D.C. (1200 17th St., N.W., 20036), February 1968, 63 pp. Paper.

People; Slum-Ghettos; Money; Power; Environment.

56. Lowe, Jeanne, *Cities in a Race with Time,* New York (Random House), 1967, 601 pp. [66-21478]

Dead Hand of the Past; Urban Reconstruction—the Early Leaders: The Man Who Got Things Done for New York, The New Coalition, Frustrating Business of Rebuilding; Emerging Social Problems: What About the People?, Who Lives Where?, End of the Line; More Comprehensive Approaches: Survival through Planning, What Urban Renewal Can and Cannot Do; Urban Landscape, Looking Backward and Ahead.

57. Lubove, Roy, *The Urban Community—Housing and Planning in the Progressive Era,* Englewood Cliffs, N.J. (Prentice-Hall), 1967, 148 pp. Paperback. [67-10119]

Roots of Urban Planning; Conservation and Community: Land Colonization and Rural Credits (State of California), Employment and Natural Resources (Benton MacKaye); Landscape Architecture and Park Planning: Recommendations for the Establishment of a Park and Boulevard System (Kansas City, Missouri), Regional Planning (Boston Metropolitan Park Commissioners); Housing Reform—Restrictive Legislation: Housing Evils and Their Significance (Lawrence Veiller); City Beautiful: City of the Future—Prophecy (John Brisbane Walker), Report on a Plan for San Francisco (Daniel H. Burnham); Garden City: Garden Cities of England (Frederick C. Howe); Emergence of Professional Planning (Frederick Law Olmsted, Jr.), Report on Building Districts and Restrictions (City of New York); Constructive Housing Legislation: State Aid for Housing (Massachusetts Homestead Commission), Federal World War I Housing (United States Housing Corporation); Community Planning—Regional Planning Association of America: Architect and the City (American Institute of Architects); Suggested Readings, pp. 145-148. ". . . a volume in the American Historical Sources Series . . ."

58. McKelvey, Blake, *The Urbanization of America—1860-1915,* New Brunswick, N.J. (Rutgers University), 1963, 370 pp. [62-21248]

"Rise of the City"—1860-1910; Building and Governing the Cities; Urban Institutions and Social Tensions; Forging an Urban Culture—1860-1910; Culmination of a Half-Century of Urbanization—1910-1915; Bibliography, pp. 333-357.

59. Meyerson, Martin; Banfield, Edward C., *Boston: The Job Ahead,* Cambridge, Mass. (Harvard University), 1966, 121 pp. [66-14449]

"Far from being a catastrophe, the American metropolis is a great achievement . . . if judged from its success in giving scores of millions of people what they want."
Introduction—A Point of View; Power to Govern; City and Suburbs; Tax Tangle; Traffic and Transit; Freight; Time Again for Enterprise; Which Way Downtown?; Housing; Schools; Youth; Law and Order; Beauty In the City.

60. Meyerson, Martin, *Face of the Metropolis,* The Building Developments That Are Reshaping Our Cities and Suburbs, New York (Random House), 1963, 249 pp. [61-13841]

Photographs with text. Changing Cityscape; Living and Working in the Center City, Middle City, Outer City; Constraints and Possibilities; Suggested Readings.

61. Moholy-Nagy, Sibyl, *Matrix of Man—An Illustrated History of Urban Environment,* New York (Praeger), 1968, 317 pp. [68-11320]

Geomorphic and Concentric Environments; Orthogonal Environment; Greek Wave; Orbit of Rome; Orthogonal Variations—Linear Merchant Cities; Clusters and the End of Origins; Options—A Conclusion.

62. Mumford, Lewis, *The City In History*—Its Origins, Its Transformations, and Its Prospects, New York (Harcourt, Brace), 1961, 657 pp. [61-7689]

Sanctuary, Village, and Stronghold; Crystallization of the City; Ancestral Forms and Patterns; Nature of the Ancient City; Emergence of the Polis; Citizen Versus Ideal City; Hellenistic Absolutism and Urbanity; Megalopolis and Necropolis; Cloister and Community; Medieval Urban Housekeeping; Medieval Disruptions, Modern Anticipations; Structure of Baroque Power; Court, Parade, and Capital; Commercial Expansion and Urban Dissolution; Paleotechnic Paradise; Capetown; Suburbia—and Beyond; Myth of Megalopolis; Retrospect and Prospect. Bibliography, pp. 579-634.

63. Mumford, Lewis, *The Culture of Cities,* New York (Harcourt, Brace), 1938, 586 pp. [38-27277]

Protection and the Medieval Town; Court, Parade and Capital; Insensate Industrial Town; Rise and Fall of Megalopolis; Regional Framework of Civilization; Politics of Regional Development; Social Basis of the New Urban Order. Bibliography, 508-552.

64. Mumford, Lewis, *The Urban Prospect,* New York (Harcourt, Brace, World), 1968, 255 pp. [68-20631]

". . . most of the essays and papers in this book were written during the last half-dozen years. . . ."
California and the Human Horizon; Planning for the Phases of Life; Quarters for an Aging Population; Neighborhood and Neighborhood Unit; Landscape and Townscape; Highway and City; Disappearing City; Yesterday's City of Tomorrow;

Megalopolis as Anti-City; Beginnings of Urban Integration; Social Complexity and Urban Design; Megalopolitan Dissolution vs. Regional Integration; Home Remedies for Urban Cancer; Brief History of Urban Frustration; Postscript—Choices Ahead.

65. National Advisory Commission on Civil Disorders, *Report of the National Advisory Commission on Civil Disorders,* New York (Bantam), 1968, 659 pp. Paperback.

What Happened?: Profiles of Disorder, Patterns of Disorder, Organized Activity, Basic Causes, Rejection and Protest—Historical Sketch, Formation of the Racial Ghettos, Unemployment, Family Structure, and Social Disorganization, Conditions of Life in the Racial Ghetto, Comparing Immigrant and Negro Experience; What Can Be Done?: Community Response, Police and the Community, Control of Disorder, Administration of Justice Under Emergency Conditions, Damages—Repair and Compensation, News Media and the Disorders, Future of the Cities, Recommendations for National Action; Supplement and Appendices.

66. National Commission on Urban Problems, *Building the American City,* 91st. Congress, 1st. Session, House Document No. 91-34, Washington, D.C. (Superintendent of Documents), 1968, 504 pp. Paper.

Urban Setting—Population, Poverty, Race; Housing Programs; Codes and Standards; Government Structure, Finance, and Taxation; Reducing Housing Costs; Improvement of the Environment.

67. Olsen, Donald J., *Town Planning in London,* New Haven, Conn. (Yale University Press), 1964, 245 pp.

"A scholarly study of the ways in which the great landowners of London managed their properties during the Georgian and Victorian eras. Ground landlords and their agents engaged in continuous planning throughout the history of their estates, creating enclaves of sophisticated urbanism within the generally chaotic and haphazard growth of the great city. The study concentrates on the Duke of Bedford estates (Russell, Tavistock, and Brunswick Squares, etc.), and the Foundling Hospital estates (Oakley and Harrington Squares)."

68. The President's Council on Recreation and Natural Beauty, *From Sea to Shining Sea,* Report on the American Environment—Our Natural Heritage, Washington, D.C. (U.S. Government Printing Office), 1968, 304 pp. Paperback [68-67300]

The Environment: Urban Areas—Neighborhood, Downtown, City, Metropolitan Region, Rural Areas—Countryside, Water and Waterways, Recreation and Wildlands, Transportation; Sharing Responsibility for Action; Government Action, Education, Research, Private Action; Summary, Keys to Action: Books and Pamphlets, Periodicals, Films, Local Agencies, State Agencies, Federal Agencies, Private Organizations.

69. Reps, John W., *The Making of Urban America, A History of City Planning in the United States,* Princeton, N.J. (Princeton University), 1965, 574 pp. [64-23414]

European Planning on the Eve of American Colonization; Spanish Towns of Colonial America; Towns of New France; Town Planning in the Tidewater Colonies; New

Towns in a New England; New Amsterdam, Philadelphia, and Towns of the Middle Colonies; Colonial Towns of Carolina and Georgia; Pioneer Cities of the Ohio Valley; Planning the National Capital; Boulevard Baroque and Diagonal Designs; Gridiron Cities and Checkerboard Plans; Cemeteries, Parks, and Suburbs—Picturesque Planning in the Romantic Style; Cities for Sale—Land Speculation in American Planning; Towns by the Tracks; Towns the Companies Built; Cities of Zion—Planning of Utopian and Religious Communities; Minor Towns and Mutant Plans; Chicago Fair and Capital City—Rebirth of American Urban Planning; Bibliography, pp. 545-562.

70. Robson, William A., *The Heart of Greater London*, London, England (London School of Economics and Political Science), 1965, 40 pp.

". . . enumerates the adverse influences marring the central area of London and ruining its prospects. . . . Suggestions include moving certain functions and activities that could well be operated elsewhere; concentrating certain businesses and professions (such as book publishers) in one area; diverting traffic from the central area and improving the underground system; and making the future planning and development of the heart of London the responsibility of a single authority, the Greater London Council."

71. Rotkin, Charles E., *Europe: An Aerial Close-Up*, Philadelphia, Pa. (Lippincott), 1962, 222 pp. [62-14807]

England; Scotland; Netherlands; Belgium; Denmark; Sweden; France; Monaco; Germany; Austria; Switzerland; Spain; Greece; Italy; Vatican City.

72. Saalman, Howard, *Medieval Cities*, New York (Braziller), 1968, 127 pp. [68-24702]

Definition; Background; Carolingian Interlude; Eleventh Century Beginnings; Premises; Faubourgs; Private Space—Public Space; Institutions and Urban Scale; City and Anti-City.

73. Schneider, Wolf, *Babylon Is Everywhere: The City as Man's Fate*, New York (McGraw-Hill), 1963, 400 pp. [63-17490]

". . . starts with the first settlements along the banks of the Euphrates and ends with the cities of today . . . no special emphasis on city planning, but vividly describes cities as a place to live. . . . 'The city is the focal point of power and the source of decline.'"

74. Schnore, Leo F. *The Urban Scene, Human Ecology and Demography*, New York (Free Press), 1965, 374 pp. [65-13067]

Human Ecology and Demography: Scope and Limits; Metropolitan Growth and Decentralization; Functions and Growth of Suburbs; Socioeconomic Status of Cities and Suburbs; Changing Color Composition of Metropolitan Areas; Studies in Urban Circulation.

75. Scott, Mel, *The San Francisco Bay Area—A Metropolis in Perspective*, Berkeley, Cal. (University of California), 1959, 333 pp. [59-12537]

Heritage; Mother of Cities; Plow, Iron Horse, and New Towns; Urban Rivalries; Heyday of Enterprise; Burnham Plan for San Francisco; New San Francisco; Oakland—End of the Village Tradition; Greater San Francisco Movement; Panama Canal—Stimulus to Planning; Seeds of Metropolitan Regionalism; Fred Dohrmann and the Regional Plan Association; Prosperity and Projects; Progress in Troubled Times; Crisis in an Arsenal of Democracy; Postwar Planning; Regional Metropolis; Bibliography, pp. 317-323.

76. Scott, Stanley (Editor), *The San Francisco Bay Area—Its Problems and Future,* Berkeley, Cal. (Institute of Governmental Studies, University of California), 1966, Vol. 1, 277 pp., Vol. 2, 314 pp. Paperback. [66-7975]

Vol. 1. Financing Local Governments . . . (Malcolm M. Davisson); Social Dependency . . . Today and Tomorrow (Margaret Greenfield); Economic Trends . . . (Orville F. Poland); Minority Groups and Intergroup Relations . . . (Wilson Record); Problems in Public Education (Theodore L. Reller); Partnership in the Arts—Public and Private Support of Cultural Activities . . . (Mel Scott).
Vol. 2. Future Demographic Growth . . . (Kingsley Davis; Eleanor Langlois); City and Regional Planning . . . (T. J. Kent, Jr.); Air Resource Management . . . (John A. Maga); Managing Man's Environment . . . (Frank M. Stead); Geography and Urban Evolution . . . (James E. Vance, Jr.); Housing and the Future of Cities . . . (Catherine Bauer Wurster); Urban Transportation . . . (Richard M. Zettel).

77. Sobin, Dennis P., *Dynamics of Community Change—The Case of Long Island's Declining "Gold Coast,"* Port Washington, N.Y. (Friedman), 1968, 205 pp. [68-8249]

Background and Approach; Emergence of a Gold Coast; Residences for the Rich; Community in Transition; Patterns of Living Past and Present; Politics of Preservation; Social Conditions and Community Survival; Economics of Change; Personal Adjustment and Property Ownership; Community Change—Process and Prospects; References and Selective Bibliography, pp. 183-193.

78. Spreiregan, Paul D. (Editor), *The Modern Metropolis: Its Origins, Growth, Characteristics, and Planning,* Selected Essays by Hans Blumenfeld, Cambridge, Mass. (M.I.T.), 1967, 377 pp. [67-13391]

Modern Metropolis—Origins, Growth, and Form: Form and Function in Urban Communities, Theory of City Form—Past and Present, Alternative Solutions for Metropolitan Development, Urban Pattern, Modern Metropolis; Metropolitan and Regional Planning: Metropolitan Area Planning, Regional Planning, Some Lessons for Regional Planning from the Experience of the Metropolitan Toronto Planning Board, Hundred-Year Plan—Example of Copenhagen; Transportation: Experiments in Transportation—For What?, Transportation in the Modern Metropolis, Transportation in San Francisco, Montreal's Subway, Monorail for Toronto?, Why Pay to Ride?; Residential Areas: Residential Densities, Comments on the Neighborhood Concept, "The Good Neighbor," Urban Renewal, Problems of Urban Renewal, City Ouvriere of Mulhouse, France; Urban Design: Universal Dilettante, Scale in Civic Design, Scale in the Metropolis, Design with the Automobile—Metropolitan Region, Continuity and Change in Urban Form, A Visitor Looks at the Montreal Exposition—"Expo 1967"; Methodology of Planning: Science and Planning, Conceptual Framework of Land Use, Projection and Planning of Transportation and Land Use, Limitation of Simulation of Future Behavior, Are Land Use Patterns

Predictable?, Economic Base of the Metropolis—Critical Remarks on the "Basic-Nonbasic" Concept.

79. Starr, Roger, *Urban Choices: The City and Its Critics*, Baltimore, Md. (Penguin), 1967, 284 pp. Paperback. [66-10429]

Introduction: One City; Easy Assumptions: A Hundred Cities—A Hundred Critics, Rotisseries and Retrogression, Lambs' Day in the Lions' Club, Money Madness, Good Night—Lincoln Steffens; Hard Choices: On With the Dig, The Holes in the Wall, They Don't Build Palaces Like Versailles Any More, Wheels Within Wheels, The Income Gap, The Inadvertent System, Automating the Political Machine, The Living End.

80. Steiner, Oscar H., *Downtown USA—Some Aspects of the Accelerating Changes Sweeping Our Nation*, Dobbs Ferry, N. Y. (Oceana), 1964, 198 pp. [64-19111]

Cities: Yesterday, Man From Mars, Changing Scene, What Is a City?, A City Is People, A City's People, Beyond the City; People Within: What Is a Home?, A Few Hard Facts, Arithmetic and Taxes, Drama in Color; Revolution of Rising Expectations: Explosion, American Dilemma, Quick Look at Tomorrow, Individual States or United States?, Conflict Within Conflict, Big Business—Big Labor, Big Government; Day After Today: Where Are We Going?, Slums and City Planning, Business Principles in Government, English Example, Back to Fundamentals, Grand Design; Postlude.

81. Tietze, Frederick J.; McKeown, James E. (Editors), *The Changing Metropolis*, Boston (Houghton Mifflin), 1964, 210 pp. Paper. [64-4379]

Embattled City: Are Cities Dead? (Robert Moses), Are Expressways Worth It? (Charles Remsburg), Decline of Railroad Commutation (George W. Hilton), Chicago Area Employment Declined from 1957 to 1961 (Illinois State Employment Service), White-Collar Corps (Edgar M. Hoover; Raymond Vernon), Appraising the Home (Arthur A. May), Violence in the City Streets (Jane Jacobs), World of the Metropolitan Giants (Robert C. Wood), Future of Retailing in the Downtown Core (George Sternlieb), Education and Metropolitan Change (Norton E. Long). City Resurgent: Teachers for Urban Schools (Matthew J. Pillard), The New "Gresham's Law of Neighborhoods"—Fact or Fiction (Charles Abrams), Urban Renewal in a University Neighborhood (Editors of *House and Home*), Great Debate in Chicago (Ulysses B. Blakely; Charles T. Leber, Jr.), Faculty Report Studies Woodlawn Area, Urges U.C. Action (John T. Williams), Must Our Cities Be So Ugly? (John E. Burchard), What Makes a Good Square Good? (Grady Clay), San Francisco Rebuilds Again (Allan Temko), Values of Urban Living (Margaret Mead).

Shifting Populations: Historical Background (Oscar Handlin), Consequences of Population Distribution (Morton Grodzins), The First Negro Family: A Strategic Aspect of Nonwhite Demand (Chester Rapkin; William G. Grigsby), Puerto Ricans in New York City (Morris Eagle), Last White Family on the Block (Marvin Caplan), Growing Negro Middle Class in Chicago (Chicago Commission on Human Relations), How Marynook Meets the Negro (Terry Sullivan), Tells Marynook Plan: Integration Solved by "Decision" (Southeast Economist).

Suburbia and Beyond: The New Masses (Daniel Seligman), Suburban Dislocation (David Riesman), Urban Sprawl (William H. Whyte), Automotive Transport in the United States (Wilfred Owen), As If Los Angeles Didn't Have Enough Trouble (Wesley Marx), Prometheus Unbound (Jean Gottmann); Bibliography, pp. 198-207.

82. Tunnard, Christopher; Reed, Henry Hope, *American Skyline, The Growth and Form of Our Cities and Towns,* New York (New American Library), 1956, 224 pp. Paperback. [55-6553]

Introduction—Temple and the City; Colonial Pattern 1607-1776; Young Republic 1776-1825; Romantic Era 1825-1850; Age of Steam and Iron 1850-1880; The City as a Way of Life—Expanding City I 1880-1910; City Beautiful—Expanding City II 1880-1910; City of Towers 1910-1933; Regional City 1933-; Seven Eras of the American City.

83. Warner, Sam Bass, Jr., *Planning for a Nation of Cities,* Cambridge, Mass. (M.I.T.), 1966, 310 pp. Paperback. [66-21355]

Federal Responsibility: National Planning for Healthy Cities—Two Challenges to Affluence (Gunnar Myrdal), Public and Private Rationale for a National Urban Policy (John W. Dyckman), Urban Constraints and Federal Policy (Sam Bass Warner, Jr.), Federal Resources and Urban Needs (Murray L. Weidenbaum); Work and Quality of Urban Life: Role of Work in a Mobile Society (Marc A. Fried), Work and Identity in the Lower Class (Lee Rainwater), Tomorrow's Workers—Prospects for the Urban Labor Force (Stanley Lebergott); Responsive Physical Planning: Suburbs, Subcultures, and the Urban Future (Bennett M. Berger), Rising Demand for Urban Amenities (Jean Gottmann), Planning Inventory for the Metropolis (Joseph R. Passonneau), Regional Planning in Britain (John R. James), Legal Strategy for Urban Development (Daniel R. Mandelker); Responsive Urban Services: Toward a Framework for Urban Public Management (Wilbur R. Thompson), Urban Education in Long-Term Crisis (Judson T. Shaplin), Urban Politics and Education (Robert H. Salisbury), Reinstitutionalization of Social Welfare (Ralph E. Pumphrey).

84. Warner, Sam Bass, Jr., *Streetcar Suburbs: The Process of Growth in Boston, 1870-1900,* Cambridge, Mass. (Harvard University), 1962, 208 pp. [62-17228]

"A historian provides a refreshing change of pace from the often pedantic, jargon-filled books that planners must read. His treatment of the late nineteenth century growth of three Boston suburbs is scholarly, yet very readable."

85. Weber, Adna F., *The Growth of Cities in the Nineteenth Century: A Study in Statistics,* Ithaca, N.Y. (Cornell University), Second Printing, 1965, 495 pp. [62-22217]

Introduction; History and Statistics of Urban Growth; Causes of the Concentration of Population; Urban Growth and Internal Migration; Structure of City Populations; Natural Movement of Population in City and in County; Physical and Moral Health of City and County; General Effects of the Concentration of Population; Tendencies and Remedies.

86. Weber, Max, *The City,* New York (Free Press), 1958, 242 pp. Paperback. [58-6492]

Prefatory Remarks—Theory of the City; Nature of the City; Occidental City; Patrician City in Antiquity and the Middle Ages; Plebean City; Ancient and Medieval Democracy; Selective Bibliography, pp. 231-232.

87. Weimer, David R. (Editor), *City and Country in America,* New York Appleton-Century-Crofts), 1962, 399 pp. Paperback. [62-12332]

Early Ambiguities: J. Hector St. John de Crevecoeur; Unambiguous Vision—Plan of Washington, D.C.: Pierre Charles L'Enfant; Into Nature—Imaginative Energy: Thomas Jefferson, Ralph Waldo Emerson, Henry David Thoreau, Henry George, Frederick Jackson Turner; Into the City—Imaginative Control: Edward Bellamy, Daniel H. Burnham and Edward H. Bennett; Europe to America—Ameliorative Revolutions: John Ruskin, Ebenezer Howard, Petr Aleksùevich Kropotkin, Patrick Geddes, Andrew Jackson Downing, Frederick Law Olmsted and Calvert Vaux, Frederick Law Olmsted, Benton MacKaye, Clarence Arthur Perry, Clarence S. Stein, Lewis Mumford, Paul and Percival Goodman; From Europe—Bold Conservatisms: Camillo Sitte, Le Corbusier; The New Centrifuge: Ralph Borsodi, Twelve Southerners, Andrew Nelson Lytle, Frank Lloyd Wright, Baker Brownell, Louis Bromfield; The New Centripety: Joseph Hudnut, Robert Moses, Christopher Tunnard, Henry Hope Reed, Jr., Victor Gruen; Guides to Study.

88. White, Morton; White, Lucia, *The Intellectual Versus the City—From Thomas Jefferson to Frank Lloyd Wright,* Cambridge, Mass. (Harvard University), 1962, 270 pp. [62-17229]

Opening Theme; Irenic Age—Franklin, Crèvecoeur, and Jefferson; Metaphysics Against the City—Age of Emerson; Bad Dreams of the City—Melville, Hawthorne, Poe; Displaced Patrician—Henry Adams, Visiting Mind—Henry James; Ambivalent Urbanite—William Dean Howells; Disappointment in New York—Frank Norris, Theodore Dreiser; Pragmatism and Social Work—William James, Jane Addams; Plea for Community—Robert Park, John Dewey; Provincialism and Alienation—Josiah Royce, George Santayana; Architecture Against the City—Frank Lloyd Wright; Legacy of Fear; Outlines of a Tradition; Romanticism Is Not Enough; Ideology, Prejudice, and Reasonable Criticism.

89. Willbern, York, *The Withering Away of the City,* University, Ala. (University of Alabama), 1963, 139 pp. [63-19642]

"To the challenges posed by the new urbanization, [the author] foresees 'no sweeping rational dramatic response.' Urban renewal and mass transit may slow the decline of downtown, but they will neither stem the suburban tide nor extend the city's boundaries. The repeated failures of proposals for metropolitan government underscore the absence of widespread dissatisfaction with the fragmented status quo . . . urban America has met sprawl with diffusion. With 'ad hoc, piecemeal, partial, and pluralistic devices and adjustments,' the concept and reality of self-sufficient local government also has withered away."

90. Wilson, James Q. (Editor), *The Metropolitan Enigma: Inquiries into the Nature and Dimensions of America's "Urban Crisis,"* Washington,

D.C. (Chamber of Commerce of the United States), 1967, 338 pp. [67-21826]

Distribution and Movement of Jobs and Industry (John F. Kain); Urban Transportation (John R. Meyer); Financing Urban Government (Dick Netzer); Pollution and Cities (Roger Revelle); Race and Migration to the American City (Charles Tilly); Housing and National Urban Goals—Old Policies and New Realities (Bernard J. Frieden); Design and Urban Beauty in the Central City (John Burchard); Urban Crime (Marvin E. Wolfgang); Schools in the City (Theodore R. Sizer); Poverty in Cities (Daniel P. Moynihan); Urban Problems in Perspective (James Q. Wilson).

91. Wissink, G. A., *American Cities in Perspective—With Special Reference to the Development of Their Fringe Areas,* Assen, Netherlands (Royal Vangorcum), 1962, 320 pp. Paper. [62-48903]

General Framework: What Is a City?, The City and Its Relation to Surrounding Areas, City Structure and City Growth in the United States, Structure and Growth of Some Non-American Cities; Fringe: Significance, Brief Description, History, Fringe and City Structure, Internal Differentiation, Delineation; Why the Fringe Is As It Is: Point of View, "Rush to the Suburbs," Past in Concrete, Why and How of Low Densities, Why of the Overall Arrangement of Land Uses, Inside Story of Residential Subdivisions, City Unique; Some Important Conclusions and Implications; Bibliography, pp. 309-320.

92. Wycherley, R. E., *How the Greeks Built Cities,* London, England (Macmillan), Second Edition, 1962, 235 pp.

Growth of the Greek City; Greek Town-Planning; Fortifications; Agora; Shrines and Official Buildings; Gymnasium, Stadium and Theatre; Greek Houses; Fountain Buildings.

Bibliographies

93. Bicker, William; Brown, David; Malakoff, Herbert; Gore, William J., *Comparative Urban Development — An Annotated Bibliography,* Papers in Comparative Public Administration, Special Series, No. 5, Washington, D.C. (American Society for Public Administration), 1965, 151 pp. Paper. [65-15606]

94. Financial Executives Institute, *Major Disasters: Policies and Plans,* New York (50 West 44th. St., 10036), 1962, 122 pp. Paperback. Bibliography, pp. 113-122.

Selected References: General—Planning, Industry Publications, Protection of Personnel, Protection of Records, Training; Other References: General, Industry Material; Civil Defense Films; Films, Film Lists.

95. Glaab, Charles N., "The Historian and the American City: A Bibliographic Survey," in: Hauser, Philip M.; Schnore, Leo F. (Editors), *The Study of Urbanization,* pp. 53-80. [65-24223]

Development of Urban History; Conceptual Framework of Urban History; Urban Histories; Conclusions; Notes.

96. Glaab, Charles N.; Brown, A. Theodore, *A History of Urban America*, Further Reading, pp. 309-318.

97. Hall, Peter, *The World Cities*, Bibliography, pp. 246-254.

98. McKelvey, Blake, *The Urbanization of America—1860-1915*, Bibliography, pp. 333-357.

99. Mumford, Lewis, *The City in History*, Bibliography, pp. 579-634.

100. Mumford, Lewis, *The Culture of Cities*, Bibliography, pp. 508-552.

101. Reps, John, *The Making of Urban America, A History of City Planning in the United States*, Bibliography, pp. 545-562.

102. Stewart, Ian R., "Nineteenth Century American Public Landscape Design," Exchange Bibliography No. 68, Council of Planning Librarians, Monticello, Ill. (P.O. Box 229, 61856), February 1969, 20 pp. Mimeo.

103. Tietze, Frederick J.; McKeown, James E. (Editors), *The Changing Metropolis*, Bibliography pp. 198-207.

104. U.S. Department of Housing and Urban Development, Library, *Housing and Planning References*, New Series 17 & 18, Washington, D.C. (Superintendent of Documents), March/April/May/June 1968, 151 pp. Paper.

Selection of publications and articles on housing and planning received during the period.

105. Wissink, G. A., *American Cities In Perspective—With Special Reference to the Development of Their Fringe Areas*, Bibliography, pp. 309-320.

2. PROCESS

Articles, Periodicals

106. Bolan, Richard S., "Emerging Views of Planning," *Journal of the American Institute of Planners*, Vol. XXXIII, No. 4, July 1967, pp. 233-245.

Classical Model Under Attack; Environment for Decision: Historical Environmental Factors, Factors in the Decisionmaking Environment, Factors in the Dynamics of Decisionmaking; Nature of the Public Agenda: Attributes of Decisions to Be Made, Generating Forces Behind Public Issues; Alternative Variations in a Planning

System: Variations in Planning Strategy—Probabilistic Programming, Informal Coordinator-Catalyst, Disjointed Incrementalist, Advocacy and Plural Planning, Adaptive and Contingency Planning; Variations in Planning Methods: Systems Analysis and Simulation, Cost Effectiveness and Program Planning, Quasi-Keynesian Planning, Ad Hoc Opportunism; Alternatives in Planning Content, in Positioning the Planning System; Sythesis—Adaptations and Convergence: Implications for the Planning Process, Research Directions Suggested by the Analysis.

107. Branch, Melville C.; Boelter, L. M. K., "Civil Engineering in City and Regional Planning," *Journal of the Urban Planning and Development Division*, Proceedings of the American Society of Civil Engineers, Vol. 92, No. UP1, Proceedings Paper No. 4827, May 1966, pp. 53-78.

Introduction; City Planning: Present Situation, Development and Trends, Opportunities, Requirements for Participation, Potentialities, Research; Regional Planning: Present Situation, Development and Trends, Opportunities; Conclusions.

108. Dror, Yehezkel, "The Planning Process: a Facet Design," *International Review of Administrative Sciences*, Vol. XXIX, No. 1, 1963, pp. 46-58.

"In this paper an effort is being made to deal with one of the first phases of a systematic study of planning, namely a preliminary concept analysis—or . . . facet design." Introduction; Methodology; Concept of Planning; Facets of Planning.

109. Drucker, Peter F., "Long-Range Planning: Challenge to Management Science," *Management Science*, Vol. V, No. 3, April 1959, pp. 238-249.

"Long-range planning as the organized process of making entrepreneurial decisions . . . what long-range planning is and what it is not; why it is needed; and what is needed to do it."

110. Dyckman, John W., "Social Planning, Social Planners, and Planned Societies, "*Journal of the American Institute of Planners*, Vol XXXII, No. 2, March 1966, pp. 66-76.

Finding Appropriate Remedies—Diagnosis of the Client; Caretakers and Long-Run Client Interests; Social Planning: and Social Leadership, Administrative Efficiency, Radical Reform; Social Planning, Social Science, and Societal Goals.

111. Frieden, Bernard J., "The Changing Prospects for Social Planning," *Journal of the American Institute of Planners*, Vol. XXXIII, No. 5, September 1967, pp. 311-323.

State of the Profession: Social Versus Physical Planning, Planning for Redistribution or Efficiency, Public Versus Private Interests; Changing Political Base: Public Interest in Redistribution, Broader Definition of Urban Problems; Governmental Setting: Local Cures for Slum Problems, Intergovernmental Relations and Metropolitan Planning; Challenge to Professional Development: Increasing the Sensitivity of Physical Planning, Extending the Scope of Planning, Health Service Planning; Unresolved Issues.

112. Institute of Management Sciences, College on Planning, *Management Technology*, Vol. 4, No. 2, December 1964, entire issue, pp. 83-195.

Planning, Anticipating and Managing (Harold F. Smiddy); Inner-Directedness in

Planning (Adrian J. Grossman); Influence of Department of Defense Practices on Corporate Planning (Donald J. Smalter); International Corporate Planning—How Is It Different?; Interrelation between Planning and Acquisition (Franc M. Ricciardi); Acquisitions Planning Panel (William H. Bokum, Neil Kirkpatrick, A. M. McKinnon, Robert D. Stillman); Marketing Planning Panel (Edward P. Gunther, David B. Learner); Research and Development Planning Panel (Carl B. Barnes, Julius Hannock, R. Hufnagel, Robert S. Rose, Jr.).

113. *Journal of the American Institute of Planners,* City Planning in Europe, Vol. XXVIII, No. 4, Special Issue, November 1962.

New Towns Program in Britain (John Madge); Building a City and a Metropolis: Planned Development of Stockholm; Urban Renewal in European Countries (Leo Grebler); Soviet City Planning: Current Issues and Future Perspectives (Robert J. Osborn; Thomas A. Reiner); Planning the City of Socialist Man (Jack C. Fisher); Images of Urban Areas: Their Structure and Psychological Foundations (Derk De Jonge); European Motherland of American Urban Romanticism (John W. Dyckman).

114. Kaufman, Jerome L., "Trends in Planning," *ASPO Newsletter,* Vol. 35, No. 3, March 1969, pp. 49-52. (American Society of Planning Officials)

Decentralization; Reduction of Local Control; Neighborhood Control; Demise of the Master Plan; Planning-Budget Mergers; Conclusion.

115. Lindbloom, Charles E., "The Science of 'Muddling Through,'" *Public Administration Review,* Spring 1968, pp. 79-88.

By Root Or By Branch; Intertwining Evaluation and Empirical Analysis; Relations Between Means and Ends; Test of "Good" Policy; Non-Comprehensive Analysis; Succession of Comparisons; Theorists and Practitioners; Successive Comparison As a System.

116. Mason, Edward S., "The Planning of Development," *Scientific American,* Vol. 209, No. 3, September 1963, pp. 235-236, 238, 240, 243-244.

"Future development will not strictly parallel classical industrial revolutions. Technology, economics and ideology all make it likely that governments will play a central role in directing the process."

117. Mitchell, Robert B. (Editor), Planning and Development in Philadelphia, *Journal of the American Institute of Planners,* Special Issue, Vol. XXVI, No. 3, August 1960, pp. 155-241.

Preface (Robert B. Mitchell); Renaissancemanship (David A. Wallace); Physical Development Plan (Arthur Row); Transportation Planning (John Rannells); Citizen Participation (Aaron Levine); PENJERDEL: A Partnership (John Bodine); Water Resources: Delaware Basin (Walter M. Phillips); Industrial Development Planning (Paul A. Wilhelm); Case Study in Urban Design (Edmund N. Bacon); Capital Programming Process (Robert E. Coughlin; Charles A. Pitts).

118. Mocine, Corwin R., "Interpretation: Urban Physical Planning and the 'New Planning,'" *Journal of the American Institute of Planners,* Vol. XXXII, No. 4, July 1966, pp. 234-241.

Process and Product; Field of Urban Physical Planning; Some Improvements in Urban Physical Planning; Physical Planning as a Component of the Total Planning Process.

119. Perin, Constance, "A Noiseless Secession from the Comprehensive Plan," *Journal of the American Institute of Planners,* Vol. XXXIII, No. 5, September 1967, pp. 336-347.

Introduction; Summary of the Reformulation; Secession from Obsession; Past Construction of Comprehensive Planning; New Form for Comprehensive Planning; Focus on Legislation; Plan as Preamble; Continuous Planning; Community Renewal Programming; Future Demands on City Planning; Conclusion.

120. Roepcke, L. A.; Rafert, W.; Benedict, R. P., "Army Material Command Planning Principles and Philosophy," *IEEE Transactions on Engineering Management,* Vol. EM-15, No. 4, December 1968, pp. 150-178.

R & D Planning Environment: Introduction, Uncertainty in Planning, Problem, General Management Criteria, Role of Planning, Time Frames; R & D Planning Philosophy: General, Planning Phases, Role of Decision Philosophy, Short-Range Planning Philosophy, Long-Range Planning Philosophy; Short- and Long-Range Planning: Comparison, R & D Technical Planning Process; Principles and Elements of Planning: Principles of R & D Plan Formulation, Sources of Task Generation, Correlation of Tasks, Planning and Basic Research; Research and Development Planning Structure: General, Internal Technical Planning Processes, R & D Integrated Technical Plan; Objectives for Technology: General, Role of Objectives, Derivation of Objectives, Analysis Matrix, Types of Warfare, Environment of Operations, Operational Functions, Use of Analysis Matrix, Quantification; Technological Forecasting: General, Forecasts in Planning and Decision Making, Forecast Methodology, Basic Data, Data Interpretation, Uses of Technical Forecasts, Army Long-Range Technological Forecast, AMC Forecast-in-Depth; Epilogue; Bibliography, pp. 176-178.

121. Seeley, John R., "What Is Planning? Definition and Strategy," *Journal of the American Institute of Planners,* Vol. XXVIII, No. 2, May 1962, pp. 91-97.

The Professions of Planning; Planning a Profession; In Clouded Clarity; Planning As Art.

122. Sidenbladh, Göran, "Stockholm, A Planned City," *Scientific American,* Vol. 213, No. 3, September 1965, pp. 107-110, 112, 114, 116, 118; also in: Scientific American (Periodical), *Cities,* New York (Knopf), 1965, 211 pp. Paperback. [65-28177]

"The concept of planning the development of a city came late to most cities. Stockholm is an exception: its growth has been planned since the establishment of a city planning office more than 300 years ago."

123. Tugwell, Rexford G., "The Fourth Power," *Planning and Civic Comment,* April-June 1939, Part II, 31 pp.

An early and historically significant discussion of the importance of directive planning in American government.

124. Wiener, Norbert, "Short-Time and Long-Time Planning," in: American Society of Planning Officials, *Planning 1954*, Chicago, Ill. (1313 East 60th St., 60637), 1954, pp. 4-11. Paperback. [39-3313]

Books, Reports, Pamphlets

125. Branch, Melville C., *Planning: Aspects and Applications*, New York (Wiley), 1966, 333 pp. [66-13523]

Context and Present State of Planning: Overview; Project Planning—Planning Environment for Research and Development: Program Formulation and Planning Process, Physical-Spatial Design, Component Plan and Effectuation; City Planning: Rome and Richmond—A Case Study In Topographical Determinism, Physical Aspects of City Planning, Simulation, Mathematical Models, and Comprehensive City Planning, Toward City Planning of Ocean Environment; Corporate Planning: Corporate Planning Process and Procedure, Psychological Factors in Business Planning, Conceptualization in Business Planning and Decision Making; Military Planning: Speculations on Some Aspects of Planning a United States Intercontinental Ballistic Missile Force; Comprehensive Planning Process: Concerning a Rationale for Comprehensive Planning, Comprehensive Planning As a Field of Study.

126. Chapin, F. Stuart, Jr., *Urban Land Use Planning*, Urbana, Ill. (University of Illinois), Second Edition, 1965, 498 pp. [64-18666]

Land Use Determinants: Land Use Perspectives, Toward a Theory of Urban Growth and Development; Tooling Up for Land Use Planning: Urban Economy, Employment Studies, Population Studies, Study of Urban Activity Systems, Urban Land Studies, Transportation and Land Use; Land Use Planning: Plan and Its Analytical Framework, Location Requirements, Space Requirements, Land Use Plan.

127. Fisher, Jack C. (Editor), *City and Regional Planning in Poland*, Ithaca, New York (Cornell University Press), 1966, 491 pp. [65-23997]

City Planning. History of Urban Development and Planning (Waclaw Ostrowski); Urban Planning Theory: Methods and Results (Boleslaw Malisz); Development of the General Plan of Warsaw (Stanislaw Dziewulski); Main Urban Planning Problems in the Silesian-Krakow Industrial Region (Czeslaw Kotela); City Planning in the Gdansk-Gdynia Conurbation (Wieslaw Gruszkowski); Postwar Housing Development in Poland (Adam Andrzejewski); Sociological Implications of Urban Planning (Janusz Ziolkowski); View of Architectural Theory (Bohdan Pniewski). Regional Planning. Postwar Changes in the Polish Economic and Social Structure (Kazimierz Dziewonski, Stanislaw Leszczcki); Regional Planning in Poland: Theory, Methods, and Results (Jozef Laremba); Regional Planning: in the Upper Silesian Industrial District (Ryszard Szmitke, Tadeuz Zielinski), in the Krakow Voivodship (Jerzy Kruczala), in the Bialystok Voivodship (Boguslaw Welpa); Water Economy in Poland (Aleksander Tuszko); Rural Planning in Poland (Marian Benko); Research Activity of the Committee for Space Economy and Regional Planning (Antoni R. Kuklinski). National Economic Planning. Comments on Economic Planning in Poland (Josef Pajestka); Location Policy and the Regional Efficiency of Investments (Andrzej Wrobel, Stanislaw M. Zawadzki); Long-Term Plan for Polish Expansion, 1961-1980 (Kazimierz Secomski); Role of Science in the Development of Socialist Society with Special Regard to the Science of Economics (Oskar Lange).

128. Ford Foundation and Harvard University, *Design for Pakistan—A Report on Assistance to the Pakistan Planning Commission*, New York (Ford Foundation), February 1965, 35 pp. [65-16714]

Evolution of Planning: Significant Difference—Activation, Test of Performance, Pre-Plan Period, Planning Art, First Plan, Second Plan; Role of the Advisory Group; Planning and Government: Financial Management of Development, Facts and Figures, Location in Government, Planning in the Provinces; Project Results: Staff Training, Advisors' Qualifications, Role of Harvard University, Role of the Foundation; Selected Bibliography.

129. Friend, J. K.; Jessop, W. N., *Local Government & Strategic Choice— An Operational Research Approach to the Processes of Public Planning*, Beverly Hills, Cal. (Sage), 1969. [75-83398]

Appraisal of Planning in a Major Local Authority; Planning—Process of Strategic Choice; Towards a Technology of Strategic Choice; Organizational Challenge.

130. Gallion, Arthur B.; Eisner, Simon, *The Urban Pattern—City Planning and Design*, Princeton, N. J. (Van Nostrand), Second Edition, 1963, 435 pp. [63-24088]

City of the Past; Industrial City; Planning Process; City Today; New Horizons; Bibliography, pp. 401-423.

131. Giedion, Sigfried, *Space, Time and Architecture*, The Growth of a New Tradition, Cambridge, Mass. (Harvard University), Fifth Edition, 1967, 897 pp. [67-17310]

Architecture of the 1960's: Hopes and Fears; History a Part of Life; Our Architectural Inheritance; Evolution of New Potentialities; Demand for Morality in Architecture; American Development; Space-Time in Art, Architecture, and Construction; City Planning in the Nineteenth Century; City Planning as a Human Problem; Space-Time in City Planning; In Conclusion.

132. Goodman, William I.; Freund, Eric C. (Editors), *Principles and Practice of Urban Planning*, Municipal Management Series, Washington, D.C. (International City Managers' Association), Fourth Edition, 1968, 621 pp. [67-30622]

Introduction; Context of Urban Planning: Antecedents of Local Planning, Intergovernmental Context of Urban Planning; Basic Studies for Urban Planning: Population Studies; Economic Studies, Land Use Studies, Transportation Planning, Open Space, Recreation, and Conservation, Governmental and Community Facilities; Special Approaches to Planning: City Design and City Appearance, Quantitative Methods in Urban Planning, Social Welfare Planning; Implementation— Policies, Plans, Programs: Defining Development Objectives, Comprehensive Plan, Programming Community Development; Implementation—Regulation and Renewal: Zoning, Land Subdivision, Urban Renewal; Urban Planning Agency: Local Planning Agency—Organization and Structure, Internal Administration, Planning and the Public; Selected Bibliography, pp. 585-604.

133. Johnson-Marshall, Percy, *Rebuilding Cities,* Chicago (Aldine), 1966, 390 pp. [64-21388]

Largely physical-architectural in approach and content. Introduction; Historical Background of Growth; Components of Planning; Visions and Designs; New Planning Legislation and Techniques; Comprehensive Development in London; Coventry; Rotterdam; City of the Twenty-First Century; Work in Progress.

134. Kaufman, Jerome L. (Study Director), *Planning for Puerto Rico,* Chicago, Ill. (American Society of Planning Officials), December 1968, 150 pp. Paper.

A Planning System for Puerto Rico; Over-all Development Policy; Resource Allocation; Resource Utilization; Underlying Conditions for an Effective Planning System; Land Planning and Controls.

135. Keeble, Lewis, *Principles and Practice of Town and Country Planning,* London (Estates Gazette), Third Edition, 1964, 382 pp. [BNB 64-15631]

Definition, Scope and Operation of Town and Country Planning; Continuity of the Planning Process and the Planner's Skills; Visual Presentation for Planning Purposes; Regional Planning; Urban Regions; Regional Surveys; Preparation of the Regional Plan; Planning of Towns; Local Survey; Preparation of the Town Plan; Space Standards and Detailed Planning Design: Transport System, Town Centre, Industrial Area, Residential Neighborhood Unit, Open Spaces, Large Establishments and Schools, Neighborhood Centres and Sub-Centres, Residential Accommodation, Density and Layout, Village Plan, Country Planning; Development Control and Planning Organization: Development Control, Planning Inquiries, the Planning Machine in Britain; Ministry Notations for Planning Maps, Development and Building Costs, Control of Height and Spacing of Buildings to Secure Adequate Daylight, Typical Conditions and Grounds of Refusal in Connection with Applications for Planning Permission, Relative Sizes of Familiar Objects; Bibliography, pp. 373-376.

136. League of Women Voters of the United States, *Planning in the Community,* Washington, D.C. (1026 17th St., N.W., 20036), 1964, 33 pp.

"Prepared for League members studying local planning programs, this booklet includes one of the best checklists for evaluating programs. . . ."

137. Mayer, Albert, *The Urgent Future—People, Housing, City, Region,* New York (McGraw-Hill), 1967, 184 pp., illus.

Crisis and Opportunity; Public Housing as Key to Community; Urban Renewal as Creative Catalyst?; Underlying Dynamics of Social and Physical Development—Existing, Entrenched—Embryonic, Potential; New Towns and Fresh In-City Communities; Restructuring the City, Sub-City or District, Neighborhood Reconsidered, Decentralization of Excellence; Central City Center—Phenomenon of Giantism, Competition with Housing-Community; From City to City-Region—Metropolitan

Planning Federated Metropolitan Governments; Twentieth-Century Pioneering—Reconstructed Regions and New Regions—Rational City Scales, Regional City; Synthesis and Sublimation—Organic Architecture; Epilogue and Prologue—The Urgent Future.

138. Perloff, Harvey S. (Editor), *Planning and the Urban Community*, Pittsburgh, Penn. (University of Pittsburgh), 1961, 235 pp. [60-16547]

Understanding the Urban Community: Economic Functions and Structure of the Metropolitan Region (Edgar M. Hoover), Evolving Metropolis of New Technology (Richard L. Meier), Culture and Esthetics of the Urban Community (Norman Rice), Sociological Research and Urban Planning (Donald J. Bogue), Note on "Community" (Perry L. Norton); Metropolitan Politics and Organization for Planning: Politics of the Metropolis (Frederick Gutheim), Planning Organization and Activities Within the Framework of Urban Government (Henry Fagin), General Planning and Planning for Services and Facilities (James A. Norton); Nature of Planning and Planning Education: City Planning as a Social Movement, a Governmental Function, and a Technical Profession (John T. Howard), Urban Design and City Planning (G. Holmes Perkins), Decision Making and Planning (Herbert Simon), Comprehensive Planning as a Field of Study (Melville C. Branch), Planning Education and Research—A Summary Statement (Harvey S. Perloff).

139. Scott, Brian, W., *Long-Range Planning in American Industry*, New York (American Management Association), 1965, 288 pp. [65-16484]

Long-Range Planning—Introduction and Background: Nature of Business Planning, Meaning of Long-Range Planning, Historical Evolution of Long-Range Planning; Mission of Strategic Long-Range Planning: Some Basic Tenets . . ., Corporate Self-Appraisal, Establishing Objectives, Assumptions About the Future, Anticipation of Technological Change, Choice of Strategy; Organization for Long-Range Planning: Role of Top Management . . ., Staff Planning Unit; Conclusion: Implications for Top Management; Appendix; Bibliography, pp. 265-279.

140. Smith, Herbert H., *The Citizen's Guide to Planning*, West Trenton, N.J. (Chandler-Davis), 1961, 106 pp. Paper.

Reasons for Planning; Development of Planning; Function of Planning; Planning Board; Master Plan; Relationship of Zoning to Planning; Regulation of Land Subdivision; Capital Improvements Program; Care and Feeding of Planning Professionals; Planning and the School Board; Planning and the Urban Renewal Program; Community Relations Problem; Conducting Public Hearings; Planning Action and the Citizen. Appendixes: Suggested By-Laws for Planning Boards, Sources of Information, Suggested References.

141. Town and Country Planning Association, *The Intelligent Voter's Guide to Town and Country Planning*, London, England (Planning Bookshop), 1964, 39 pp.

"Aimed directly at the public, this booklet sets out to strengthen public understanding of the issues involved in town and country planning. The views of the three major parties—some in the form of party statements and others simply extracts from the speeches of party leaders—are given after the Association's general summary and basic recommendations in the areas of: general planning aims and machinery; land use and development; new towns and regional expansion; urban renewal; traffic and transport; and green belts, national parks and rural planning."

142. Waterston, Albert, *Development Planning—Lessons of Experience*, Baltimore, Md. (Johns Hopkins), 1965, 706 pp. [65-26180]

Development Planning Process: Many Meanings of Planning, Spread of Development Planning, Stages of Development Planning, Development Plans, Basic Data for Planning, Budget's Role in Planning, Administrative Obstacles to Planning, Implementation of Plans; Organization of Planning: Planning Machinery Priorities, Distribution of Planning Functions, Central Planning Agency—Function and Role, Location, Types, Organization; Subnational Regional and Local Planning Bodies; Programming Units in Operating Organizations; Appendices.

143. Webster, Donald H., *Urban Planning and Municipal Public Policy*, New York (Harper & Row), 1958, 572 pp. [58-5111]

Governmental Framework of Planning: Planning Within the Framework of American Government, Government at the Local Level, Planning Agency and Planning Administration; Subject Matter of Planning: Physical Planning and Community Development, Comprehensive Planning of Municipal Services and Programs; Means of Implementation: Plan Implementation, Public Improvements Program and Financial Planning, Zoning, Subdivision of Land and Platting, Urban Redevelopment and Renewal; Future of Planning: Guidelines to Planning Progress.

Bibliographies

144. Gallion, Arthur B.; Eisner, Simon, *The Urban Pattern—City Planning and Design*, Bibliography, pp. 401-423.

145. Keeble, Lewis, *Principles and Practice of Town and Country Planning*, Bibliography, pp. 373-376.

146. Scott, Brian W., *Long-Range Planning in American Industry*, New York (American Management Association), 1965, Bibliography, pp. 265-279. [65-16484]

147. Violi, Louis C., *State Planning in the United States*, Council of Planning Librarians Exchange Bibliography 71, Monticello, Ill. (P.O. Box 229, 61856), March 1969, 27 pp. Paper.

Introduction; 1920-1932; 1933-1941; 1942-1959; 1960-1968; List of State Planning Agencies.

3. THEORY

Articles, Periodicals

148. Banfield, Edward C., "Ends and Means in Planning," *UNESCO International Social Science Journal*, Vol. XI, No. 3, 1959, pp. 361-368.

Develops "common conceptions to provide a simple theory of planning, one which is essentially a definition." Discusses "the argument that the procedures of organizations do not in fact even roughly approximate to those described in the theoretical model. . . ." The author then considers "the question of why it is that organizations do so little planning and rational decision making."

149. Blumenfeld, Hans, "Form and Function in Urban Communities," *Journal of the American Society of Architectural Historians,* Vol. 3, No. 1-2, January-April 1943, pp. 11-21.

What IS a City?; Planning from Inside Out and from Outside In; Vicissitudes of the Rectangular Pattern; Vicissitudes of the Concentric Plan; Planners Are Always Late; Yesterday and Tomorrow.

150. Blumenfeld, Hans, "The Modern Metropolis," *Scientific American,* Vol. 213, No. 3, September 1965, pp. 64-74; also in: Scientific American (Periodical), *Cities,* New York (Knopf), 1965, 211 pp. Paperback. [65-28177]

"The urban revolution that began in the latter half of the 19th century has culminated in a qualitatively new kind of human settlement: an extended urban area with a dense central city."

151. Carrothers, Gerald A. P., "An Historical Review of the Gravity and Potential Concepts of Human Interaction," *Journal of the American Institute of Planners,* Vol. XXII, No. 2, Spring 1956, pp. 94-102.

Introduction; Basic Concept of Interaction; Early Formulations of the Concept; Formalization of the Concepts; Mapping Measures of Gravity and Potential; Modification of the Distance Factor; Modification of the Population Factor; Adaptations of the Basic Concepts; Conclusions. Selected Bibliography, pp. 100-102.

152. Chapin, F. Stuart, Jr., "Activity Systems and Urban Structure—A Working Schema," *Journal of the American Institute of Planners,* Vol. XXXIV, No. 1, January 1968, pp. 11-18.

"With a focus on households as users of city space, activity systems of urban residents are seen both to shape and to be shaped by the spatial organization of the metropolitan area . . . the outcome of the optimization process in which choice is made on the basis of a suboptimal combination of satisfactions with each activity in the routine selected with respect to security, achievement, status, and other needs appropriate to that stage of the life cycle and income level."

153. Chapin, F. Stuart, Jr.; Hightower, Henry C., "Household Activity Patterns and Land Use," *Journal of the American Institute of Planners,* Vol. XXXI, No. 3, pp. 222-231.

"Most land uses and transportation routes exist not for their own sake, but because they are means for accomplishment of some desired activity. Whereas movements are the key component of transportation planning, *activities* are fundamental to land use planning and constitute the link between the city and the people. Study of activity patterns which are recurrent in time and space may ultimately yield better theoretical explanations for the urbanist and improved predictive tools for the planner. Some problems and potentialities of activity analysis are explored using data from interviews in two census tracts of a small city."

154. Creese, Walter, "The Planning Theories of Sir Raymond Unwin, 1863-1940," *Journal of the American Institute of Planners,* Vol. XXX, No. 4, November 1964, pp. 295-304.

Influence of World War I; Recurring Themes; Tradition and Creativity.

155. Dakin, John, "An Evaluation of the 'Choice' Theory of Planning," *Journal of the American Institute of Planners,* Vol. XXIX, No. 1, February 1963, pp. 19-27. Also: Davidoff, Paul; Reiner, Thomas A., "A Reply to Dakin," pp. 27-28.

"The core of the theory is that the 'exercise of choice is its (planning) characteristic intellectual act' . . . examines this theory of planning by stating some requirements any theory of planning must meet and testing the 'choice' theory against them." Total Planning Process Is to Be Included; It Must Not Depend on a Particular Political Ideology or System; Theory Must Be Related to Certain Physical and Social Theory; Must Account for Non-Rational Elements; Unclassified Comments: Composite Theory, Ideal-Real Theory, Process Theory; Conclusions.

156. Davidoff, Paul; Reiner, Thomas A., "A Choice Theory of Planning," *Journal of the American Institute of Planners,* Vol. XXVIII, No. 2, May 1962, pp. 103-115. Dakin, John, "An Evaluation of the 'Choice' Theory of Planning," *ibid.,* Vol. XXIX, No. 1, February 1963, pp. 19-27; Davidoff, Paul; Reiner, Thomas A., "A Reply to Dakin," *ibid.,* pp. 27-28.

Planning Defined; Environment Surrounding Planning; Planning's Purposes; Planning Characteristics; Planning Process; Value Formulation: Fact and Value, Responsibility, Clients, Analysis of Values, Evaluation of Values; Time Perspective of Plans; Means Identification; Effectuation; Conclusions; Bibliography, p. 115.

157. Doxiadis, C. A., "Man's Movement and His City—Cities Are Systems Created By Man's Need and Ability to Move," *Science,* Vol. 162, No. 3851, 18 October 1968, pp. 326-334.

Before the Era of Cities; Cities; Capital Cities; Dynamic Cities; Present Situation; Future; Goals; Practical Steps.

158. Duncan, Beverly; Sabagh, Georges; Van Arsdol, Maurice D., Jr., "Patterns of City Growth," *The American Journal of Sociology,* Vol. LXVII, No. 4, January 1962, pp. 418-429.

"Residential construction is proposed as the mechanism whereby city-wide population growth is translated into population redistribution from mature to relatively undeveloped areas . . . the rapidity of redistribution varies directly with both the rate of population growth and the ratio of new dwellings to incremental population."

159. Durden, Dennis; Marble, Duane F., "The Role of Theory in CBD Planning," *Journal of the American Institute of Planners,* Special Issue:

Planning the City's Center, Vol. XXVII, No. 1, February 1961, pp. 10-16.

Need for Theory; Theory in Professional Planning Practice; Excursions Overseas and Into the Past; Concern for Traditional American Forms; Geometers; Social Mechanics; Consequences of Inadequate Perspective; Project Approach to CBD Planning; Conflicting Project Objectives; Lack of an Adequate Regional Context; Low Planning Efficiencies; Applicable Contemporary Social Science Theory; Location Theory as a Specific Case; Some Areas of Theoretical Work; CBD and the Region; Internal Structure of the CBD; Interregional Variations; Limitations on the Utility of CBD Theory.

160. Dyckman, John W., "The Changing Uses of the City," *Daedalus*, Vol. 90, No. 1, Winter 1961, pp. 111-131.

Introduction: Forces Influencing the Changing City; Late Industrial Technology and the Physical City; Technology and Population Distribution; Centralization, Decentralization, and Density; Citizens of a Changing City; Alienation in the City; Leveling of Urban Culture; Autonomous Growth of Organization; Educative City; Open City; Conclusion.

161. Dyckman, John W., "Planning and Decision Theory," *Journal of the American Institute of Planners*, Vol. XXVII, No. 4, November 1961, pp. 335-345.

Rational Action and Economics; Extended Rationality and Strategy; Strategy of Scientific Inquiry; Condition of the Decision Maker; Time and Decision; Requirements of Social Choice: The End of the Plan?; Turning Point in the Literature; Outside of Efficiency; Burden of Planning Decisions, Traditions of Public Decision; Efficiency and Planning Decision; Special Stance of the Planner; Selected Bibliography, pp. 343-345.

162. Forrester, Jay W., "Industrial Dynamics: A Major Breakthrough for Decision Makers," *Harvard Business Review*, Vol. 36, No. 4, July-August 1958, pp. 37-66.

Toward a Theory; Tools of Progress; Production and Distribution; Improving Control; Effect of Advertising; Complete Analysis; Shape of the Future. Explanatory Charts.

163. Friedmann, John, "A Conceptual Model for the Analysis of Planning Behavior," *Administrative Science Quarterly*, Vol. 12, No. 2, pp. 225-252.

"A conceptual model is presented and hypotheses are derived as a means for ordering the data of empirical research into planning processes. Four modes of planning are distinguished: developmental and adaptive, allocative and innovation . . . forms of thought relevant to planning, institutions for political guidance and conflict resolution, and types of implementation procedures are discussed in terms of their proper level and position within a comprehensive system."

164. Guttenberg, Albert Z., "Urban Structure and Urban Growth," *Journal of the American Institute of Planners*, Vol. XXVI, No. 2, May 1960, pp. 104-110.

Urban Structure: Elements, Structure and Form, Hierarchical Aspects, Whole and Part, Density; Urban Structure and Urban Growth: Spatial Extent of a Community in Relation to Its Structure, Growth and Its Consequences; Translocation: Meaning, Size and Density Changes.

165. Haight, Frank A., "The Future of the Traffic Flow Theory," *Traffic Quarterly*, Vol. 17, No. 4, pp. 516-527.

Need for Statistical Sophistication; Bridge from Theory to Engineering; Traffic Flow Theory and Applied Mathematics; Suggestions for Useful General Research Areas.

166. Handler, A. Benjamin, "What Is Planning Theory?", *Journal of the American Institute of Planners*, Vol. XXIII, No. 3, 1957, pp. 144-150.

Reviews some of the thoughts about planning theory of Walter Isard, Hans Blumenfeld, Douglas Carroll, Robert Mitchell, and Harvey Perloff. Author proposes a broad concept of "capital" and "efficiency criteria" as a basis for the development of a theory of planning, and as a unifying concept in planning education.

167. Harris, Britton, "Some Problems in the Theory of Intra-Urban Location," *Operations Research*, Vol. 9, No. 5, September-October 1961, pp. 695-721.

Theory and Practice; Theories of Urban Form and Intra-Urban Location: Design of Urban Form, Description of Urban Form, Explanation of Urban Form, Analogy with Inter-Regional Models, Differences in Time Scales, Role of Land, Role of Capital, Household Sector, External Economies and Diseconomies of Scale, Summary; Classes of Units and Their Interaction: Analyzing the Environment, Analysis of Decision Behavior; Structure of Models: Simulation in Models, Stochastic, Probabilistic, and Deterministic Models, Static and Dynamic Models, A Possible Model; Some Technical Considerations: Sources of Data, Computer Technology, Systems Approach; Conclusions.

168. Herbert, Gilbert, "The Organic Analogy in Town Planning," *Journal of the American Institute of Planners*, Vol. XXIX, No. 3, August 1963, pp. 198-209.

"This paper is concerned with postulating the validity of the organic concept; with attempting a more precise definition of the term; and with examining the organic connotations of some of the more significant town planning theories in the light of the concept as here defined." Philosophy of Organism; Organic Analogies in Town Planning: Cosmological, Nature, Systemic, Ecological, Cellular; Toward an Organic Town Planning Theory.

169. Hill, Forest G., "Wesley Mitchell's Theory of Planning," *Political Science Quarterly*, Vol. LXXII, No. 1, March 1957, pp. 100-118.

Introduction; Theoretical Framework; Institutional and Psychological Analysis; The Planning Process; Welfare Analysis; Conclusion.

170. Koontz, Harold, "Making Sense of Management Theory," *Harvard Business Review*, Vol. 40, No. 4, July-August 1962, pp. 24-26, 31-32, 34, 36, 38, 40, 42, 46.

". . . scholars and practitioners alike have succeeded for the most part only in muddying the waters still further . . . The Result? A jungle of confusion and conflict . . . a distinguished authority on management, conquers confusion by deft division."

171. Lessinger, Jack, "The Case For Scatteration—Some Reflections on the National Capital Region *Plan for the Year 2000*," *Journal of the American Institute of Planners*, Vol. XXVIII, No. 3, August 1962, pp. 159-169.

Compaction and Scatter Defined; Model 1—Compact Radial Corridors; Model 2—Corridors with Scattered Development; Compaction and Scatter in the Washington, D.C., Region; Compaction-Scatter and Residential Slums; Compaction Enforces Substandard Concentration; Implications of Scatter and Compaction; Conclusions.

172. Lynch, Kevin, "The Pattern of the Metropolis," *Daedalus*, Vol. 90, No. 1, Winter 1961, pp. 79-98.

Critical Aspects of Urban Form; Dispersed Sheet; Galaxy of Settlements; Core City; Urban Star; Ring; Objectives of Metropolitan Arrangement; Relation of Forms to Goals; Dynamic and Complex Forms.

173. Lynch, Kevin; Rodwin, Lloyd, "A Theory of Urban Form," *Journal of the American Institute of Planners*, Vol. XXIV, No. 4, 1958, pp. 201-214.

Possible Analytical Approaches; Criteria for Analytical Categories of Urban Form; Proposed Analytical System; Example of the Analytical System; Problems of Goal Formulation; Criteria for the Choice of Goals; Goal Form Interaction; Complex Form and Goal Relationships; Evaluation.

174. Maccoby, Michael, "The Social Psychology of Utopia: The Writings of Paul Goodman," *Journal of the American Institute of Planners*, Vol. XXVIII, No. 4, November 1962, pp. 282-292.

American Society; Utopia vs. Old Frontier; Three Alternatives; Man or Goodman; Social and Human Change; Conclusion: Decline of Utopian Thought.

175. Odum, Eugene P., "The Strategy of Ecosystem Development," *Science*, Vol. 164, No. 3877, 18 April 1969, pp. 262-270.

"An understanding of ecological succession provides a basis for resolving man's conflict with nature."

Definition of Succession; Bioenergetics of Ecosystem Development; Comparison of Succession in a Laboratory Microcosm and a Forest; Food Chains and Food Webs; Diversity and Succession; Nutrient Cycling; Selection Pressure—Quantity versus Quality; Overall Homeostasis; Relevance of Ecosystem Development Theory to Human Ecology; Pulse Stability; Prospects for a Detritus Agriculture; Compartment Model.

176. Ostrom, Vincent; Tiebout, Charles M.; Warren, Robert, "The Organization of Government in Metropolitan Areas: A Theoretical Inquiry," *American Political Science Review*, Vol. LV, No. 4, December 1961, pp. 831-842.

Nature of Public Goods and Services; Scale Problems in Public Organization; Public Organization in Gargantua; Public Organization in a Polycentric Political System.

177. Riesman, David, "Some Observations on Community Plans and Utopia," *The Yale Law Journal*, December 1947, pp. 172-200.

"A revival of the tradition of utopian thinking [is] one of the important intellectual tasks. . . . Since we live in a time of disenchantment, such thinking, where it is rational in aim and method and not mere escapism, is not easy; it is easier to concentrate on programs for choosing among lesser evils, even to the point where these evils can scarcely be distinguished, one from the other."

178. Rydell, C. Peter; Schwarz, Gretchen, "Air Pollution and Urban Form —A Review of Current Literature," Review Article, *Journal of the American Institute of Planners*, Vol. XXXIV, No. 2, March 1968, pp. 115-120.

Designing Buildings and Streets; Locating Industrial Sites; Transporting People and Goods; Allocating Open Space; Land Use Planning; Air Use Planning.

179. Spencer, Milton H., "Uncertainty, Expectations, and Foundations of the Theory of Planning," *Journal of the Academy of Management*, Vol. 5, No. 3, December 1965, pp. 197-206.

"The purpose . . . is to outline some fundamental theoretical concepts which form an important part of the basis of any general theory of planning."

180. Tan, T., "Road Networks In an Expanding City," *Operations Research*, Vol. 14, No. 4, July-August 1966, pp. 607-613.

"Effect of growth in the number of persons living in the suburbs of an idealized circular city and working in a central business district of constant size is investigated in the case of five alternative basic road systems. . . . The total city size, relative to the size of its central business district, can increase only up to a critical value beyond which a major change in the basic road network is necessary to ensure efficient travel. Critical values for the ratio of the density of work places to homes are then obtained. The criteria used to evaluate the road networks are distance traveled, road space required, and travel time."

181. Weintraub, Sidney, "Theoretical Economics," in: Mitchell, Robert B. (Special Editor), Urban Revival: Goals and Standards, *Annals of the American Academy of Political and Social Science*, Vol. 352, March 1964, pp. 152-164.

". . . recent trends in theoretical economics . . . survey of the literature . . ."
Origins of Modern Theory; Microeconomics: Revealed Preferences, Hick's Revision, Numerical Utility, Theory of the Firm, Market Classifications and Doctrines of Competition, Duopoly and Games, Multiple-Product Firms, Expectations and Decision-Making; General Equilibrium; Macroeconomics: Capital and Growth, Inflation and Growth, Monetary Theory, Theory of Interest and Distribution, Welfare Economics; Concluding Note.

182. Weiss, Shirley F.; Kaiser, Edward J., "A Quantitative Evaluation of

Major Factors Influencing Urban Land Development in a Regional Cluster," *Traffic Quarterly*, Vol. XXII, No. 1, January 1968, pp. 109-115.

Relationship With Previous Analyses; Selection of Variables for Explanation and Measurement.

183.   Wilson, James Q., "An Overview of Theories of Planned Change," in: Morris, Robert (Editor), *Centrally Planned Change: Prospects and Concepts*, New York (National Association of Social Workers), 1964, 149 pp. [64-25690]

Definition; Does Centrally Planned Change Exist?; Conditions for Its Existence or Absence; Circumstances Permitting Change; What of the Future?

Books, Reports, Pamphlets

184.   Alexander, Christopher, *Notes on the Synthesis of Form*, Cambridge, Mass. (Harvard University), 1964, 216 pp. [64-13417]

Introduction: Need for Rationality; Goodness of Fit, Source of Good Fit, Unself-conscious Process, Selfconscious Process; Program, Realization of the Program, Definitions, Solution; Epilogue; Appendix 1—Worked Example; Appendix 2—Mathematical Treatment of Decomposition.

185.   Alonso, William, *Location and Land Use: Toward a General Theory of Land Rent*, Cambridge, Mass. (Harvard University), 1964, 204 pp. [63-17193]

". . . concerns the relation of land values to land uses in the city. The major focus is on residential land. Conclusions differ from those based on agricultural rent theories. Relationships in urban areas are more complex, involving variations in size of site, tastes, income, profits and other costs. While the analysis is largely mathematical, the author has attempted to present his theories in words and diagrams for readers without a mathematical background."

186.   Bellamy, Edward, *Looking Backward*, New York (New American Library), 1960, 222 pp. Paperback.

". . . novel about a young Boston gentleman . . . mysteriously transported from the nineteenth to the twenty-first century—from a world of war and want to a world of peace and plenty. Translated into more than twenty languages and the most widely read novel of its time . . . it is a blueprint of the 'perfect society,' a guidebook that stimulated some of the greatest thinkers of our age."

187.   Bertalanffy, Ludwig Von; Rapoport, Anatol (Editors), *General Systems*, Yearbook of the Society for the Advancement of General Systems Theory, Vol. 1, Ann Arbor, Mich. (University of Michigan, Mental Health Research Institute), 1956, 162 pp.

Introduction: General System Theory (Ludwig von Bertalanffy), General System Theory—The Skeleton of Science (Kenneth Boulding), Definition of System (A. D. Hall, R. E. Fagen); Exploration of Mathematical Models: Mathematical Models in the Social Sciences (Kenneth J. Arrow), The Diffusion Problem in Mass Behavior

(Anatol Rapoport), Behavior—Imbalance in a Network of Chemical Transformations (D. F. Bradley, M. Calvin), Toward a General Theory of Growth (Kenneth Boulding), Principle of Allometry in Biology and the Social Sciences (Raoul S. Naroll, Ludwig von Bertalanffy), On the Parallel between Learning and Evolution (J. W. S. Pringle), In Search of New Foci of Integration: The Hypothesis of Cybernetics (J. O. Wisdom), Topology and Life—In Search of General Mathematical Principles in Biology and Sociology (N. Rashevsky), Dynamic Systems, Psychological Fields, and Hypothetical Constructs (David Krech), Dynamic Systems of Open Neurological Systems (David Krech), Biologist's View of Society (R. W. Gerard), Rights of Man—Biological Approach (R. W. Gerard).

188. Beshers, James M., *Urban Social Structure,* New York (Free Press), 1962, 207 pp. [62-11844]

". . . extension of present social theory is needed to cope with urban facts . . . social classes, ethnic groups and racial groups are integral parts of American urban social structure that cannot be meaningfully treated in isolation of each other . . . historical research, census tract research, survey research, and observational research must be integrated in an effective analysis of urban social structure . . . rigorous development of theory, specifically mathematical formulation, can clarify issues presently confused."

Values, Policy, and Urban Sociology; Ecology and Functionalism; Urban Social Structure; Historical Processes; Social Structure and Residential Areas; Consequences of Spatial Distribution; Theoretical Model for Urban Social Structure; Urban Social Organization; Appendix—Mathematical Aspects.

189. Carver, Humphrey, *Cities in the Suburbs,* Toronto, Canada (University of Toronto), 1962, 120 pp.

"Can something be done to lend variety, color and meaning to our spreadout regional cities? . . . The author's search for the answers includes a lively examination of cities, their changing nature, and the idea of leading planners to what they should be. His own proposal . . . a 'Town Centre' that gives focus to the social, political and cultural life of the suburbanite. . . . Because local governments in the fringe areas are traditionally weak . . . there must be a special public body with powers to buy the community land, to make plans for the Town Centre, and to start building its structures and preserving its open space."

190. Chapin, F. Stuart, Jr.; Weiss, Shirley F. (Editors), *Urban Growth Dynamics in a Regional Cluster of Cities,* New York (Wiley), 1962, 484 pp. [62-19867]

Introduction (F. Stuart Chapin, Jr.); Economic Orientations to Urbanization (Lowell D. Ashby); Patterns of Economic Interaction in the Crescent (Ralph W. Pfouts); Labor Mobility in the Crescent (Robert L. Bunting); Industrial Development Trends and Economic Potential (Lowell D. Ashby); Agricultural Adjustment to Urban Growth (William N. Parker; David G. Davies); Leadership, Decision Making, and Urban Growth (Frederic N. Cleaveland); Political Decision Making in Arcadia (Benjamin Walter); Roles of the Planner in Urban Development (Robert T. Doland; John A. Parker); Roles of Top Business Executives in Urban Development (E. William Noland); Negro Political Participation in Two Piedmont Crescent Cities (Bradbury Seasholes; Frederic N. Cleaveland); Social Correlates of Urban

Growth and Development (John Gulick); Newcomer, Enculturation in the City, Attitudes and Participation (John Gulick; Charles E. Bowerman; Kurt W. Back); Livability of the City, Attitudes and Urban Development (Robert L. Wilson); Occupational Careers and Mobility (Richard L. Simpson); Patterns of Urban Development (F. Stuart Chapin, Jr.); Land Development Patterns and Growth Alternatives (F. Stuart Chapin, Jr.; Shirley F. Weiss); Policy Implications of Research Findings (F. Stuart Chapin, Jr.).

191. Christaller, Walter, *Central Places in Southern Germany*, Englewood Cliffs, N. J. (Prentice-Hall), 1966, 230 pp. [66-14747]

"Although conditions have changed and the specific results of Christaller's study are now somewhat out of date, his contribution to the method and procedures for investigating the organized relationships and functions of an economic system from its spatial point of view has not diminished."

Introduction; Economic-Theoretical Foundations of Town Geography: Fundamental Meanings, Static Relations, Dynamic Processes, Results; Application of the Theory of Location to the Actual Geography of Settlements: Method of Determination of Central Places, Preliminary Results; Number, Sizes, and Distribution of the Central Places in Southern Germany: L-System of Munich; Conclusion: Verification of the Theory, Methodological Results of the Geography of Settlements; Appendix; Bibliography, 227-230.

192. Doxiadis, Constantinos A., *Ekistics—An Introduction to the Science of Human Settlements*, New York (Oxford), 1968, 527 pp. [68-1162]

Introduction; The Subjects and Their Study; Facts; Theory; Action; Epilogue.

193. Forrester, Jay W., *Urban Dynamics*, Cambridge, Mass. (M.I.T.), 1968, 256 pp.

". . . models (or theory) selectively simulate processes that affect the life cycle of the city: growth equilibrium, decay, and renewal . . . include variables such as different kinds and levels of employment, housing, and industry."

194. Gerlough, D. L.; Capelle, D. G. (Editors), *An Introduction to Traffic Flow Theory*, Highway Research Board Special Report 79, Washington, D. C. (Highway Research Board), 1964, 149 pp. Paper.

Not for the mathematically uninitiated.

Hydrodynamic Approaches (L. A. Pipes; M. J. Lighthill; G. B. Whitham); Car Following and Acceleration Noise (E. W. Montroll; R. B. Potts); Queueing Theory Approaches (D. E. Cleveland; D. G. Capelle); Simulation of Traffic Flow (D. L. Gerlough); Some Experiments and Applications (R. S. Foote); Bibliography on Theory of Traffic Flow and Related Subjects, pp. 139-145.

195. Gutkind, E. A., *The Twilight of Cities*, New York (Free Press of Glencoe), 1962, 211 pp. [62-15341]

"[The author's] main thesis is that there is no longer need for the great concentration of human beings and their activities in the centers of metropolises that we now have, let alone in the megalopolises that we are building. He calls for urban settlement on a 'general scattering out' plan, which, to judge from his sketches, would result in a giant checkerboard with squares of urban development alternating with squares of park, forest, or agriculture. . . . Being a designer, Gutkind is convinced of the primacy of design, and would assign an infallibility to planning that is not widely accepted."

196. Howard, Ebenezer, *Garden Cities of To-Morrow,* Cambridge, Mass. (M.I.T.), 1965, 168 pp. Paperback. [65-10521]

". . . The book holds a unique place in town planning literature, is cited in all planning bibliographies, stands on the shelves of the more important libraries, and is alluded to in most books on planning; yet most of the popular writers on planning do not seem to have read it—or if they have read it, to remember what it says."
Town-Country Magnet; Revenue of Garden City, and How It Is Obtained—The Agricultural Estate; Revenue of Garden City: Town Estate, General Observations on Its Expenditure; Further Details of Expenditure on Garden City; Administration; Semi-Municipal Enterprise, Local Option, Temperance Reform; Pro-Municipal Work; Some Difficulties Considered; Unique Combination of Proposals; Path Followed Up; Social Cities; Future of London; Select Book List, pp. 160-163.

197. Isard, Walter, *Location and Space-Economy,* A General Theory Relating to Industrial Location, Market Areas, Land Use, Trade, and Urban Structure, Regional Science Studies Series 1, Cambridge, Mass. (M.I.T.), 1956, 350 pp. [56-11026]

Introduction—Posing the Location and Regional Problem; Some General Theories of Location and Space Economy; Some Empirical Regularities of the Space Economy; Transport Inputs and Related Spatial Concepts; Locational Equilibrium of the Firm: Transport, Orientation, Labor and Other Orientation; Market and Supply Area Analysis and Competitive Locational Equilibrium; Agglomeration Analysis and Agricultural Location Theory; Some Basic Interrelations of Location and Trade Theory; Aspects of General Location Theory—Mathematical Formulation; Partial Graphic Synthesis and Summary.

198. Le Breton, Preston P.; Henning, Dale A., *Planning Theory,* Englewood Cliffs, N. J. (Prentice-Hall), 1961, 359 pp. [61-6610]

Not a theoretical construct in the strict or scientific sense of the term.
Role of Planning; Dimensions of a Plan; Theory of: Need Determination, Choice, Data Collection and Processing, Testing; Organization For Planning—Role of: Top-Level Units, Committees, Specialists, Communication Theory in Planning, Persuasion Theory in Planning; Summary and Conclusions.

199. MacKaye, Benton, *The New Exploration: A Philosophy of Regional Planning,* Urbana, Ill. (University of Illinois), 1962, 243 pp. [62-17516]

"Some of the data and illustrations [of this essay, first appearing in 1928] are no longer applicable, but the philosophy and all the salient points of the arguments for regional planning are as fresh as they ever were. Lewis Mumford's delightful introduction tells much about the life of [the author]."

200. Maier, Henry W., *Challenge to the Cities, An Approach to the Theory of Urban Leadership,* New York (Random House), 1966, 210 pp. Paperback. [66-19852]

Background and Backdrop: Background City, Gap in Research; D-Stepp Formula; S for Strategy, T for Tactics; S for Strategy, T for Tactics—Community-Wide Programs: Forming the Social Development Commission, Establishing the Milwaukee Department of City Development, Initiating the Community Renewal Program; S for Strategy, T for Tactics—Programs within the Mayor's Office: Division of Economic Development, Tactics of the Executive Budget Message; Enrollment and Power: Discussion of Power, Case of the Letters of Resignation, Enrollment, Milwaukee Idea; Nondeferable Decision: 72nd Street Library Case, Brown Deer Case; Philosophy of Local Leadership: Further Definition of Institutional Leadership, Involvement with Strategy, Involvement with Tactics—Relationship with Groups, Enrollment—People a Mayor Depends On, Involvement with Power, Conclusion. Bibliography, pp. 193-194.

201. Mairet, Philip, Pioneer of Sociology: The Life and Letters of Patrick Geddes, London, England (Lund Humphries), 1957, 226 pp. [57-59160]

"The ideas of one of the most original thinkers in the modern planning movement."

202. Meier, Richard L., *A Communications Theory of Urban Growth*, Cambridge, Mass. (M.I.T.), 1962, 184 pp. [62-20480]

"The author examines in detail the relation between urban growth and increased communication capacity, the effects of imbalance, and regulatory possibilities."
Urban Communications Systems; The Civic Bond; Uses of Time and Space; Transaction Capacity; Public Controls for Communications; Socio-Cultural Growth; Information Flows and Human Channel Capacity; Conservation Principles; Growth in Newly Developing Areas; Priorities for Further Study.

203. Olsson, Gunnar, *Distance and Human Interaction: A Review and Bibliography*, Philadelphia, Pa. (Regional Science Research Institute), 1965, 112 pp.

"Both concepts suggested by the title have received extensive treatment in regional science literature. . . . In reviewing much of this material, the author suggests some possibilities for connecting central place theory with spatial models of migration and diffusion."

204. Planning Research Corporation, *Urban Form As a Passive Defense Variable*, PRC D-1341, Los Angeles, Cal. (1100 Glendon Ave., 90024), 16 January 1967, 36 pp. Paper.

". . . summarizes three studies conducted for the Office of Civil Defense."
Civil Defense Implications of Urban Configurations; Economic Feasibility of Decentralized Metropolitan Regions; Implementation of the Ordered Sprawl Configuration; Perspectives.

205. Pred, Allen, *Behavior and Location—Foundations for a Geographic and Dynamic Location Theory*, Part I, Lund, Sweden (Gleerup), 1967, 128 pp. Paper.

Introduction; Behavioral Matrix; Real-World Deviations from Economic Location Theory; Literature Cited, pp. 122-128.

206. Reiner, Thomas A., *The Place of the Ideal Community in Urban Planning*, First Volume of City Planning Series, Philadelphia, Penn. (University of Pennsylvania), 1963, 194 pp. [61-11264]

Introduction; Physical Planning Principles in Ideal Communities; Prospect of Ideal Communities [examples]; Physical Planning Content of Ideal Communities; Summary Conclusions; Bibliography, pp. 175-192.

207. Reissman, Leonard, *The Urban Process—Cities in Industrial Societies*, New York (Free Press), 1964, 255 pp. [64-20301]

Urban Achievement; Typology of Urban Studies; Visionary—Planner for Urban Utopia; Empiricists—Classifiers of Cities; Ecologists—Analysts of Urban Patterns; Theoreticians—Developers of the Urban Concept; Scope of an Urban Theory; City in Industrial Society.

208. Saarinen, Eliel, *The City—Its Growth, Its Decay, Its Future*, Cambridge, Mass. (M.I.T.), 1965, 385 pp. Paperback. [Original Publication: 43-9034]

Reprint of original 1943 publication.

Introduction; Past: Mediaeval Case, Decline of the City, Civic Rehabilitation; Toward the Future: Problems of Today, Concentration, Organic Decentralization, Revaluation, Relegislation, Urban Population; Epilogue—Town-Design.

209. Simon, Herbert A., *The Sciences of the Artificial*, Cambridge, Mass. (M.I.T.), 1969, 144 pp.

"Engineering, medicine, business, architecture, and painting are concerned . . . not with how things are but with how they might be—in short, with design. The possibility of creating a science or sciences of design is exactly as great as the possibility of creating any science of the artificial."
Understanding the Natural and Artificial Worlds; Psychology of Thinking—Embedding Artifice in Nature; Science of Design—Creating the Artificial; Architecture of Complexity.

210. Sitte, Camillo, *City Planning According to Artistic Principles*, Columbia University Studies in Art History and Archaeology, No. 2, New York (Random House), 1965, 205 pp. Paper. [64-17102]

Relationship between Buildings, Monuments, and Their Plazas; That the Center of Plazas Be Kept Free; That Public Squares Should Be Enclosed Entities; Size and Shape of Plazas; Irregularities of Old Plazas; Plaza Groupings; Streets; Layout of Public Squares in the North of Europe; Meager and Unimaginative Character of Modern City Plans; Modern Systems; Artistic Limitations of Modern City Planning; Improvements in the Modern System; Example of an Urban Arrangement According to Artistic Principles; Conclusion; Appendix: Greenery Within the City (1900), To Our Readers (1904).

211. Webber, Melvin M.; Others, *Explorations into Urban Structure*, Philadelphia (University of Pennsylvania), 1963, 246 pp. [63-15009]

An Approach to Metropolitan Spatial Structure (Donald L. Foley); The Urban Place and the Nonplace Urban Realm (Melvin M. Webber); Public and Private

Agents of Change in Urban Expansion (William L. C. Wheaton); The Tactical Plan (Albert Z. Guttenberg); Summary: Planning and Metropolitan Systems (John W. Dyckman).

212. Willhelm, Sidney M., *Urban Zoning and Land-Use Theory*, New York (Free Press), 1962, 244 pp. [62-15355]

Introduction; Materialistic Approaches to Ecology; Voluntaristic Approaches to Ecology; Means and End in the Zoning Process; Value Orientations and Zoning Data; Value Orientations in the Zoning Process; Value Contentions in the Zoning Process.

213. Williams, John D., *The Compleat Strategyst: Being a Primer on the Theory of Strategy*, New York (McGraw-Hill), Revised Edition, 1966, 288 pp. [65-25049]

Introduction; Two-Strategy Games; Three-Strategy Games; Four-Strategy Games and Larger Ones; Miscellany; General Methods of Solving Games.

Bibliographies

214. Anderson, Stanford, *Planning for Diversity and Choice*, Bibliography: Planning Techniques and Theory, pp. 333-334.

215. Berkman, Herman G., *Planning Theory*, Bibliography No. 33, Council of Planning Librarians, Monticello, Ill. (Exchange Bibliographies, P.O. Box 229, 61856), 1967, 14 pp.

216. Berry, Brian J. L.; Pred, Allen, *Central Place Studies: A Bibliography of Theory and Applications with Supplement through 1964*, Philadelphia, Penn. (Regional Science Research Institute), 1965, 202 pp.

"Materials published in the 1961 bibliography on central place theory are retained in this new edition and supplemented by nearly 300 annotated entries covering research in the field from 1961 to 1964. Among the sections in the new supplement are: internal business structure of the city, urban spheres of influence, and measurement of retail trade areas."

217. Carrothers, Gerald A. P., "An Historical Review of the Gravity and Potential Concepts of Human Interaction," Bibliography, pp. 100-102.

218. Dyckman, John W., "Planning and Decision Theory," Selected Bibliography, pp. 343-345.

219. Gerlough, D. L.; Capelle, D. G. (Editors), *An Introduction to Traffic Flow Theory*, Bibliography on Theory of Traffic Flow and Related Subjects, pp. 139-145.

Prepared by Highway Research Board Committee on Theory of Traffic Flow. Hydrodynamic Analogies and Kinematic Waves; Traffic Dynamics (Car Following); Probabilistic Flow Models; Simulation of Traffic Flow; Studies of Mechanics and Geometry; Experimental Studies of Capacity, Speed, and Flow-Concentration Relationships; Intersection Situations; Miscellaneous.

220. Howard, Ebenezer, *Garden Cities of To-Morrow*, Select Book List, pp. 160-163.

221. Howard, William A., *Concept of an Optimum Size City, A Selected Bibliography*, Bibliography No. 52, Council of Planning Librarians, Monticello, Ill. (Exchange Bibliographies, P.O. Box 229, 61856), 1968, 5 pp.

Books, Monographs, Reports and Theses; Periodicals.

222. Pred, Allen, *Behavior and Location—Foundations for a Geographic and Dynamic Location Theory*, Literature Cited, pp. 122-128.

223. Reiner, Thomas A., *The Place of the Ideal Community in Urban Planning*, Bibliography, pp. 175-192.

4. INFORMATION, COMMUNICATION

Articles, Periodicals

224. Barraclough, Robert E., "Mapping and EDP," *Planning 1965*, Chicago (American Society of Planning Officials), 1965, pp. 313-318. [39-3313]

Today's Maps; Computers and Mapping; Geographic Identification of Data; Conclusion.

225. Beged-Dov, Aharon G., "An Overview of Management Science and Information Systems," *Management Science*, Vol. 13, No. 12, August 1967, pp. B—817-831.

Applied Study of Business Problems—A Perspective; Systems Approach; Planning—Man/Machine Configuration; Value of Information—Theory and Practice; Information Systems—Purpose and Concepts; Approach to Information Systems Design; Information Systems in 1970; Impact on Management Practice; References.

226. Colwell, Robert N., "Remote Sensing of Natural Resources," *Scientific American*, Vol. 218, No. 1, January 1968, pp. 54-69.

"Several devices are now available to supplement the aerial camera in detecting natural resources from airplanes and spacecraft. They include radar, gamma ray, detectors and sensors of infrared energy."
Remote-Sensing Equipment; Related Devices; Analysis of Data; Some Applications; Additional Applications; A Prospect.

227. Evans, Marshall K.; Hague, Lou R., "Master Plan for Information Systems," *Harvard Business Review*, Vol. 40, No. 1, January-February 1962, pp. 92-103.

Need for Action; Long-Range Objectives; Analysis of Present System; Short-Range Improvements; Time & Responsibility; Accomplishing the Plan; Conclusion.

228. Hearle, Edward F. R., "Electronic Data Processing in Planning: A Framework of Alternatives," in: *Planning 1965*, Chicago (American Society of Planning Officials), 1965, pp. 301-305. [39-3313]

Extent of Integration with Non-Planning Agencies; Geographic Scope of System; Depth Versus Breadth of Data; Areal Unit; Type of Computer Programming Language; Conclusion.

229. Herbert, Evan, "Information Transfer," *International Science and Technology*, No. 51, March 1966, pp. 26-37.

"New ways of manipulating data can give instant access to networks of files but retrieval of information still hinges on the transfer of meaning."

What's in a File?; The Creamed File; Decentralization Debate; Negotiating the Question; Thread of Relevance; Form of an Answer; Coded Data; Image Analogs; Tiny Pictures; Electronic Images; Remote Access to Text; Library of the Future; Economics of Information.

230. Hodge, Gerald, "Use and Mis-Use of Measurement Scales in City Planning," *Journal of the American Institute of Planners*, Vol. XXIX, No. 2, May 1963, pp. 112-121.

Measurement, Numbers, and Scales: Nominal, Ordinal, Interval, Ratio; Measuring Housing Quality—Pitfalls and Potential: APHA Appraisal Method, APHA Variations and Other Housing Scales, Housing Quality Measurement and Program Decisions; Outlook on City Planning Measurement: Potential of Index Numbers, Ordinal Scales and Scalogram, Aids in Defining Dimensions and Units of Measurement; Challenge of Measurement.

231. Johnson, David L.; Kobler, Arthur L., "The Man-Computer Relationship—The Potential Contributions of Computers Crucially Depend Upon Their Use by Very Human Beings," *Science*, Vol. 138, No. 3543, 23 November 1962, pp. 873-879.

Basic Parameters; "Routine" and "Special" Problems; Parameters of Value; Results of Human Limitations; Machine-Learning Systems; The Man-Computer Problem; Conclusions.

232. Moore, Eric G.; Wellar, Barry S., "Urban Data Collection by Airborne Sensor," *Journal of the American Institute of Planners*, Vol. XXXV, No. 1, January 1969, pp. 35-43.

Role of Remote Sensors in Data Collection; Relation of the Remote Sensor to an Urban Data System: Characteristics of Data Collected by Remote Sensor, Potential Uses of Remote Sensors; Multiband Camera; Thermal Infrared Sensors; Imaging Radars: Conclusion.

233. Pierce, John R., "Communication," *Daedalus*, Vol. 96, No. 3, Summer 1967, pp. 909-921.

234. Profet, Karen, "Information to Manage Public Programs," *SDC Magazine*, Vol. 11, No. 1, Winter 1968, pp. 1-31; *SDC Magazine*, "A Centralized Management Information System: The Experience of the

'Largest Organization in the Free World,'" Vol 11, No. 9, October 1968, pp. 2-16.

Need for Public Management Information Systems; Development of Management Information Systems—Three Case Studies; Human Element in Management Information Systems; Data Element Standardization.

Answers for the White House, What's Going On, Perspective on ADEPT, Answering the Unexpected Question, Tracking Decisions for Management Control, Keeping Track of a Hundred Billion Dollars, Day-to-Day Business, Surprising Economy, Close to an Integrated System, How Much Information Is Enough?

235. Schmedel, Scott R., "Computers Gain Use Among Cities, States Trying to Cut Costs," *The Wall Street Journal*, Pacific Coast Edition, Vol. LXXI, No. 70, 16 October 1964, p. 1.

"They Help Washington Record Liquor Sales; Data Center Gets Facts on Chicago Pupils." Savings in Wisconsin; Pioneering in Planning.

236. *Scientific American*, Information, Vol. 215, No. 3, September 1966, Entire Issue, pp. 64-260.

Information (John McCarthy); Computer Logic and Memory (David C. Evans); Computer Inputs and Outputs (Ivan E. Sutherland); System Analysis and Programming (Christopher Strachey); Time-Sharing on Computers (R. M. Fano; F. J. Corbató); Transmission of Computer Data (John R. Pierce); Uses of Computers in: Science (Anthony G. Oettinger), Technology (Steven Anson Coons), Organizations (Martin Greenberger), Education (Patrick Suppes); Information Storage and Retrieval (Ben-Ami Lipetz); Artificial Intelligence (Marvin L. Minsky).

237. Sessions, Vivian S., "The City Planning and Housing Library: An Experiment in Organization of Materials," *Municipal Reference Library Notes*. Vol. XXXVII, No. 9, November 1963, pp. 269-283.

Presents the classification of the City Planning and Housing Library of the New York municipal government—"one of the few centers in the country that treats urban planning as a depth problem in public administration rather than as an adjunct to architecture or another discipline."

238. Sessions, Vivian S., "Document Retrieval and Planning, "Planning 1965, Chicago (American Society of Planning Officials), 1965, pp. 330-335. [39-3313]

Brief description and discussion of the URBANDOC project in document retrieval.

239. Vazsonyi, Andrew, "Automated Information Systems in Planning, Control and Command," *Management Science*, Vol. 11, No. 4, February 1965, pp. B—2-41.

Introduction; Problems of Automation; On-Line-Real-Time Information Systems; Control of Space Travel, of Airline Reservations; Military Command and Control; Planning and Control of Research and Development; Where Computers Excel; Where Men Excel; How Man Works; Heuristic Problem Solving; Man-Machine Communication Consoles; Solution of Differential Equations; Computer Aided

Design; Solution of Partial Differential Equations and Integral Equations; Automated Program Evaluation and Review Technique; Automated Teaching of Languages; Approach for Automation; References, pp. B–39-41.

240. Webber, Melvin M., "The Roles of Intelligence Systems In Urban-Systems Planning," *Journal of the American Institute of Planners.* Vol. XXXI, No. 4, November 1965, pp. 289-296.

"In response to the growing demand for good information that might support rational development decisions, 'intelligence centers' are proposed. . . ." New Images of Urban Systems: From Stocks to Flows, Merging Actors; Strategies for Planning: Present Goal-Setting Methods, In Search of a Programming Strategy, Interim Programming Strategy; Politics of Information; Conclusion.

Books, Reports, Pamphlets

241. American Society of Planning Officials, *Threshold of Planning Information Systems*, Chicago, Ill. (1313 East 60th. St., 60637), 1967, 108 pp. Paper.

Threshold of Planning Information Systems: State of the Art of Planning Information Systems (Edgar M. Horwood), Initiating the Data Bank Concept Into Planning Operations (Robert C. Jacobson), Development of an Information System for the Ottawa National Capitol Commission (D. L. McDonald); Explorations in Municipal Information Systems Research: Introductory Comment (W. H. Mitchel), Cost-Benefit Ratio of EDP—Political Challenge (Myron E. Weiner), Decisions, Data Needs, and Rationality (John K. Parker), Integrated Information Systems—Problem In Psychology (Robert E. Graetz), Anatomy of a Municipal Information and Decision System (William H. Mitchel); Geographic Implications of Urban Information Systems: Computer Graphic Techniques for Governmental Boundary Analysis (Welden E. Clark; Donald B. Gutoff), Urban Geocoding Systems and Their Utility (William L. Clark); Data Processing for Planning: Urban Information Systems and Data Banks—Better Prospects With an Environmental Model (Fred J. Lundberg), LOGIC—Santa Clara County Government Information System and Its Relationship to the Planning Department (Robert A. Clark), Time-Sharing Applications of Regional Data Handling (Claude Peters; Eugene Kozik).

242. Backstrom, Charles H.; Hursh, Gerald D., *Survey Research*, Evanston, Ill. (Northwestern University), 1963, 192 pp. Paperback. [63-18013]

Planning a Survey; Drawing the Sample: Sample Size, Representativeness of the Sample, Information Needed, Mechanics of Drawing the Sample, Interviewer Assignments; Writing the Questions: Kinds, Form, Content, Sequence; Designing the Questionnaire; Going into the Field: Setting Up Field Work, Coordinating Field Work; Processing the Data; Index.

243. Becker, Joseph; Hayes, Robert M., *Information Storage and Retrieval: Tools, Elements, Theories*, New York (Wiley), 1963, 448 pp. [63-12279]

Tools: Librarian and Recorded Knowledge, Documentalist and the Development of New Techniques, Information Framework and the User, Printed Data and the Creation of a Machine Language, Analysis, Logical Processing, and the Computer, Indexes, Documents, and the Storage of Data; Elements: Interdisciplinary Character of Information Systems, Elements of—Usage, Organization, Equipment, Parameters and Implementation; Theories: Role, of File Organization, of System Design; Periodicals and Journals Relating to Information Storage and Retrieval.

244. Bolt Beranek and Newman, Inc., *Man/Computer Processing of Graphical Urban Data for Geographical Reference*, Proposal P66-LA-29, Van Nuys, Cal. (15808 Wyandotte St., 91406), 3 February 1966, 42 pp.

Introduction and Summary; Objectives and Techniques: Importance of Geographical Reference Data—City-wide Coordinate System, Parcel Coordinates, City Block Coordinates, Block Face Coordinates, Gross and Net Parcel Area, Parcel Frontage, Street Segments and Intersections, Street and Alley Width, Lot Position, Capability of Deriving Additional Geographic Data for APOF, Utility Network File, Other Graphic Input/Output Capabilities; Approach—System Design and Data Acquisition, Typical Input Procedure, Block Face Input, Special Input Situations, Editing, Output; Statement of Proposed Work: Data to Be Generated, Scope of Proposed Work; Cost.

245. Canning, Richard G.; Sisson, Roger L., *The Management of Data Processing*, New York (Wiley), 1967, 124 pp. [67-25643]

Past Is Prologue; Choosing the Direction; Organizing the Effort; Staffing the Operation; Changing Technology—Its Implications; Managing a Future EDP System; Bibliography, pp. 115-117.

246. Crowley, Thomas H., *Understanding Computers*, New York (McGraw-Hill), 1967, 142 pp. Paperback. [67-14891]

". . . [assumes] no knowledge of mathematics beyond arithmetic or any concepts from physics or engineering."

Introduction; Basic Computer Functions; What Is Being Processed?; Interpretation of Symbols; Memory; Input-Output Operations; Symbol-Processing Operations; Control Process; Stored-Program Computers; Programming; Applications of Computers; Computer "Priests"; What Does the Future Hold?; 1984?; Selected General Bibliography, pp. 137-139.

247. Davis, Jeanne M., *Uses of Airphotos for Rural and Urban Planning*, Agricultural Handbook No. 315, Economic Research Service, U.S. Department of Agriculture, Washington, D.C. (Superintendent of Documents), June 1966, 40 pp.

Summary; Introduction; Types of Airphotos in General Use; Uses of Airphotos in Planning; Photo-Interpretation Problems in Planning; Steps in Use of Airphotos for Interpretation; Availability of Airphotos; How to Obtain Aerial Photography; Selected References.

248. Duke, Richard D. (Editor), *Automatic Data Processing: Its Application to Urban Planning*, East Lansing, Mich. (Institute for Community Development and Services, Michigan State University), 1961, 112 pp. Paper. [61-63704]

Automatic Data Processing and Urban Planning Education (Myles G. Boylan); Means: Punched-Card Equipment (Francis B. Martin), Electronic Computer (Gerard P. Weeg), Data Retrieval Procedures (James E. Skipper); Representative Applications: Utilization of Data Processing in Transportation Research (Roger L. Creighton), Utilization of the Electronic Computer in Economic Studies (Richard C. Henshaw, Jr.); Overview: Data Processing and Planning (Robert S. McCarger). Bibliography pp. 101-112.

249. Hearle, Edward F. R.; Mason, Raymond J., *A Data Processing System for State and Local Governments*, Englewood Cliffs, N. J. (Prentice-Hall, September 1963, 176 pp. [63-21039]

Introduction; Data Processing Equipment; State and Local Governments; The Data; Unified Organization System: Organization; Unified Information System: Activities; Unified Information System: Costs and Benefits; Steps Toward Implementation; Appendices.

250. Hirsch, Werner Z. (Editor), *Elements of Regional Accounts*, Baltimore, Md. (Johns Hopkins), 1964, 221 pp. [64-16309]

". . . nine essays provide a methodology for organizing regional and metropolitan area data to facilitate private and public decision-making. Several deal with techniques useful in investigating short- and long-range regional changes, and in understanding the effects of zoning, land values and circulation within metropolitan areas. Others concentrate on local government policy effects, and on households as the core of the manpower and consumer factors. A final group of essays presents ways of integrating information on income and output flows with that on assets and wealth."

251. Howe, Robert T., *Fundamentals of a Modern System of Land Parcel Records*, Cincinnati, Ohio (University of Cincinnati), May 1968, 51 pp. Paper.

Need for a Modern System of Land Parcel Records; Problems of Identifying Points on the Surface of the Earth; Systems for Identifying Land Parcels; Deed Descriptions and Public Records Relating to Land Title; A Modern, Coordinate-Based System of Land Records; Summary and Conclusions; References, pp. 50-51.

252. Janowitz, Morris, *The Community Press in an Urban Setting*, The Social Elements of Urbanism, Chicago (University of Chicago), Second Edition, 1967, 275 pp. Paperback [67-21391]

Communications and the Urban Metropolis; Growth and Organization of the Community Press; Image of the Community—Content for Consensus; Attributes of Readership—Family, Status, and Community; Impact of Readership—Interest, Penetration, and Imagery; Social Role of the Community Publisher—Agent of Personal and Mass Communications; Social Dimensions of the Local Community; Future of the Community Press; Postscript—Communication and Community (Scott Greer).

253. Keane, Mark E.; Arnold, David S.; Havlick, J. Robert; Howe, George F., *The Municipal Year Book, 1968*, International City Managers'

Association, Washington, D.C. (1140 Connecticut Ave., N.W., 20036), 1968, c. 600 pp.

Government Setting; Small Cities; Personnel Administration; Municipal Finance; Municipal Activities; References; Directories.

254. Kish, Leslie, *Survey Sampling*, New York (Wiley), 1965, 643 pp. [65-19479]

Fundamentals of Survey Sampling: Introduction; Basic Concepts of Sampling; Stratified Sampling; Systematic Sampling; Stratification Techniques; Cluster Sampling and Subsampling; Unequal Clusters; Selection with Probabilities Proportional to Size Measures; Economic Design of Surveys. Special Problems and Techniques: Area Sampling; Multistage Sampling; Sampling from Imperfect Frames; Some Selection Techniques. Related Concepts: Biases and Nonsampling Errors; Some Issues of Inference from Survey Data; Appendices.

255. McLuhan, Marshall, *Understanding Media: The Extensions of Man*, New York (New American Library), 1964, 318 pp. Paper. [64-16296]

The Medium Is the Message, Media Hot and Cold, Reversal of the Overheated Medium, Gadget Lover—Narcissus as Narcosis, Hybrid Energy—Les Liaisons Dangereuses, Media as Translators, Challenge and Collapse—Nemesis of Creativity; Spoken Word—Flower of Evil, Written Word—Eye for an Ear, Roads and Paper Routes, Number—Profile of the Crowd, Clothing—Our Extended Skin, Housing—New Look and New Outlook, Money—Poor Man's Credit Card, Clocks—Scent of Time, Print—How to Dig It, Comics—Mad Vestibule to TV, Printed Word—Architect of Nationalism, Wheel, Bicycle and Airplane, Photograph—Brothel-without-Walls, Press—Government by News Leak, Motorcar—Mechanical Bride, Ads—Keeping Upset with the Joneses, Games—Extensions of Man, Telegraph—Social Hormone, Typewriter—Into the Age of Iron Whim, Telephone—Sounding Brass or Tinkling Symbol?; Phonograph—Toy That Shrank the National Chest, Movies—Reel World, Radio—Tribal Drum, Television—Timid Giant, Weapons—War of the Icons, Automation—Learning a Living; Further Reading for Media Study, pp. 315-318.

256. National Science Foundation, Office of Science Information Service, *Current Research and Development in Scientific Documentation, No. 14*. NSF-66-17, Washington, D.C. (Superintendent of Documents), 1966, 662 pp.

Information Needs and Uses; Document Creation and Copying; Language Analysis; Translation; Abstracting, Classification, Coding, and Indexing; System Design; Analysis and Evaluation; Pattern Recognition; Adaptive Systems.

257. Passonneau, Joseph R.; Wurman, Richard S., *Urban Atlas: 20 American Cities*, A Communication Study Notating Selected Urban Data at a Scale of 1:48,000, Cambridge, Mass. (M.I.T.), 1966, 118 pp. plates. [Map 66-36]

". . . a pattern of colored geometrical symbols is overprinted on [U.S. Geological Survey] maps, on separate plates. . . . The symbols representing the range of one variable are then overprinted on the symbols representing the range of others, allowing a visual correlation. . . ."

Population Density Calculations; Total Income Calculations; Industrial Data; Land Use Data; Church Data; Recording and Printing Data; Condition of Housing.

258. Security First National Bank of Los Angeles, Research Department, *The Growth and Economic Stature of the South Coast Area of Los Angeles County, 1963,* Los Angeles, Cal. (561 S. Spring St., 90054), 1963, 64 pp.

Highlights; Pattern of Population Growth; Construction Volume; Homebuilding; Land Zoning; Harbor Activities; Oil Industry; Employment; Family Income; Bank Debits; Retail Trade; Freeways; Educational Facilities; Tourism and Recreation; In Conclusion.

259. Ruesch, Jurgen; Kees, Weldon, *Nonverbal Communication—Notes on the Visual Perception of Human Relations,* Berkeley, Cal. (University of California), 1956, 205 pp. [55-11228]

". . . an attempt to . . . explore the informal and often spontaneous methods of communication that, when considered in verbal and, particularly, in abstract terms, tend grossly to distort actual events."

Frame of Reference; Message Through Nonverbal Action; Message Through Object and Picture; Language of Disturbed Interaction; Summary; References, pp. 197-201.

260. Tulsa Metropolitan Area Planning Commission, *Land Use: A Procedural Manual For Collection, Storage, & Retrieval of Data,* Tulsa, Okla. (600 Kennedy Bldg., 74103), 12 March 1965, 95 pp. Paper.

Preface; Two Concepts; Past Land Use Information Programs; Pre-Inventory Analysis and Development; Activity Inventory Procedures; Data Processing Procedures; Continuing Land Use Information Flow; Appendix.

261. (Project) URBANDOC, *Urbandoc Thesaurus,* New York (City University), Field Test Edition, May 1967, 190 pp. Paper.

". . . designed mainly for a computerized system of bibliographic information . . . the electronic equivalent of a library catalogue, but greatly augmented . . . the thesaurus contains words which may be used . . . to analyze and retrieve documents in urban planning, renewal, and housing . . ."

Introduction; Alphabetic Listing; Permuted Listing. Prepared under Urban Renewal Demonstration Project No. N.Y. D-9.

262. U.S. Bureau of the Census, *Statistical Abstract of the United States, 1965,* Washington, D.C. (Superintendent of Documents), 1965, 1,050 pp.

"Presents data for the most recent year or period available during the early part of 1965. Included in the new material are statistics from the 1963 Census of Transportation, Mineral Industries, Manufacturers, and Business. The new data appear in the sections dealing with Public Lands, Parks, Recreation and Travel; Transportation—Land; Mining and Mineral Products; Manufacturers; and Distribution and Services."

263. U.S. Housing and Home Finance Agency, *Metropolitan Data Center*

Project, Washington, D.C. (U.S. Department of Housing and Urban Development), February 1966, 112 pp.

Report Summary; Introduction; Information Systems Program; Capital Improvement Programming; Appendices.

264. U.S. Housing and Home Finance Agency, Urban Renewal Administration, Urban Renewal Service, *Using Computer Graphics in Community Renewal,* Community Renewal Program Guide No. 1, Washington, D.C. (Superintendent of Documents), January 1963, c. 250 pp. Paper.

Introduction and Glossary; Card Mapping Program; Array Program; BCD to Binary Tape Transfer Program; Distribution Program; Tape Mapping Program; Appendices; Data System Design for the Spokane CRP, Principles of Multiphasic Screening, Model Constructs in Urban Analysis, Block Diagrams of the Computer Program.

265. Warren, Roland L., *Studying Your Community,* New York (Free Press), First Paperback Edition, 1965, 385 pp. [55-7727]

How To Use This Book; Your Community, Its Background and Setting; Your Community's Economic Life; Government, Politics, and Law Enforcement; Community Planning Associations; Community Organization; Organizing a Community Survey; Aids to the Survey; Some Important Aspects of the Community; List of Agencies.

266. Wiener, Norbert, *The Human Use of Human Beings—Cybernetics and Society,* Garden City, N. Y. (Doubleday), 1954, 199 pp. Paperback.

Cybernetics in History; Progress and Entropy; Rigidity and Learning—Two Patterns of Communicative Behavior; Mechanism and History of Language; Organization as the Message; Law and Communication; Communication, Secrecy, and Social Policy; Role of the Intellectual and the Scientist; First and Second Industrial Revolution; Some Communication Machines and Their Future; Language, Confusion, and Jam.

267. Yavitz, Boris; Stanback, Thomas M., Jr., *Electronic Data Processing in New York City—Lessons for Metropolitan Economics,* New York (Columbia University), 1967, 159 pp. [67-29876]

Introduction; Challenge of Metropolitan Economics; EDP and the Computer Revolution; EDP As a Business Service; Service Bureau—Firm and Industry; Manpower and Employment—Current Picture; Forces of Change; Expected Impacts of the Computer Revolution on Headquarters, EDP Services, and Employment; Competition Among Cities; Mobilizing Resources—Taking the Lead.

Bibliographies

268. Aronoff, Leah, *Current Information Sources for Community Planning,* Bibliography No. 25, Council of Planning Librarians, Monticello, Ill. (Exchange Bibliographies, P.O. Box 229, 61856), 1967, 55 pp.

269. Baerwald, Diane A., "Survey of Planning Information In Standard Reference Books," Exchange Bibliography No. 62, Council of Planning Librarians, Monticello, Ill. (P.O. Box 229, 61856), 1968, 34 pp. Mimeo.

Defining the Field; The Librarian and the Reference Question; Literature Search: Biographies, Encyclopedias, Government Documents, Indexes, Dictionaries, Atlases, Yearbooks, Bibliographies, Directories; Conclusion; Bibliography.

270. Clark, Robert A., *Data Bank or Information Systems Publications— With Emphasis on Land Use,* Bibliography No. 59, Council of Planning Librarians, Monticello, Ill. (Exchange Bibliographies, P.O. Box 229, 61856), 1968, 10 pp. Mimeo.

271. Duke, Richard D. (Editor), *Automatic Data Processing: Its Application to Urban Planning,* Bibliography, pp. 101-112.

272. Goodman, Edith H., *Information Systems Bibliographic Index,* Detroit, Mich. (American Data Processing, Inc.), 1965, 202 pp [62-21678]

". . . second volume in a series of bibliographic indices on Information Systems . . . Taken cumulatively . . . this index covers virtually the entire span of 'the computer age.' . . . In the earlier period articles tended mostly toward actual and projected applications of the earlier computer systems. Most recently published information has related more toward . . . 'think type' material . . . the more conjectural, theoretical, or abstract . . . often titles do not faithfully describe the content of articles . . . it may appear at times that articles listed in this index are misplaced or sound general. . . ."

273. Howe, Robert T., *Fundamentals of a Modern System of Land Parcel Records,* References, pp. 50-51.

274. McLuhan, Marshall, *Understanding Media: The Extensions of Man,* Further Reading for Media Study, pp. 315-318.

275. Tucker, Dorothy, "Computers and Information Systems in Planning and Related Governmental Functions," Exchange Bibliography No. 42, Council of Planning Librarians, Monticello, Ill. (P.O. Box 229, 61856), 21 pp. Mimeo.

Introduction; General Discussion—Feasibility; Statewide EDP Systems; EDP in Local Governments; Intergovernmental EDP Systems; EDP in the Planning Agency; Data Banks, Information Systems; Computer Graphics; Information Retrieval; Time-Sharing, User-Oriented Systems; List of Periodicals Devoted to EDP; List of Other Periodicals Cited; Agency Addresses Not Given in Bibliographic Listing.

276. Vazsonyi, Andrew, "Automated Information Systems in Planning. Control and Command," References, pp. B—39-41.

277. Voos, Henry, *Organizational Communication: A Bibliography,* New Brunswick, N.J. (Rutgers University), 1967, 251 pp. [67-63681]

Literature Survey: General, Decision Making, Upward Communication, Downward Communication, Persuasion, Horizontal Communication, Feedback, Cognitive Dissonnance, Networks—Circle, Chain, Wheel, Completely Connected, Characteristics of the Literature, Future Work; References; Supplementary References; Indexes—Author, Subject.

5. RESEARCH ANALYSIS

Articles, Periodicals

278. ABT Associates Inc., "Survey of the State of the Art: Social, Political, and Economic Models and Simulations," in: National Commission on Technology, Automation, and Economic Progress, *Technology and the American Economy*, Appendix Vol. V, Washington, D.C. (Superintendent of Documents), February 1966, pp. V-209-239.

Introduction; Survey; Limitations of the Selected Survey Content; Typology of Models Surveyed; Economic and Managerial Models; Social and Political Models; Computer Simulations; Manual Simulations—Human Player Games; Staff, Time, and Money Requirements; Conclusions and Recommendations; Selected Bibliography for Economic Models, pp. V-247-248 . . . for Social and Political Models, pp. V-249-250.

279. Adelson, Marvin, "The System Approach—A Perspective," *SDC Magazine*, Vol. 9, No. 10, October 1966, pp. 1-9.

What Is the System Approach?; Major Characteristics of the Systems Approach: Organized, Creative, Empirical, Theoretical, Pragmatic; Caveat; System Analysis Performance: State of California, National Information System, Air Force Personnel Data System, New York City, Airport Transportation.

280. Alonso, William, "Predicting Best with Imperfect Data," *Journal of the American Institute of Planners*, Vol. XXXIV, No. 4, July 1968, pp. 248-255.

How Error Cumulates; Modeling Strategies: Value of Better Data, Simple and Complex Models, An Example; Weak Models and Robust Theorems; Ripening and Using Complex Models.

281. Ansoff, H. Igor; Brandenburg, Richard C., "A Program of Research In Business Planning," *Management Science*, Vol. 13, No. 6, February 1967, pp. B-219—B-239.

"The purpose of this paper is to outline a program of research which is needed to improve the state of the art of business planning."

Taxonomy of Science; Planning As Part of the Decision Process; Planner As a Specialist Within Management Science; Program of Research for Planners: Concepts and Methodology, Social Behavior, Physical and Information Resource Conversion, Applied Theory of the Firm, Design of the Planning Process; Methodology for Research on Planning; References, pp. B-237—B-239.

282. Black, Guy, "Systems Analysis in Government Operations," *Management Science,* Vol. 14, No. 2, October 1967, pp. B—41-58.

Current Scene; Brief Description of Systems Approach: Systems Analysis, Systems Engineering, Systems Management, Tools of Systems Analysis; Phases of Systems Analysis: General, Stating Objectives, Jamming Model, Models as a Means for System Design, Input-Output Relationships, Benefit Functions, Cost Functions, Optimization, Quantification of Parameters; Relation of Systems Approach to Program Budgeting: PPBS and Systems Approach, Suboptimization and the Locus of Analysis; Conclusion; References.

283. Branch, Melville C., "Simulation, Mathematical Models, and Comprehensive City Planning," *Urban Affairs Quarterly,* Vol. 1, Issue 3, March 1966, pp. 15-38; also Chap. 7 in: _____, *Planning: Aspects and Applications,* New York (Wiley), 1966, pp. 146-174. [66-13523]

Basic Requirements for Simulation in Comprehensive Planning; Potentialities of Mathematical Models; The Decision-Making Interface; The Judgmental Process.

284. Burns, Leland S., "A Programming Model for Urban Development," in: Thomas, Morgan D. (Editor), *Regional Science Association Papers,* Vol. 11, 1963, pp. 195-210. Paper.

"The methodology is an econometric programming model, which is linear and dynamic. By substituting planning priorities for the traditional market mechanism, the model schedules the allocation of production factors such that target capacity is gained in minimal time: thus the model is normative. It is also operational, for this study represents one of the first attempts at empirically testing a dynamic linear program conceived for a spatially delimited area. The test uses data generated from the reconstruction experience of a principal European city severely damaged during World War II: Rotterdam."

285. Chase, Lawrence, "Progress Report on Harnessing the H-Bomb: Vast New Power Source Seems Nearer Realization," *Princeton Alumni Weekly,* Special Issue With "University" Magazine, Vol. LXIX, No. 21, 18 March 1969, pp. 7-11, 50-54; also: *University,* A Princeton Quarterly, No. 40, Spring 1969, pp. 2-7, 30-34.

". . . if the forward strides continue, the H-bomb may be controlled well before the year 2000. Then the world may witness widespread construction of thermonuclear reactors capable of turning hydrogen isotopes from ordinary water into commercial power—at a cost of about half that of today's coal-fed or atomic (nuclear) power sources. This could avert a certain future shortage of fossil fuels (coal, oil), help reduce atmospheric pollution, raise standards of developing nations around the globe."

286. Cohen, John, "Subjective Probability," *Scientific American,* Vol. 197, No. 5, November 1957, pp. 128-138.

"In the course of their daily lives men must constantly measure the probability of events. How these personal judgments conform to the actual laws of probability is investigated by experiment."

287. Crecine, John P., "A Computer Simulation Model of Municipal Budgeting," *Management Science*, Vol. 13, No. 11, July 1967, pp. 786-815.

Positive, Empirical Theory of Municipal Budgeting; Overview of Municipal Budgeting; Formal Model of Municipal Budgeting; Model Tests; Analysis of Model Residuals; Summary.

288. Czamanski, Stanislaw, "Effects of Public Investments on Urban Land Values," *Journal of the American Institute of Planners*, Vol. XXXII, No. 4, July 1966, pp. 204-217.

". . . an attempt to develop a model for identifying and assessing the impact of alternative development policies on urban land values. The statistical analysis carried out suggests that the most important factor determining the values of urban land is accessibility to the central urban functions. Age of existing structures and zoning regulations are secondary factors. Since future development of central urban functions can be projected with the help of existing models and since accessibility of places (particularly by means of public transportation) and zoning regulations can be influenced by public authorities, indirect public actions considerably influence urban land values."

289. *Daedalus*, Toward the Year 2000: Work in Progress, Vol. 96, No. 3, Summer 1967, pp. 639-994. [12-30299]

The Year 2000—Trajectory of an Idea (Daniel Bell); Working Session I; Some Specific Problems: The Next Thirty-Three Years—Framework for Speculation (Herman Kahn; Anthony J. Wiener), Can Social Predictions Be Evaluated? (Fred Charles Iklé), Forecasting and Technological Forecasting (Donald A. Schon), Information, Rationality, and Free Choice in a Future Democratic Society (Martin Shubik), Planning and Predicting—Or What to Do When You Don't Know the Names of the Variables? (Leonard J. Duhl), Modernizing Urban Development (Harvey S. Perloff), Relationship of Federal to Local Authorities (Daniel P. Moynihan), Need for a New Political Theory (Lawrence K. Frank), University Cities in the Year 2000 (Stephen R. Graubard), Educational and Scientific Institutions (Harold Orlans), Biological Man and the Year 2000 (Ernest Mayr), Deliberate Efforts to Control Human Behavior and Modify Personality (Gardner C. Quarton), Religion, Mysticism, and the Institutional Church (Krister Stendahl), Memorandum on Youth (Erik H. Erikson), Life Cycle and Its Variations—Division of Roles (Margaret Mead), Problems of Privacy in the Year 2000 (Harry Kalven, Jr.), Some Psychological Perspectives in the Year 2000 (George A. Miller), Notes on Meritocracy (David Riesman), Communication (John R. Pierce), Thinking About the Future of International Society (Eugene V. Rostow), Political Development and the Decline of the American System of World Order (Samuel P. Huntington), International System in the Next Half Century (Ithiel De Sola Pool); Working Session II; Coda—Work in Further Progress (Daniel Bell).

290. Duke, Richard D., "Gaming Urban Systems," in: *Planning 1965*, Chicago (American Society of Planning Officials), 1965, pp. 293-300. [39-3313]

Brief, general description of urban games, including METROPOLIS I and II.

291. Dyckman, John W., "The Scientific World of the City Planners," *The American Behavioral Scientist*, Vol. 6, No. 6, February 1963, pp. 46-50.

"There are strong trends within city planning in the direction of making it more scientific. The comprehensive viewpoint of city planners can be of real assistance to urban studies in the development of a science of metropolitan organization and behavior, and planning method is essential in the application of urban knowledge."

292. Feldt, Alan G., "Operational Gaming in Planning Education," *Journal of the American Institute of Planners*, Vol. XXXII, No. 1, January 1966, pp. 17-23.

"A general description of the Cornell Land Use Game is provided, together with an interpretation of its use by student, faculty, and professional groups. Some suggestions for further improvements of games of this nature are offered."
Operational Gaming As a Teaching Device; Cornell Land Use Game; Experience With the Game; Effectiveness of the Game in Teaching; Further Modifications of the Game; Conclusion.

293. Fishburn, Peter C., "Methods of Estimating Additive Utilities," *Management Science*, Vol. 13, No. 7, March 1967, pp. 435-453.

"Additive utility formulations for risky and nonrisky multiple-factor decision situations are reviewed. Twenty-four methods of estimating additive utilities are listed and classified."

294. Geisler, Murray A.; Steger, Wilbur A., "The Combination of Alternative Research Techniques In Logistics Systems Analysis," *Management Technology*, Vol. 3, No. 1, May 1963, pp. 68-77.

Introduction; System Characteristics; Purpose of Systems Analysis; Classification of Systems-Analysis Techniques and Their Attributes; Classification of Simulation Techniques; Example of the Evolution of a Simulation; Conclusion.

295. Gerwin, Donald, "A Process Model of Budgeting in a Public School System," *Management Science*, Vol. 15, No. 7, March 1969, pp. 338-361.

Review of Relevant Theories; Outline of the Model; Description of the Model: Preliminary Allocations to the Major Subunits, General Salary Increases, New Debt Service, Deficit Removal Processes, Surplus Distribution Processes; Analysis of the Model: Testing the Model; Requests; Discussion: Normative Implications.

296. Greniewski, Henryk, "Intention and Performance: A Primer of Cybernetics of Planning," *Management Science*, Vol. 11, No. 9, Series A, July 1965, pp. 763-782.

"Cybernetics of planning is concerned with planning in any field. . . ."
Task; Initial Assumptions; Plan and Payoff Matrix; Correction of the Plan; Feedback; Logical Construction of the Plan; Types of Models; Conclusion.

297. Gross, Donald; Ray, Jack L., "A General Purpose Forecast Simulator," *Management Science*, Vol. 11, No. 6, April 1965, pp. B-119-135.

"A simulator is developed which enables a variety of forecasting procedures to be tested on any given time series . . . determines the standard deviation of the forecast error for each of the procedures tried, indicates the procedure which is optimal in the sense of lowest standard deviation of forecast error, and presents projections and graphs of actual versus predicted values. The time series under study can be either a real time series read into the simulator or a time series generated by the simulator."

298. Harmon, Leon D.; Knowlton, Kenneth C., "Picture Processing by Computer," *Science*, Vol. 164, No. 3875, 4 April 1969, pp. 19-29.

"Computers which process graphical material are new, powerful tools for science, engineering, education, and art."

Equipment and Processes; Picture-to-Picture; Picture-to-Abstraction; Abstraction-to-Picture; Summary.

299. Harris, Britton, "Plan or Projection: An Examination of the Use of Models in Planning," *Journal of the American Institute of Planners*, Vol. XXVI, No. 4, November 1960, pp. 265-272.

Varying Scales of Planning; Major Planning Goals; Metropolitan System; Scientific Prediction; What Planners Must Project; Concluding Notes.

300. Harris, Britton (Guest Editor), Urban Development Models: New Tools for Planning, *Journal of the American Institute of Planners*, Entire Issue, Vol. XXXI, No. 2, May 1965, 182 pp.

Urban Development Models—New Tools for Planning (Britton Harris); Opportunity-Accessibility Model for Allocating Regional Growth (George T. Lathrop; John R. Hamburg); Land Use Plan Design Model (Kenneth J. Schlager); Growth Allocation Model for the Boston Region (Donald M. Hill); Model for Simulating Residential Development (F. Stuart Chopin, Jr.); Simulation Model for Renewal Programming (Ira M. Robinson; Harry B. Wolfe; Robert L. Barringer); Retail Market Potential Model (T. R. Lakshmanan; Walter G. Hansen); Pittsburgh Urban Renewal Simulation Model (Wilbur A. Steger); Retail Component of the Urban Model (Brian J. L. Berry); Urban Transportation Planning Models in 1975 (W. L. Garrison). Review Articles: Short Course in Model Design (Ira S. Lowry); Review of Analytic Techniques for the CRP (Wilbur A. Steger).

301. Harris, Britton, "The Uses of Theory in the Simulation of Urban Phenomena," *Journal of the American Institute of Planners*, Vol. XXXII, No. 5, September 1966, pp. 258-273.

Theory and Practice; Scientific Method; Social Sciences; Induction and Deduction; Generalization; Models; Land Use Systems and Transportation Systems; Communicating Mechanisms; Behavior of Decision Units; Implications for Planning.

302. Hermann, Cyril, "Systems Approach to City Planning," *Harvard Business Review*, Vol. 44, No. 5, September-October 1966, pp. 71-80.

"San Francisco applies management techniques to urban redevelopment."

Challenge & Opportunity; Organizing to Do the Job: "VP for Planning," Stress on Private Business; Analytical Approach: City as a System, Purposes of the Model, Construction & Operation, Impact of Public Actions; Stimulus to Progress: Goals Identified, Program of Action; Problems & Solutions: Remedial Programs, Improved Investment; Conclusion.

303. Heuston, M. C.; Ogawa, G., "Observations on the Theoretical Basis of Cost-Effectiveness," *Operations Research,* Vol. 14, No. 2, March-April, 1966, pp. 242-266.

". . . describes the results of continuing research to develop a comprehensive and rigorous description of the important elements of cost-effectiveness as used by the aerospace industry for military and commercial systems planning. The primary objective is to utilize basic mathematical and statistical theory to construct the rules, properties, and hypotheses . . . needed to satisfy the contractual requirements imposed by various governmental customers."

304. Highway Research Board, Land Use Forecasting Concepts, *Highway Research Record,* No. 126, Washington, D. C. (2101 Constitution Ave., 20418), 1966, pp. 1-87.

Uses of Theory in the Simulation of Urban Phenomena (Britton Harris); Difficult Decisions in Land Use Model Construction (W. L. Garrison); Recursive Programming Theory of the Residential Land Development Process (Kenneth J. Schlager); Test of Some First Generation Residential Land Use Models (Carl N. Swerdloff; Joseph R. Stowers); Verification of Land Use Forecasting Models: Procedures and Data Requirements (David E. Boyce; Roger W. Cote); Methodology for Developing Activity Distribution Models by Linear Regression Analysis (Donald M. Hill; Daniel Brand); Sampling Technique for Updating a Quantitative Land Use Survey (Sam M. Hadfield; John D. Orzeske).

305. Isenson, Raymond S., "Technological Forecasting in Perspective," *Management Science,* Vol. 13, No. 2, October 1966, pp. B-70—B-83.

"The purpose of this paper is to review generally the various, apparently useful forecasting techniques, to offer a rationale basis for undertaking a technological forecast, and to place technological forecasting in perspective with regard to other planning inputs."

306. Johnson, Richard A.; Kast, Fremont E.; Rosenzweig, James E., "Systems Theory and Management," *Management Science,* Vol. 10, No. 2, January 1964, pp. 367-384.

"General systems theory is reviewed for the reader. Next, it is applied as a theory for business, and an illustrative model of the systems concept is developed to show the business application. Finally, the systems concept is related to the traditional functions of a business, i.e., planning, organizing, control, and communications."

307. Journal of the American Institute of Planners, An Issue devoted to the Science of Planning, Vol. XXII, No. 2, Spring 1956, pp. 58-102.

Building the Middle-Range Bridge for Comprehensive Planning (Martin Meyerson); Home Owners and Attitude Toward Tax Increase, Flint Metropolitan Area (Basil G. Zimmer; Amos H. Hawley); Fringe Living Attitudes (Thomas B. Brademas); Estimating Daytime Populations (Robert C. Schmitt); Financial Variables in Rental Housing (Leon Pollard); Historical Review of the Gravity and Potential Concepts of Human Interaction (Gerald A. P. Carrothers).

308. Kilbridge, Maurice D.; O'Block, Robert P.; Teplitz, Paul V., "A Conceptual Framework for Urban Planning Models," *Management Science*, Vol. 15, No. 6, February 1969, B-246–B-266.

Conceptual Framework; Subject of the Model: Land Use, Transportation, Population, Economic Activity, Degree of Aggregation; Model's Function: Projection, Allocation, Derivation, Model Diagrams; Model's Underlying Theory: Behavioral, Growth Forces; Operational Method of the Model: Econometric Forms, Mathematical Programming, Simulation; Other Analytic Forms; Problems and Prospects; Annotated Bibliography of Models . . ., pp. B-262–B-264; General Bibliography of Urban Models, B-264–B-266.

309. Kirby, Robert M., "A Comparison of Short and Medium Range Statistical Forecasting Methods," *Management Science*, Vol. 13, No. 4, December 1966, pp. B-202-210.

Introduction, Statistical Projection Methods Tested, Actual and Constructed Series Used in Tests, Simulation Test Method, Caused Noise, Forecasting Trend Dominated Series, Forecasting Cycle Dominated Series, Short Versus Intermediate Range Forecasts, Seasonal Adjustment, Choice of Parameter Values, Other Forecasting Tools, Observations on Exponential Smoothing and Moving Average, Conclusion.

310. Krasnov, Howard S.; Merikallio, Reino A., "The Past, Present, and Future of General Simulation Languages," *Management Science*, Vol. 11, No. 2, November 1964, pp. 236-267.

Introduction—The Past: Services Provided by General Simulation Languages; The Present: Characteristics of Some Simulation Languages, Comparison of Some Simulation Languages; The Future.

311. Krutch, Joseph Wood, "What the Year 2000 Won't Be Like," *Saturday Review*, 20 January 1968, pp. 12-14, 43.

"Prediction is hazardous. But rereading previous forecasts convinces a prominent social critic of one thing—even in a computer age, nobody can foresee the future."

312. Kuhn, Alfred, "Science, Models, and Systems," in: . . . . *The Study of Society—A Unified Approach*, Homewood, Ill. (Irwin and Dorsey), 1963, pp. 27-54. [63-16897]

Things, Versus Information About Things: "Amplification" of Information; Science: Observations and Classifications: Observation Stage, Classification Stage; Science: Functional Relations and Causes; Science and Models; Science and Systems: Simple Action Systems, or Nonfeedback Systems, Feedback Systems, Uncontrolled Feedback Systems, Cybernetic Systems, or Controlled Feedback Systems, Oscillation, Systems and Subsystems, Boundaries of Systems; Is This a System?: Learning in Systems; Concluding Observations.

313. Leibowitz, Martin A., "Queues," *Scientific American*, Vol. 219, No. 2, August 1968, pp. 96-103.

"Waiting in line is done not only by people but also by such things as freight cars, airplanes, telephone calls and computer routines. Mathematical analysis of queues suggests ways to shorten them."

314. Lichfield, Nathaniel; Margolis, Julius, "Benefit-Cost Analysis as a Tool in Urban Government Decision Making," in: Schaller, Howard G. (Editor), *Public Expenditure Decisions in the Urban Community,* Washington, D. C. (Resources for the Future), 1963, pp. 118-146. [63-22774]. Also: reprint by Resources for the Future, Washington, D. C., 1963.

Elements of Decision-Making and Cost-Benefit Analysis; Special Aspects of Urban Government; Logic of Benefit-Cost Analysis in Urban Government: Identifying Goals, Constraints, and Criteria; Methodological Comments: Social Evaluation of Benefits, Discount Rate, Prices, "With and Without," Interdependencies and Externalities, Parties—Balance Sheet or Index; Typical Analyses Suggested by Benefit-Cost Considerations: Constraint Upon Choice—Cost-Revenue and Density, Criterion for Choice—Fire Prevention, Balance Sheet of Benefits and Costs— Preservation of a Historic Building, Balancing Conflicting Goals—Alternative Plans for a Central Area.

315. Lichfield, Nathaniel, "Cost-Benefit Analysis in City Planning," *Journal of the American Institute of Planners,* Vol. XXVI, No. 4, November 1960, pp. 273-279.

Need for a Welfare Test in City Planning; Welfare Test Should Be of Cost-Benefit Type; Method of Cost-Benefit Analysis for Testing a City Plan: Definition of Goals, Program of Projects, "With and Without" Comparison of Projects, Nature of Costs and Benefits, Measurement of Costs and Benefits, Summation, Planning Decision.

316. Lichfield, Nathaniel, "Cost-Benefit Analysis in Plan Evaluation," *The Town Planning Review,* Vol. XXXV, No. 2, July 1964, pp. 159-169.

"Economics can make its contribution throughout the planning process. . . . This paper is concerned with one limited aspect; how a particular tool of economic analysis, cost-benefit analysis, which in planning circles has come to be popularly called the 'Planning Balance Sheet,' should and could be used in one particular kind of plan evaluation."
Evaluation in Planning: Economists' Approach to Evaluation; What Is Cost-Benefit Analysis?; Toward Cost-Benefit Analysis in Planning; The Approach Can Be Used Today.

317. Lichfield, Nathaniel, "Implementing Buchanan: Problem of Social Choice," *Traffic Engineering and Control,* Vol. 6, No. 8, December 1964, pp. 498-500.

Social Choice; Decision Making Model; Goal; Using the Model; Professional Advisor; Levels of Choice in Implementing Buchanan; Applying the Decision Model; Immediate Possibilities.

318. Lowry, Ira S., "A Short Course in Model Design," *Journal of the American Institute of Planners,* Vol. XXXI, No. 2, Special Issue, May 1965 pp. 158-166.

Uses of Models: Descriptions, Predictions, Planning; Theories and Models; Strategy of Model Design: Level of Aggregation, Treatment of Time, Concept of Change,

Solution Methods; Fitting a Model: Variables, Parameters; Testing a Model; Evaluation; References, pp. 165-166.

319. Meier, Richard L.; Duke, Richard D., "Gaming Simulation for Urban Planning," *Journal of the American Institute of Planners*, Vol. XXXII, No. 1, January 1966, pp. 3-17.

"A game employing the roles of planner-administrator, politician, and businessman for three to fifteen players has been developed and played more than thirty times. It employs a reduction in time scale of more than one in ten million and other necessary simplifications. That model is now being hybridized so as to become a 'teaching machine' for developing a metropolis of a quarter million population."
Undertaking Simulation; Styles of Simulation; Designing Computer Simulations; Urban Developers; Hybridization Process; Summary.

320. Miller, George A., "Thinking Machines: Myths and Actualities," *The Public Interest*, No. 2, Winter 1966, pp. 92-112.

Can Machines Think?; Automatic Desk Calculator; Automatic Filing Cabinet; Automatic Voltmeter; Cheap Memories; Giant Brains; Cybernetic Approach; An Example—Noise, Subroutines, Recursive Subroutines; Psychological Experiment; Implications for the Future; Will Our Computers Destroy Us?; Case for Optimism.

321. Nadler, Gerald, "An Investigation of Design Methodology," *Management Science*, Vol. 13, No. 10, June 1967, pp. B-642—B-655.

Shortcomings Of the Research Approach In Design; Toward a New Approach; Suggested Design Methodology; Additional Research; Appendices: Interview With an Engineer, Suggested Research Topics For Phase 1, Five Interviews Arranged By Decision Points in Professional's Approach, Suggested Research Topics For Phase 2.

322. Nelson, Eldred C.; Branch, Melville C., "Techniques of Analysis," Chapter V in: Branch, Melville C., *The Corporate Planning Process*, New York (American Management Association), 1962, pp. 121-179. [62-15083]

Mathematical-Statistical Method: Probability and Statistics, Models, Control Theory, Theory of Games, Information Theory, Mathematical Programming, Systems Engineering; Subjective Judgment: Identification, Development; Synthesis: Conceptualization, Institutional Mechanisms of Conceptualization, The Corporate Representation, Performance of People; Summary.

323. Peterson, William, "On Some Meanings of 'Planning'," *Journal of the American Institute of Planners*, Vol. XXXII, No. 3, May 1966, pp. 130-142.

Deductive Planning; Utopian Planning; Inductive Planning; Facts; Evaluation.

324. Pfouts, Ralph W., "An Empirical Testing of the Economic Base Theory," *Journal of the American Institute of Planners*, Vol. XXIII, No. 2, [1957], pp. 64-69. Harris, Britton, "Comment on Pfouts' Test of the Base Theory," *ibid.*, Vol. XXIV, No. 4, 1958, pp. 233-237; Pfouts, Ralph W., "Reply to Harris on Testing the Base Theory," *ibid.*, pp. 238-243;

Tiebout, Charles M., "Harris versus Pfouts: A Third Party Note," *ibid.*, pp. 244-246; Andrews, Richard B., "Comment Re Criticisms of the Economic Base Theory," *ibid.*, Vol. XXIV, No. 1, pp. 36-40.

". . . the results of a statistical testing of the economic base theory of urban development. Since the statistical tests do not support the economic base theory . . . alternative directions of inquiry [are suggested] that may lead to an economic theory and an apparatus of analysis that would be of use to city planners and others interested in urban growth and development."

325. Pollard, Leon, "The Interrelationships of Selected Measures of Residential Density," *Journal of the American Institute of Planners,* Vol. XX, No. 2, Spring 1954, pp. 87-94.

"Various attempts to determine desirable densities for residential development have resulted in standards for several of these ratios. . . . The planner should be aware of those factors in the situation under consideration which would tend toward the qualification of standards, or more precisely, toward the variation of ratios."

326. Prest, A. R.; Turvey, R., "Cost-Benefit Analysis: A Survey," *The Economic Journal,* Vol. LXXV, No. 300, December 1965, pp. 683-735.

Introduction; General Principles: Preliminary Considerations—Statement of the Problem, A General Issue, Main Questions—Enumeration of Costs and Benefits, Valuation of Costs and Benefits, Choice of Interest Rate, Relevant Constraints, Final Considerations—Investment Criteria, Second-Best Matters; Particular Applications: Water Projects—Irrigation, Flood Control, Hydro-Electric Power Schemes, Multi-Purpose Schemes, Transport Projects—Roads, Railways, Inland Waterways, Land Usage—Urban Renewal, Recreation, Health, Education, Other Fields; Conclusions; Bibliography, pp. 731-735.

327. Rogers, Andrei, "Matrix Methods of Population Analysis," *Journal of the American Institute of Planners,* Vol. XXXII, No. 1, January 1966, pp. 40-44.

Mortality; Fertility; Mortality, Fertility, and the Interregional System; Growth Model for California Region; Conclusion.

328. Rowe, Alan J., "Simulation: A Decision-Aiding Tool," in: *Proceedings of International Conference, Institute of Industrial Engineers,* New York, 23-25 September 1963, pp. 135-144.

Introduction; Organizational Decisions; Management Decision Making and the Computer; Simulation Technique; Uses of Simulation; Problems of Simulation; Successful Applications of Simulation; Case Study; Future of Simulation; Appendix and References.

329. *Scientific American,* Mathematics in the Modern World, Vol. 211, No. 3, September 1964, Entire Issue, pp. 40-216.

Mathematics in the Modern World (Richard Courant); Number (Philip J. Davis); Geometry (Morris Kline); Algebra (W. W. Sawyer); Probability (Mark Kac); Foundations of Mathematics (W. V. Quine); Mathematics in the Physical Sciences (Freeman J. Dyson), Biological Sciences (Edward F. Moore), Social Sciences

(Richard Stone); Control Theory (Richard Bellman); Computers (Stanislaw M. Ulam).

330. *SDC Magazine*, "System Development for Regional, State and Local Government," Vol. 8, No. 10, October 1965, pp. 1-27.

Education: Rockland County, N. Y., Province of Quebec; Civic Improvement: State of California, Bay Area Transportation System; Economic Development: Appalachia; Administration of Justice: Los Angeles Police Department, New York State Identification and Intelligence System, New York City Police Department.

331. Shubik, Martin, "Gaming: Costs and Facilities," *Management Science*, Vol. 14, No. 11, July 1968, pp. 629-660.

Postscript and Introduction; Uses and Scope of Gaming: Teaching, Operational Gaming, Experimental Gaming; Administration, Facilities, and Costs: General Comments, The Computer, Laboratory or Other Facilities and Supplies, Game Construction, "Debugging and Dress Rehearsals," Debriefing and Analysis, Documentation and Replication; Conclusions and Suggestions: Need for Joint Usage, Making Gaming Pay for Itself; Appendix I: Questions on the Costs and Economics of Gaming.

332. Silvers, Arthur L.; Sloan, Allan K., "A Model Framework for Comprehensive Planning in New York City," *Journal of the American Institute of Planners*, Vol. XXXI, No. 3, August 1965, pp. 246-251.

Policy Issues; Components of the Modeling Framework: An Investor Behavior Model.

333. Smalter, Donald J., "The Influence of Department of Defense Practices on Corporate Planning," *Management Technology*, Vol. 4, No. 2, December 1964, pp. 115-138.

Lessons for the Corporate Planner: Planning by Missions, Application of Operations Research Principles to Complex Strategy Research, Stepwise Thought Process in Problem Solving, Logic Sequence Network Diagrams, Decision-Making Room.

334. Smith, David C., "Detailed Plant Models Help More Firms Cut Cost of Construction," *The Wall Street Journal*, Pacific Coast Edition, Vol. LXXI, No. 53, 14 September 1964, pp. 1 & 12.

Refinery Miniature Saves Fluor $25,000 on Drawings, Time; DuPont Makes Own Models. Other Advantages, Too; Popular for "Process" Plants.

335. Smith, Robert D.; Greenlaw, Paul S., "Simulation of a Psychological Decision Process in Personnel Selection," *Management Science*, Vol. 13, No. 8, April 1967, pp. B-409—B-419.

". . . computer simulation of the decision processes of a psychologist dealing with the ill-structured problem of analyzing psychological test scores and other data concerning individuals being considered for various types of clerical and clerical-administrative positions."

336. Sonenblum, Sidney; Stern, Louis H., "The Use of Economic Projections

in Planning," *Journal of the American Institute of Planners,* Vol. XXX, No. 2, May 1964, pp. 110-123.

View of Economic Planning; Private Planning; Economic Policy Planning; Program Planning; Planning and Projections; Logic of Planning; National Economic Projections; Private Planning by Business; Difference between the Problem at the National and State Level; Economic Policy Planning at the State Level; Program Planning; Summary; Some Components of an Informational System for Planning; Determinants of Regional Characteristics; Informational System and National and State Planning; Informational System and Program Planning; Conclusion.

337. Starr, Martin K., "Planning Models," *Management Science,* Vol. 13, No. 4, December 1966, pp. B-115–B-141.

Definition of the Planning Function, Basic Building Blocks—Unit Decisions; Plans vs. Policies, More Formal Definition of Planning; Types of Planning Models, Fully-Constrained Planning Systems: Extensive Networks of Determinate Form, Non-Search Methods, Search Methods—Linear Programming, Dynamic Programming, Simulation Methods, Competitive Systems; Partially-Constrained Planning Systems: Stochastic Analog of a PERT Network, In Summary; Threshold-Constrained Planning Systems: Action Limits, Sequential Decisions, Size and Scale of Available Resources, Planning Horizons, In Summary; Appendices.

338. Stegman, Michael A., "Accessibility Models and Residential Location," *Journal of the American Institute of Planners,* Vol. XXXV, No. 1, January 1969, pp. 22-29.

"Many residential location models have been developed within the context of long-range transportation planning programs and tend to explain housing consumer behavior largely in terms of minimizing the journey to work. This article questions the preeminence of accessibility in the residential location process and offers empirical evidence that neighborhood considerations are more important to residential locators than accessibility to place of work. It concludes with some recommendations for future modeling activity."

339. System Development Corporation, "Simulation: Managing the Unmanageable," *SDC Magazine,* Vol. 8, No. 4, April 1965, pp. 1-25.

Simulating National Policy Formation (Robert Boguslaw; Robert H. Davis): Simulation Vehicle, Pilot Experiment One, Pilot Experiment Two, Results and Conclusions; Leviathan (Beatrice and Sidney Rome): Introduction to Method, Conceptual Approach—Social Hierarchy, Simulation, 1963 Experiments, 1964 Experiments, Conclusion.

340. Voorhees, Alan M. (Special Editor), Land Use and Traffic Models: A Progress Report, *Journal of the American Institute of Planners,* Special Issue, Vol. XXV, No. 2, May 1959, pp. 55-104.

Nature and Uses of Models in City Planning (Alan M. Voorhees); Are Land Use Patterns Predictable? (Hans Blumenfeld); Predicting Chicago's Land Use Pattern (John R. Hamburg; Roger L. Creighton); How Accessibility Shapes Land Use (Walter G. Hansen); Economic Forces Shaping Land Use Patterns (Arthur Row; Ernest Jurrat); Forecasting Traffic for Freeway Planning (William B. Colland); Method for Predicting Urban Travel Patterns (Howard W. Bevis); Transit vs. Auto

Travel in the Future (James Booth; Robert Morris); Data Processing for City Planning (Roger L. Creighton; J. Douglas Carroll, Jr.; Graham S. Finney). Selected Bibliography on Land Use and Traffic Models, p. 104.

341. Vorhaus, Alfred H., "General Purpose Display System," *SDC Magazine*, Vol. 8, No. 8, August 1965, pp. 1-15.

"Is there a display system that contributes to the user-computer dialogue? Yes."

342. Wendt, Paul F., "Forecasting Metropolitan Growth," *California Management Review*, Vol. IV, No. 1, Fall 1961, pp. 27-34.

"There is a pattern to urban growth which makes it possible to forecast booms in industrial areas and spot high rent residential areas in the making. Underlying principles are topography, transportation, and the uneasy dynamics of the high income class seeking better, more exclusive residential sites. Time maps, sector analyses, input-output forecast are tools used to predict change in the megalopolis."

343. Wheaton, William L. C., "Operations Research for Metropolitan Planning," *Journal of the American Institute of Planners*, Vol. XXIX, No. 4, November 1963, pp. 250-258.

Washington Plan; Denver Plan; Planners' Biases; Problems in Metropolitan Analysis; Area Data Systems; Some Attainable Goals for Operations Research Analysis.

344. Wolfe, Harry B., "Model of San Francisco Housing Market," *Socio-Economic Planning Sciences*, Vol. 1, No. 1, September 1967, pp. 71-95.

A model is described . . . designed to predict the effects of public programs (zoning restrictions, code enforcement, taxation, subsidies, renewal and improvement projects) on the quality, quantity and location of city housing.

Model Operation; Model Elements: Housing Stock, Space Transitions, Space Pressure; Model Results; Extension of the CRP Simulation Model.

345. Wood, Marshall K., "Parm—An Economic Programming Model," *Management Science*, Vol. 11, No. 7, May 1965, pp. 619-680.

Background: Historical Development, Communication and Model Structure, Acceptable Programs, Organizational Context; Model Structure: Activities, Factor Records, Mathematical Model, Triangularization, System Organization, Production Computation, Activity Transformation; Empirical Implementation: Consumer Demand, Government, Capacity and Investment, Production, Manpower Requirements; Current Economy Applications: Need, Proposal, Approach, Model Structure, Behavioristic Relationships, Regional Relationships.

Books, Reports, Pamphlets

346. Almon, Clopper, Jr., *The American Economy to 1975, An Interindustry Forecast*, New York (Harper & Row), 1966, 169 pp. [67-11339]

". . . the first long-range forecasts of the American economy to make full use of what we now know about the supply connections among the industries of the economy." Consistent Interindustry Forecasting for Long-Range Business Planning; Personal Consumption; Capital Investment; Government Demands; Exports and Imports;

Input-Output Table; Labor Force and Productivity; Prospect; Appendix: Mathematical Method Used in Generating the Forecasts.

347. Anderson, Stanford, *Planning for Diversity and Choice,* Cambridge, Mass. (M.I.T.), 1968, 340 pp. [68-14458]

Introduction (Stanford Anderson); Utopian Thinking and the Architect (I. C. Jarvie); Long-Range Studies of the Future and Their Role in French Planning (Bernard Cazes); Parameters of Urban Planning (Leonard J. Duhl); About "Mankind 2000" (Robert Jungk); Population of the United States in the Last Third of the Twentieth Century (Herbert Moller); Natural Resources in the Changing U.S. Economy (Harold J. Barnett); Leisure—Old Patterns and New Problems (Pardon E. Tillinghast); Obsolescence and "Obsolescibles" in Planning for the Future (Bruce Mazlish); Normative Planning (Paul Davidoff); Ideology and Architecture —Dilemmas of Pluralism in Planning (Leonard J. Fein); Triumph of Technology— "Can" Implies "Ought" (Hasan Ozbekhan); Social Indicators—Or Working in a Society Which Has Better Social Statistics (Raymond A. Bauer); Telos and Technique—Models As Modes of Action (Marx Wartofsky); Outline of a Pluralistic Theory of Knowledge and Action (Paul K. Feyerabend); Environmental Conjecture—In the Jungle of the Grand Prediction (Melvin Charney); Bibliography, pp. 327-334.

348. Asimov, Morris, *Introduction to Design,* Englewood Cliffs, N.J. (Prentice-Hall), 1962, 135 pp. [62-10550]

Philosophy of Engineering Design; Engineering Design and the Environment; Morphology of Design; Feasibility Study; Preliminary Design; Detailed Design Phase; Design Process, Analysis of Needs and Activity, Decision Processes in Design; Archetypes and Computers; Techniques of Optimization; Case Study.

349. Bartelli, L. J.; Klingebiel, A. A.; Baird, J. V.; Heddleson, M. R. (Editors), *Soil Surveys and Land-Use Planning,* Madison, Wis. (Soil Science Society of America; American Society of Agronomy), 1966, 196 pp. [66-26147]

Soil Surveys for Community Planning (Charles E. Kellogg); Soil Surveys and the Regional Land Use Plan (Robert H. Doyle); Soils and Their Role in Planning a Suburban County (David B. Witwer); Use of Soil Maps by City Officials for Operational Planning (W. R. Hunter; C. W. Tipps; J. R. Coover); Use of Soils Information in Urban Planning and Implementation (John G. Morris); Applications of Soils Studies in Comprehensive Regional Planning (Kurt W. Bauer); Use of Soil Surveys by a Planning Consultant (Carol J. Thomas); Use of Soil Surveys in Subdivision Design (John R. Quay); Use of Agricultural Soil Surveys in the Planning and Construction of Highways (Thomas H. Thornburn); Use of Soil Surveys in Planning for Recreation (P. H. Montgomery; F. C. Edminister); Improving Soil Surveys Interpretations Through Research (Gerald W. Olson); Urban Soils Program in Prince William County, Virginia (Dwight L. Kaster; Oscar W. Yates, Jr.), "System" of Soil Survey, Interpretation, Education, and Use (W. J. Meyer; W. H. Bender; K. A. Wenner; J. Rinier); Use of Soil Surveys in the Equalization of Tax Assessments (Robert R. Kinney); Soil Surveys and Urban Development—An Educational Approach (Harry M. Galloway); Educational Programs to Aid Agricultural Users of Soil Survey Reports (O. W. Bidwell); Quantitative Aspects of Soil

Survey Interpretation in Appraisal of Soil Productivity (William R. Oschwald); Role of Detailed Soil Surveys in Preparation and Explanation of Zoning Ordinances (M. T. Beatty; D. A. Yanggen); Changes in the Need and Use of Soils Information (S. S. Obenshain).

350. Berry, Brian J. L.; Marble, Duane F. (Editors), *Spatial Analysis— A Reader in Statistical Geography,* Englewood Cliffs, N.J. (Prentice-Hall), 1968, 512 pp. [68-10856]

Methodology: Quantitative Revolution and Theoretical Geography (Ian Burton), Approaches to Regional Analysis—A Synthesis (Brian J. L. Berry), Identification of Some Fundamental Spatial Concepts (John D. Nystuen), Geography and Analog Theory (Richard J. Chorley), Short Course in Model Design (Ira S. Lowry); Spatial Data and Spatial Statistics: Use of Computers in the Processing and Analysis of Geographic Information (Richard C. Kao), Geographic Area and Map Projections (Waldo R. Tobler), Geographic Sampling (Brian J. L. Berry; Alan M. Baker), Statistical Analysis of Geographical Series (Roberto Bachi), Temperature Extremes in the United States (Arnold Court), Aspects of the Morphometry of a "Polycyclic" Drainage Basin (Richard J. Chorley), Interactance Hypothesis and Boundaries in Canada—Preliminary Study (J. Ross Mackay), Physics of Population Distribution (John Q. Stewart; William Warntz); Analysis of Spatial Distributions: Statistical Study of the Distribution of Scattered Villages in Two Regions of the Tonami Plain, Toyama Prefecture (Isamu Matui), Quantitative Expression of the Pattern of Urban Settlements in Selected Areas of the United States (Leslie J. King), Family of Density Functions for Lösch's Measurements on Town Distribution (Michael F. Dacey), Modified Poisson Probability Law for Point Pattern More Regular Than Random (Michael F. Dacey), Maps Based on Probabilities (Mieczyslaw Choynowski), Climatic Change As a Random Series (Leslie Curry), Trend-Surface Mapping in Geographical Research (R. J. Chorley; P. Haggett), Fourier Series Analysis in Geology (John W. Harbaugh; Floyd W. Preston), Connectivity of the Interstate Highway System (William L. Garrison), Aspects of the Precipitation Climatology of Canada Investigated by the Method of Harmonic Analysis (Michael E. Sabbagh; Reid A. Bryson); Study of Spatial Association: Distribution of Land Values in Topeka, Kansas (Duane S. Knos), Correlation and Regression Analysis Applied to Rural Farm Population Densities in the Great Plains (Arthur H. Robinson; James B. Lindberg; Leonard W. Brinkman), Mapping the Correspondence of Isarithmic Maps (Arthur H. Robinson), Regional and Local Components in the Distribution of Forested Areas in Southeast Brazil—Multivariate Approach (Peter Haggett), Maps of Residuals from Regression (Edwin N. Thomas), Multivariate Statistical Model For Predicting Mean Annual Flood in New England (Shue Tuck Wong), Monte Carlo Approach to Diffusion (Tors'ten Hågerstrand); Regionalization: Geographic Distribution of Crop Productivity in England (M. G. Kendall), Graph Theory Interpretation of Nodal Regions (John D. Nystuen; Michael F. Dacey), Synthesis of Formal and Functional Regions Using a General Field Theory of Spatial Behavior (Brian J. L. Berry); Problems in the Analysis of Spatial Series: Additional Comments on Weighting Values in Correlation Analysis of Areal Data (Edwin N. Thomas; David L. Anderson), Some Alternatives to Ecological Correlation (Leo A. Goodman), Contiguity Ratio and Statistical Mapping (R. C. Geary), Review on Measures of Contiguity for Two and K-Color Maps (Michael F. Dacey).

351. Black, Guy, *The Application of Systems Analysis to Government Operations*, New York (Praeger), 1968, 186 pp. [68-18914]

"The typical reader . . . in mind was a trained and experienced professional person, occupying a high or middle level in government, and involved in a program or activity for which systems approaches had been suggested."

Place of Systems Analysis in Government; Concepts of a System; Methods of Systems Analysis; Benefit Functions in Systems Analysis; Systems Concept and Costing; Optimization; Data Needs and Systems Analysis; After Systems Analysis; Do We Want More Systems Analysis in Government?; Appendix; Bibliography, pp. 173-181.

352. Bowman, William J., *Graphic Communication*, New York (Wiley), 1968, 210 pp. [67-29931]

Figuratively Speaking; Visual Language; Graphic Statement; To Show: What, How, How Much, Where.

353. Branch, Melville C., *Aerial Photography in Urban Planning and Research*, Harvard City Planning Studies, Vol. 14, Cambridge, Mass. (Harvard University), 1948, 150 pp. [48-7715]

Use of Aerial Photographs for Urban Planning and Research: Vertical Stereophotographs, Oblique Photographs, Photographic Mapping, Cost of Aerial Surveys; Special Knowledge Prerequisite to the Effective Use of Air Photos for Urban Planning and Research: Plain Vertical Photographs, Oblique Photographs, Stereophotographs, Photoruns and Plotting, Photo Mosaics, Flight Planning, Photographic Quality, Cameras and Film, Night Photographs, Photographic Interpretation.

354. Brown, Harrison; Bonner, James; Weir, John, *The Next Hundred Years*, New York (Viking), 1963, 193 pp. [57-8404]

Presents facts and trends relative to the availability and utilization of raw materials, population and specialized manpower, food production, energy resources, the development of knowledge.

355. California Institute of Technology, *The Next Ninety Years*, Pasadena, Cal. (Office for Industrial Associates), 1967, 186 pp. Paper. [67-7369]

Welcoming Remarks; Next Ninety Years; Beyond Survival; Man, Water, and Waste; Some Problems of Emerging Nations; Experimental Cities; Next Ninety Years—Summary Discussion.

356. Chapin, F. Stuart, Jr.; Weiss, Shirley F., *Factors Influencing Land Development*, Chapel Hill, N.C. (Institute for Research in Social Science, University of North Carolina), 1962, 101 pp.

"This monograph on the formulation of a land development forecast model takes into account the relative importance of various land development 'priming' factors such as expressway location and provision of sewer facilities. Using both dwelling density and total land in urban use as dependent variables, the model is multidimensional in allowing for the structuring effect of physical distribution and the timing effect of sequential distribution."

357. Chestnut, Harold, *System Engineering Tools*, New York (Wiley), 1965, 646 pp. [65-19484]

Systems Engineering in Industry; Energy, Materials, Information; Modeling and Simulation; Computing; Control; Probability and Statistics; Signals and Noise; Optimizing; Tolerances, Variations and Disturbances; Engineering an Information Handling System.

358. Churchman, C. West, *The Systems Approach*, New York (Delacorte), 1968, 243 pp. [68-20106]

What Is a System?; Applications of Systems Thinking; Systems Approach to the Future; Systems Approach and the Human Being; Supplementary Sections: Exercises in Systems Thinking, Suggested Readings, pp. 239-243.

359. Cochran, William G., *Sampling Techniques*, New York (Wiley), Second Edition, 1963, 413 pp. [63-7553]

"An up-to-date discussion of the techniques appropriate for handling any type of survey, whether in business, market research, opinion research, medical science, sociology, or agriculture. It explains methods suitable for small-scale and large-scale surveys."

360. Coleman, James S., *Introduction to Mathematical Sociology*, Glencoe, Ill. (Free Press), 1964, 554 pp. [64-13241]

Uses of Mathematics in Sociology; Problems of Quantitative Measurement in Sociology; Mathematics As a Language for Relations Between Variables; Mathematical Language for Relations Between Qualitative Attributes; Relations Between Attributes—Over-Time Data; Multivariate Analysis; Multiple-Level Systems and Emergent Propositions; One-Way Process With a Continuous Independent Variable; Social and Psychological Processes and Their Equilibrium States; Poisson Process and Its Contagious Relatives; Poisson Process and Its Contagious Relatives—Equilibrium Models; Social and Psychological Organization of Attitudes; Change and Response Uncertainty; Measures of Structural Characteristics; Method of Residues; Study of Local Implications; Diffusion in Incomplete Social Structures; Tactics and Strategies in the Use of Mathematics; References, pp. 531-542.

361. Cullingworth, J. B.; Orr, S. C. (Editors), *Regional and Urban Studies —A Social Science Approach*, Beverly Hills, Cal. (Sage), 1969, 280 pp. Paperback.

Participation of Social Scientists in Planning—Background to the Studies (J. B. Cullingworth; Sarah C. Orr); Regional Economic Planning and Location of Industry (Sarah C. Orr); Planning and the Labour Market (L. C. Hunter); Regional Multiplier—Some Problems in Estimation (K. J. Allen); Regional Input-Output Analysis (Edith M. F. Thorne); Population (T. H. Hollingsworth); Housing Analysis (J. B. Cullingworth); Economics and Methodology in Urban Transport Planning (R. M. Kirwin); Employment Projection and Urban Development (J. T. Hughes); Comprehensive Urban Renewal and Industrial Relocation—Glasgow Case (G. C. Cameron; K. M. Johnson).

362. Department of Scientific and Industrial Research, Road Research

Laboratory, *Research on Road Traffic*, London (Her Majesty's Stationery Office), 1965, 505 pp.

Traffic Flow; Speeds; Traffic Surveys; Theoretical Models of Traffic; Traffic Capacity; Urban Traffic and Roads; Parking; Traffic Control at Intersections; Regulation and Automatic Control; Traffic Signs and Carriageway Markings; Pedestrians; Layout of Rural Roads; Safety Fences and Kerbs; Economics; Accident Studies Before and After Road Changes.

363. Dorfman, Robert (Editor), *Measuring Benefits of Government Investments*, Washington (Brookings Institution), 1965, 414 pp. [65-18313]

A symposium representing "a sophisticated attempt to come to grips with the problems of measuring the benefits of government programs. . . ."

Outdoor Recreation (Ruth P. Mack, Sumner Myers); Urban Highway Investment (Herbert Mohring); Urban Renewal Programs (Jerome Rothenberg); Preventing School Dropouts (Burton A. Weisbrod); Syphilis Control Programs (Herbert E. Klarman); Government Research and Development Programs (Frederic M. Scherer); Civil Aviation Expenditures (Gary Fromm).

364. Dove, Donald A., *Housing and Population*, LARTS Technical Bulletin 1-6, Los Angeles (Los Angeles Regional Transportation Study, Transportation Association of Southern California), 1966, 13 pp.

"Presents a method for estimating distribution of housing units and population by traffic zones or census tracts."

Purpose; Hypothesis; Assumptions; Data Collection: Dwelling Units Inventory, Mobile Homes Inventory, Group Quarters; Making the Estimates; Maintenance.

365. Draper, Norman; Smith, Harry, *Applied Regression Analysis*, New York (Wiley), 1966, 407 pp. [66-17641]

Fitting a Straight Line by Least Squares; Matrix Approach to Linear Regression; Examination of Residuals; Independent Variables; More Complicated Models; Selecting the "Best" Regression Equation; Typical Multiple Regression Problem; Multiple Regression and Mathematical Model Building; Multiple Regression Applied to Analysis of Variance Problems; Introduction to Non-Linear Least Squares; Answers to Exercises; Bibliography, Appendices, Index.

366. Enrick, Norbert Lloyd, *Management Operations Research*, New York (Holt, Rinehart & Winston), 1965, 320 pp. [65-12740]

"This book offers the principles and methods of operations research (OR) as a management science in a simplified nonmathematical form for the practical manager and interested student."

General Nature of Operations Research as a Management Science: Introduction—Operations Research in Modern Management, Operations Research and Accounting Methods, Decision Making under Uncertainty—Case Study, Introducing Operations Research Techniques in an Organization; Planning, Programming, and Program Review: Linear Programming—Graphic Method, Matrix Method, Profitability Analysis, Simplex Programming—Case History, Linear Programming—Transportation Method, Case History, Planning of Capital Improvements and Other Long-

Term Investments, Capital Investment—Further Aspects of Discounting Future Earnings, Program Evaluation and Review Techniques (PERT), PERT in Large-Scale Systems, Line-of-Balance (LOB) Systems; Management of Inventories and Waiting Lines: Principles of Inventory Management, Economic Lot Sizes, Proper Reorder Points, Inventories to Smooth Market Fluctuations, Management of Waiting Lines (Queues), Simulation of Inventories and Queues and Other Business Problems; Sampling and Statistical Analysis: Statistical—Surveys Using Probability Samples, Reports Based on Probability Samples, Analysis Techniques Pertinent to Samples, Work Sampling as an OR Tool, Principles of Efficient Management, Experimental Design Application; Management in the Seventies.

367. Enrick, Norbert Lloyd, *Management Planning—A Systems Approach,* New York (McGraw-Hill), 1967, 217 pp. [67-27823]

"[Purpose of] this book [is to] serve the practical manager and the student alike by bringing a clear, use-oriented introduction to the quantitative Management Science methods of Mathematical Programming for operations planning."

368. Ewald, William R., Jr. (Editor), *Environment and Change—The Next Fifty Years,* Bloomington, Ind. (Indiana University), 1968, 392 pp. Paperback. [68-27345]

Introduction (William R. Ewald, Jr.); Creating the Future Environment: Future of Man (Pierre Bertaux), On Attending to the Future (Bertrand de Jouvenal), Arts in Modern Society (Harold Taylor), Role of the Spirit in Creating the Future Environment (Joseph Sittler), Life Where Science Flows (John R. Platt), Role of Technology in Creating the Environment Fifty Years Hence (Ralph G. H. Siu); A Future Filled With Change: Faustian Powers and Human Choices—Some Twenty-First Century Technological and Economic Issues (Herman Kahn; Anthony J. Wiener), How Technology Shapes the Future (Emmanuel G. Mesthene), Young Rebels (Carl Oglesby), Effective Society (Claude Brown), On Whose Behalf Is the Dream Being Dreamt? (Max Lerner), Planning *with* People (Robert Theobald); Prologue to the Future: Culture of Urban America (John Burchard), Taking Stock—Resumé of Planning Accomplishments in the United States (Carl Feiss); Future American Society and Future Role of the Individual: Development of Administrative and Political Planning in America (Robert C. Wood), Necessity and Difficulty of Planning the Future Society (Gunnar Myrdal), New Factor in American Society (David T. Bazelon), The Individual—Not the Mass (August Heckscher), Individuals in a Collective Society (Sir Geoffrey Vickers); Context of the Future—Youth, Technology, and the World: Technology and the Underdeveloped World (Renato Severino), The Hope There Is in People (Ann Schrand), People Build with Their Hands (John F. C. Turner), Operating Manual for Spaceship Earth (R. Buckminster Fuller), Epilogue—Mood for Development (William L. C. Wheaton).

369. Fite, Harry H., *The Computer Challenge to Urban Planners and State Administrators,* London, England (Macmillan), 1965, 142 pp. [65-27826]

Challenge of Technology: Computer Challenge to State and Local Government, Computers, Technology and Government, World Automation and Management Education, The Future—Big Enough for Man *and* Automation; Computers and

Urban Government: Sounder Decisions in City Government Through Computers, Automation's New Frontier—Municipal Process Control, Centralized Computer Traffic Control, Computer Traffic Control for Smaller Cities; Computers and State Government: Administrative Evolution in ADP in State Government, Automation of a State Revenue Department—Systems Approach; Selected References, pp. 131-140.

370. Gibbs, Jack P. (Editor), *Urban Research Methods,* Princeton, N.J. (Van Nostrand), 1961, 625 pp. [61-66842]

Urban Units, Their Nature and Boundaries: Influence of the Definition of the Urban Place on the Size of the Urban Population (Milos Macura), Notes on the Concepts of "City" and "Agglomeration" (G. Goudswaard), Delimitation of Urban Areas (Olaf Boustedt), "Urbanized Areas" (U.S. Bureau of the Census), Growth and Study of Conurbations (General Register Office), Standard Metropolitan Statistical Areas (U.S. Bureau of the Census), Methods and Problems in the Delimitation of Urban Units (Jack P. Gibbs); Some Basic Characteristics of Urban Units: Measurement and Control of Population Densities (William H. Ludlow), Method for Comparing the Spatial Shapes of Urban Units (Jack P. Gibbs), Measurement of Change in the Population Size of an Urban Unit (Jack P. Gibbs), Components of Population Change in Suburban and Central City Populations of Standard Metropolitan Areas, 1940-1950 (Donald J. Bogue; Emerson Seim), Methods for Describing the Age-Sex Structure of Cities (Harley L. Browning); Spatial Structure of Urban Units: City Block As a Unit for Recording and Analyzing Urban Data (Edward B. Olds), Theory and Practice of Planning Census Tracts (Calvin F. Schmid), Compatibility of Alternative Approaches to the Delimitation of Urban Sub-Areas (William H. Form, et al), Delimiting the CBD (Raymond E. Murphy; J. E. Vance, Jr.), Use of Local Facilities in a Metropolis (Donald L. Foley), Some Measures of the Spatial Distribution and Redistribution of Urban Phenomena (Jack P. Gibbs); Urban Hinterlands and Functional Types of Cities: Urban Hinterlands in England and Wales—Analysis of Bus Services (F. H. W. Green), Hinterland Boundaries of New York City, and Boston in Southern New England (Howard L. Green), Differentiation in Metropolitan Areas (Leslie Kish), Measurement of the Economic Base of the Metropolitan Area (John M. Mattila; Wilbur R. Thompson), Functions of New Zealand Towns (L. L. Pownall), Service Classification of American Cities (Howard J. Nelson), Economic Structural Inter-relations of Metropolitan Regions (Walter Isaard; Robert Kavesh); Characteristics of Urbanization: Some Demographic Characteristics of Urbanization (Jack P. Gibbs), Conventional Versus Metropolitan Data in the International Study of Urbanization (Jack P. Gibbs; Kingsley Davis), Some Measures of Demographic and Spatial Relationships Among Cities (Harley L. Browning; Jack P. Gibbs); Rural-Urban Differences: Distinction Between Urban and Rural—National Prac-tices and Recommendations (United Nations), Community Size and Rural-Urban Continuum (Otis Dudley Duncan), Traits of the Urban and Rural Populations of Latin America (Ana Casis; Kingsley Davis), Trends in Rural and Urban Fertility Rates (T. J. Woofter, Jr.), Regional Comparisons Standardized for Urbanization (Otis Dudley Duncan); Rural-Urban Interrelations: Gradients of Urban Influence on the Rural Population (Otis Dudley Duncan), Note on Farm Tenancy and Urbanization (Otis Dudley Duncan), On the Estimation of Rural-Urban Migration (Jack P. Gibbs); Bibliography, Compiled in 1959, pp. 581-621.

371. Gibson, W. L., Jr.; Hildreth, R. J.; Wunderlich, Gene, *Methods for Land Economics Research*, Lincoln, Neb. (University of Nebraska), 1966, 242 pp. Paperback. [66-19269]

Perspective on Content and Methodology of Land Economics (Walter E. Chryst; W. B. Black); Identification of Problems (R. J. Hildreth; Emery N. Castle); Hypothesis—Guides for Inquiry (Loyd K. Fischer; Howard W. Ottoson); Classification in the Research Process (W. B. Black; Gene Wunderlich); Choice of Empirical Techniques (Gene Wunderlich; W. L. Gibson, Jr.); Analysis of Variance and Covariance (Henry Tucker); Correlation and Regression Analysis (E. J. R. Booth); Nonparametric Statistics (Robert F. Boxley, Jr.); Statistical Inference: Classical and Bayesian (Albert N. Halter); Budgeting and Programming in Economic Research (Clark Edwards); Operations Research Techniques (Earl R. Swanson); The Input-Output Model as a Tool for Regional Analysis (Frank M. Goode, George S. Tolley).

372. Gilmore, John S.; Ryan, John J.; Gould, William S., *Defense Systems Resources in the Civil Sector: An Evolving Approach, An Uncertain Market*, Prepared by the University of Denver Research Institute for the U.S. Arms Control and Disarmament Agency (Contract ACDA/E-103), Washington, D.C. (Superintendent of Documents), July 1967, 201 pp. Paper.

Evolution of the Systems Approach in the Defense Community: Components, Approach, Resources; Evolving Systems Approach in the Civil Sector: Early Experience; Application of Defense Systems Resources in the Civil Sector—Future Markets: Institutional Obstacles and Imponderables, Supply of Resources, Characteristics Affecting Demand; Conclusions and Recommendations; Appendices; Bibliography, pp. 165-190.

373. Gordon, T. J.; Helmer, Olaf, *Report on a Long-Range Forecasting Study*, P-2982, Santa Monica, Calif. (RAND), September 1964, 110 pp.; also as appendix in: Olaf, Helmer, *Social Technology*, New York (Basic Books), 1966.

Intent; Subject Matter; Method; Illustration of Procedure; Substantive Outcome—Introductory Remarks; Predicted Scientific Breakthroughs; Predicted Population Trends; Automation Predictions; Predicted Progress in Space; Predictions Concerning War and Its Prevention; Predicted Weapon Systems of the Future; World of 1984; World of 2000; Conceivable Features of the World in the Year 2100; Editorial Comments on These Forecasts; Convergence of Opinions; Prediction Precision as a Function of Time; Prediction Frequency as a Function of Time; Confidence as a Function of Predicted Date; Critique of Experimental Procedure; Conclusions; Bibliography, p. 65.

374. Greenberger, Martin (Editor), *Computers and the World of the Future*, Cambridge, Mass. (M.I.T.), 1962, 340 pp. Paperback. [62-13234]

Scientists and Decision Making (C. P. Snow; Others); Managerial Decision Making (J. W. Forrester; Others); Simulation of Human Thinking (H. A. Simon; Others); Library for 2000 A.D. (J. G. Kemeny; Others); Computer in the University (A. J.

Perlis; Others); Time-Sharing Computer Systems (J. McCarthy; Others); New Concept in Programming (G. W. Brown; Others); What Computers Should Be Doing (J. R. Pierce; Others); Selected Bibliography, pp. 327-332.

375. Guttenberg, Albert Z., *New Directions in Land Use Classification,* Chicago, Ill. (American Society of Planning Officials), 1965, 30 pp. Paper.

Introduction; Conceptual Framework; Analysis—Referential Mode; Evaluation—Appraisive Mode; Control—Prescriptive Mode; Perspectives.

376. Hall, Peter, *London 2000,* London, England (Faber and Faber), 1963, 220 pp.

Posing the Problem; Guiding Growth; Building the New London; Running the New London; Living in It.

377. Hauser, Philip M. (Editor), *Handbook for Social Research in Urban Areas,* Paris, France (Educational, Scientific and Cultural Organization), 1965, 214 pp.

Social Research Data and Procedures: Areal Units for Urban Analysis (Philip M. Hauser, Judah Matras), Basic Statistics and Research (Giuseppe Parenti), Field and Case Studies (P. H. Chombart de Lauwe), Other Social Research Approaches (Judah Matras); Types of Studies: Comprehensive Urban Studies (Philip M. Hauser), Demographic Trends in Urban Areas (Judah Matras), Social Organization in an Urban Milieu (P. H. Chombart de Lauwe), Migration and Acculturation (Gino Germani), Social and Personal Disorganization (Judah Matras), Research on Urban Plant and Administration (Z. Pioro).

378. Hendrick, Thomas W., *Modern Architectural Model,* London (Architectural Press), 1957, 144 pp. [58-448]

"Textbook for professionals and amateurs in [physical architectural] model making."

379. Highway Research Board, *Urban Development Models,* Special Report 97, Publication 1628, Washington, D.C. (2101 Constitution Ave., 20418), 1968, 266 pp.

Introduction: Conference Summary and Recommendations (Britton Harris), Opening Statements (Edward H. Holmes; William B. Ross), Agency Expectations from Predictive Models (Charles J. Zwick); Planning, Decision-Making, and the Urban Development Process: Plan Evaluation Methodologies—Some Aspects of Decision Requirements and Analytical Response (Wilbur A. Steger; T. R. Lakshmanan), Activity Systems as a Source of Inputs for Land Use Models (F. Stuart Chapin, Jr.), Towards a Theory of the City (Charles Levin); Design and Construction of Models: Seven Models of Urban Development—Structural Comparison (Ira S. Lowry), Access and Land Development (Morton Schneider), Quality of Data and Choice and Design of Predictive Models (William Alonso), Construction of Models (Kenneth Schlager; Britton Harris; T. R. Lakshmanan; Boris Pushkarev); Use of Models: Survey of Planning Agency Experience with Urban Development Models, Data Processing, and Computers (George C. Hemmens), Evaluation of Land Use Patterns (John R. Hamburg; Roger L. Creighton; Robert S. Scott), Communication in the Field of Urban Development Models (David E. Boyce).

LEWIS AND CLARK COLLEGE LIBRARY
PORTLAND, OREGON 97219

380. Hirsch, Werner Z., *Introduction to Modern Statistics—With Applications to Business and Economics*, New York (Macmillan), 1957, 429 pp. [57-5774]

". . . arranged primarily to meet the needs of schools of business and departments of economics . . ."

Getting Meaning Out of a Mass of Data; "On the average"; Dispersion; Superstition, Hunch, and the Laws of Chance; Inference; Sampling Distribution of Means; Estimating Means and Totals; Biromial Distribution—Success or Failure, How Likely Are They?; Estimating Percentages; Decisions About Means; Decisions About Percentages; Index Numbers—Tying Up to the Hitching Post; Association Among Quantitative Data—Regression and Correlation Analysis; Time Series Analysis—Prophecy Galore, I and II; Association Among Qualitative Data; Decisions by Control Charts; Statistician in the Age of Electronics; Appendices; Bibliography, pp. 416-421.

381. Hogben, Lancelot, *Mathematics for the Million*, New York (Norton), 1950, 690 pp. [51-8025]

Mathematics, Mirror of Civilization; Mathematics in Prehistory; Translating Number Language; What You Can Do With Geometry; Beginnings of Arithmetic; What We Can Do With Trigonometry; How Algebra Began; Spherical Triangles; What Are Graphs?; How Logarithms Were Discovered; What the Calculus Is About; Arithmetic of Human Welfare.

382. Howlett, Bruce, *Land Use Handbook—A Guide to Understanding Land Use Surveys*, Chicago, Ill. (Northeastern Illinois Metropolitan Area Planning Commission), 1961, 35 pp.

What This Handbook Does; Idea of Land Use; What Is a Land Use Map?; Uses of Land Use Studies; Preparing a Land Use Survey; Land Use Classification System; Preparing for Field Work; Conducting the Field Survey; Field Mapping Procedure; Showing Mixed Uses; Office Procedures; Preparing Final Maps; Special Study Maps; Technical Notes; Appendices: Index of Land Uses, Abbreviations for Detailed Studies.

383. Hoyt, Homer, *The Structure and Growth of Residential Neighborhoods in American Cities*, Federal Housing Administration, Washington, D.C. (Superintendent of Documents), 1939, 178 pp.

Technique of Analysis; Structure of Residential Neighborhoods in American Cities; Ground Plan of Cities, Segregation of Land Uses, Analysis of Residential Areas, Alternative Technique in the Analysis of Residential Areas, Composition of Urban American Dwellings and Their Inhabitants, Patterns of Residential Rent Areas; Growth of Residential Neighborhoods: Influence of the Rate of City Growth, Form of City Growth, Changes in Urban Land Uses, Patterns of Movement of Residential Rental Neighborhoods; Appendix; Map Supplement.

384. Huff, Darrell, *How to Lie with Statistics*, New York (Norton), 1954, 142 pp. Paperback.

Sample with the Built-in Bias; Well-Chosen Average; Little Figures That Are Not There; Much Ado About Practically Nothing; Gee-Whiz Graph; One-Dimensional

Picture; Semiattached Figure; Post Hoc Rides Again; How to Statisticulate; How to Talk Back to a Statistic.

385. Huff, David L., *Determination of Intra-Urban Retail Trade Areas,* Los Angeles, Calif. (Graduate School of Business Administration, University of California), 1962, 47 pp. Paper.

Summary and Conclusions; Introduction: Historical Background of Gravity Models In Retailing, Limitations of Gravity Models, Purpose and Scope of This Study; Characteristics of the Model: Terms and Definitions, Basic Propositions, Empirically Measuring a Shopping Center's Utility; Empirical Test of the Model; Type and Size of Sample, Analysis of Observed and Expected Behavior, Variations in λ Among Products; Use of Model in Determining a Shopping Center's Trading Area. Appendix: Tables, Figures.

386. Isard, Walter, *Methods of Regional Analysis: An Introduction to Regional Science,* Cambridge, Mass. (M.I.T.), 1960, 784 pp. [60-11723]

Setting; Population Projection; Migration Estimation; Regional Income Estimation and Social Accounting; Interregional Flow Analysis and Balance of Payments Statements; Regional Cycle and Multiplier Analysis; Industrial Location Analysis and Related Measures; Interregional and Regional Input-Output Techniques; Industrial Complex Analysis; Interregional Linear Programming; Gravity, Potential, and Spatial Interaction Models; Channels of Synthesis; Retrospect and Prospect.

387. Janke, Rolf, *Architectural Models,* New York (Praeger), 1968, 139 pp. [68-10688]

Introduction; Designing with Models; Model Types; Model-Making; Photographing Architectural Models; Bibliography, pp. 134-135.

388. Johnson, Richard A.; Kast, Fremont E.; Rosenzweig, James E., *The Theory and Management of Systems,* New York (McGraw-Hill), 1963, 350 pp. [62-20722]

Systems Concepts and Management: Systems Theory and Management, Concepts—Planning and Systems, Organization and Systems, Control and Systems, Communication and Systems, Integration of Systems; Applications: Weapon-System Management, Rhochrematics, Automation and Numerical Control, Data Processing Systems; Implementation: Management Science, PERT/PEP Techniques of Network Analysis, Systems Design, People and Systems; The Future: Systems Management; Bibliography, pp. 330-341.

389. Kahn, Herman; Wiener, Anthony J., *The Year 2000—A Framework for Speculation on the Next Thirty-Three Years,* New York (Macmillan), 1967, 431 pp. [67-29488]

Change and Continuity; Comments on Science and Technology; Some "Surprise-Free" Economic Projections—Quantitative Scenario; Postindustrial Society in the Standard World; International Politics in the Standard World; Some Canonical Variations from the Standard World; Some Possibilities for Nuclear Wars; Other Twenty-First Century Nightmares; International System in the Very Long Run; Policy Research and Social Change.

390. Land Classification Advisory Committee of the Detroit Metropolitan Area, *Land Use Classification Manual*, Chicago, Ill. (Public Administration Service), 1962, 53 pp. Paper. [62-22018]

Developing a Standard Land Use Code; Using the Code; Reporting Results; Land Use Classification Color and Screen Guide; Land Use Classification and Code; Index to Code to 3-Digit Level.

391. Levinson, Horace C., *Chance, Luck and Statistics: The Science of Chance*, New York (Dover), 1963, 357 pp. [63-3453]

Chance, Luck and Statistics; Gamblers and Scientists; World of Superstition; Fallacies; Grammar of Chance; "Heads or Tails"; Betting and Expectation; Poker Chances; Poker Chances and Strategy; Roulette; Lotteries, Craps, Bridge; From Chance to Statistics; Chance and Statistics; Fallacies in Statistics; Statistics at Work; Advertising and Statistics; Business and Statistics.

392. Loomba, N. Paul, *Linear Programming—An Introductory Analysis*, New York (McGraw-Hill), 1964, 284 pp. [64-18902]

". . . written primarily for the beginning student in linear programming."
Linear Programming and Management; Graphical Method; Systematic Trial-and-Error Method; Matrices and Vectors; Vector Method; Simplex Method, I and II; Dual; Degeneracy; Transportation Model; Assignment Model; Meaning of Linearity; Note on Inequalities; System of Linear Equations Having a Unique Solution; System of Linear Equations Having No Solution; System of Linear Equations Having an Infinite Number of Solutions; Selected Bibliography, pp. 279-280.

393. Los Angeles, Department of City Planning, *Introduction and Proposed Program, The Mathematical Model Development Program*, Los Angeles, Cal. (City Hall, 90012), 28 September 1966, 48 pp. Paper. *Ibid., Residential Location Models*, 23 November 1966, 17 pp. Paper.

Abstract; Preface; Introduction: Mathematical Models, Initiation of the Mathematical Model Development Program, System Development Corporation's Recommended Approach, Conclusion of Phase I; Recommended Short-Range and Long-Range Programs: Short-Range, Long-Range, Discussion of Programs, Summary by Prof. Harris; Some Representative Models of Residential Location; Data Availability; Bibliography, pp. 41-48.
Introduction; University of Pennsylvania Residential Location Model; References, p. 17.

394. Lowry, Ira S., *Migration and Metropolitan Growth*, San Francisco (Chandler), 1966, 120 pp. [66-27475]

Introduction; Place-to-Place Migration Flows: Evolution of a Migration Model, Migration, Distance and Economic Opportunity, Regression Analysis, Illustrative Comparison—Albany and San Jose, Some Implications of the Model; Population Change Due to Migration: Blanco Model, Regression Analysis, Demand for Labor and Net Migration, Changes in the Residential Labor Force and Net Migration, Military Employment and Net Migration, School Enrollment and Net Migration, Income Changes and Net Migration, Parsimonious Model of Population Change Due to Migration, Final Note; Forecasting Migration: Place-to-Place Migration

Flows, Population Change Due to Migration, Errors of Estimate, Evaluating the "Independent" Variables, Generality of the Fitted Parameters, Forecasting Net Migration, 1960-63; Summary and Conclusions.

395. Lowry, Ira S., *Seven Models of Urban Development: A Structural Comparison*, Pub. No. P-3673, Santa Monica, Cal. (RAND), September 1967, 47 pp. Paper.

Introduction; Market for Urban Land; Classifying Models of Urban Development; Land Use—CATS Model, Land Use Succession—UNC Model, Location—EMPIRIC Model, Migration—POLIMETRIC Model, Hybrid—Pittsburgh Model, Market Demand—Penn Jersey Model, Market Supply—San Francisco Model; Conclusions.

396. Mace, Ruth L., *Municipal Cost-Revenue Research in the United States*, A Critical Survey of Research to Measure Municipal Costs and Revenues in Relation to Land Uses and Areas: 1933-1960, Chapel Hill, N.C. (Institute of Government, University of North Carolina), 1961, 201 pp. Paperback.

Background and Preview; Slum Losses and Redevelopment Gains; Balancing Land Uses in the Suburbs; Cost-Revenue Analysis in Connection with Annexation; The Missing "Big Picture"; Summary and Forecast.

397. McMillan, Claude; Gonzalez, Richard F., *Systems Analysis—A Computer Approach to Decision Models*, Homewood, Ill. (Irwin), 1965, 336 pp. [65-12415]

Systems and Models; Simulation; Introduction to Programming; Inventory System Under Certainty; Probability Concepts; Monte Carlo Simulation—Inventory System Under Uncertainty; Basic Queueing Concepts: Process Generators; Simulation of Queueing Systems; Management Planning Models; Large-Scale Simulation Models of the Firm; Industrial Dynamics; Study in Total Systems Simulation; Experimentation.

398. Martin, Brian V.; Memmott, Frederick W., III; Bone, Alexander J., *Principles and Techniques of Predicting Future Demand for Urban Area Transportation*, M.I.T. Report No. 3, Cambridge, Mass. (M.I.T.), June 1961, 222 pp. [64-65157]

Total Transportation Planning; Principles and Techniques: Inventories of Existing Conditions, Estimates of Future Urban Area Growth, Determining Future Travel Demand; Summary; Future Research; Appendices; Bibliography: Traffic Estimation and Assignment, pp. B1-B8.

399. Miller, Delbert C., *Handbook of Research Design and Social Measurement*, New York (David McKay), 1964, 332 pp. Paperback [64-10332]

General Description of the Guides to Research Design and Sampling; Guides to Statistical Analysis; Selected Sociometric Scales and Indexes: Social Status, Group Structure and Dynamics, Morale and Job Satisfaction, Community, Social Participation, Leadership in the Work Organization, Important Attitude Scales, Family and Marriage, Personality Measurements, Inventory of Measures Utilized in the American Sociological Review During 1951-60; Research Costing and Reporting.

400. Monkhouse, F. J.; Wilkinson, H. R., *Maps and Diagrams, Their Compilation and Construction*, New York (Dutton), Second Edition, 1963, 432 pp.

Materials and Techniques; Relief Maps and Diagrams; Climatic Maps and Diagrams; Economic Maps and Diagrams; Population Maps and Diagrams; Maps and Diagrams of Settlements; Appendix—Introduction to Numerical and Mechanical Techniques.

401. Morgan, Bruce W., *An Introduction to Bayesian Statistical Decision Processes*, Englewood Cliffs, N.J. (Prentice-Hall), 1968, 116 pp. Paperback.

". . . written by a non-technician for his fellow non-technicians."
Chapters on Bayes theorem, decision process, rectangular and normal probability functions, single-sample procedures, and sequential decision procedures. A brief, elementary introduction to Bayesian procedures.

402. Morse, Philip M. (Editor), *Operations Research for Public Systems*, Cambridge, Mass. (M.I.T.), 1967, 212 pp. Paperback. [67-27347]

Introduction (Philip M. Morse); Operational Research in Local Government (R. A. Ward); Simulation Models and Urban Planning (Harry B. Wolfe; Martin L. Ernst); Vehicular Traffic (Leslie C. Edie); Evaluation of Alternative Transportation Networks (Peter S. Loubal); Operations Research in Medical and Hospital Practice (William J. Horvath); Systems Approach to the Study of Crime and Criminal Justice (Alfred Blumstein; Richard C. Larson); Mathematical Techniques—Probabilistic Models (George P. Wadsworth); Mathematical Techniques—Mathematical Programming (John D. C. Little).

403. (Editors of) News Front, *The Image of the Future*, New York (Year), 1968, 255 pp. [68-18149]

"U.S. management in the decades ahead will need broad knowledge and long-range vision more than ever before."
53 articles on a wide variety of anticipated developments, ranging from intellectual climate and international relations outlook to miracles of medicine and infrared.

404. Norling, A. H. (Project Director), *Future U.S. Transportation Needs*, Cambridge, Mass. (United Research Corporation), October 1963, 289 pp. Paper. Typescript.

Prepared for U.S. National Aeronautics and Space Administration, Office of Aeronautical Research, under Contract No. NASw-585.

Introduction; Summary; Identification of Factors and Characteristics of Transportation Demand; Demographic and Economic Forces in the U.S. Through 1985; Intra-Urban Transportation Demand; Domestic Intercity Passenger Transportation Demand; Demand by U.S. Citizens for Foreign Travel; Future Trends in Intercity Freight Demand; Appendices; Bibliography, pp. B-1—B-10.

405. Optner, Stanford L., *Systems Analysis for Business and Industrial Problem Solving*, Englewood Cliffs, N.J. (Prentice-Hall), 1965, 116 pp. [65-16575]

Characteristics of Business Problems; Nature of Systems; Relationships Among Feedback Components; Feedback-Control in Problem Solving; Dealing with Alternatives, Assumptions, Criteria, and Risk; References, p. 109.

406. Pascal, Anthony H. (Editor), *Cities in Trouble: An Agenda for Urban Research,* Memorandum RM-5603-RC, Santa Monica, Cal. (RAND), August 1968, 159 pp. Paper.

Introduction (Adam Yarmolinsky); Housing (Ira S. Lowry; Barbara Woodfill); Urban Education (Daniel M. Weiler); Manpower Training and Jobs (Anthony H. Pascal); Public Assistance (William A. Johnson; Robert Rosenkranz); Public Order (Douglas F. Loveday; Samuel M. Genensky); Health Services (C. T. Whitehead).

407. Paterson, Robert W., *Forecasting Techniques for Determining the Potential Demand for Highways,* Columbia, Mo. (Business and Public Administration Research Center, University of Missouri), 1966, 128 pp. Paper.

Introduction; Population Estimation; Forecasts of Motor Vehicle Registrations; Traffic Forecasts; Epilogue; Bibliography, pp. 117-128.

408. Pfouts, Ralph W. (Editor), *The Techniques of Urban Economic Analysis,* West Trenton, N.J. (Chandler-Davis), 1959, 408 pp. Paperback. [59-11409]

Economic Base Theory and Its Implications: Historical Development of the Base Concept (Richard B. Andrews), Economic Base Analysis (Arthur M. Weimer; Homer Hoyt), Problem of Terminology, Classification of Base Types, Problem of Base Measurement, Base Identification—General Problem, Special Problems, Problem of Area Delimitation, Concept of Base Ratios, Base Concept and the Planning Process (Richard B. Andrews), Basic-Nonbasic Concept of Urban Economic Functions (John W. Alexander); Objections to the Economic Base Theory and An Alternative Theory: Classification Errors in Base-Ratio Analysis (James Gillies; William Grigsby), Economic Base of the Metropolis (Hans Blumenfeld), Economic Base Reconsidered (Charles M. Tiebout), Empirical Testing of the Economic Base Theory (Ralph W. Pfouts), Limitations of the Economic Base Analysis (Ralph W. Pfouts; Erle T. Curtis), Statics, Dynamics and the Economic Base (Charles E. Ferguson), Community Income Multiplier—Case Study (Charles M. Tiebout); An Alternative Methodology—Input-Output Approach: Economic Structural Interrelations of Metropolitan Regions (Walter Isard; Robert Kavesh), Planning Elements of an Inter-Industry Analysis—Metropolitan Area Approach (Abe Gottlieb), Regional and Interregional Input-Output Models—Appraisal (Charles M. Tiebout).

409. Phillips, Bernard S., *Social Research—Strategy and Tactics,* New York (Macmillan), 1966, 336 pp. [66-20823]

Theory and Method: Introduction to Research Methods, Elements of Inquiry, Process of Inquiry; Data Collection: Principles, Experiment, Interviews, Questionnaires and Surveys, Use of Documents, Observation, Simulation, Principles of Measurement and Scaling, Nominal and Ordinal Scales, Interval and Ratio Scales; Analysis of Data: One and Two Variables, Principles of Statistical Decision Making,

Sampling, Statistical Testing, and Degree of Association, Multivariate Analysis; Applications of Logic and Mathematics.

410. Prehoda, Robert W., *Designing the Future—The Role of Technological Forecasting*, New York (Chilton), 1967, 310 pp. [67-24412]

Economic Considerations; Promising Material Developments; Biological Applications.

411. Rogers, Andrei, *Matrix Analysis of Interregional Growth and Distribution*, Berkeley, Cal. (University of California), 1968, 119 pp. [68-11530]

". . . how the techniques of matrix algebra facilitate incorporation of interregional population migration into a population model . . . examines the possibilities of intervention into the population system in order to redistribute it and analyses various theories to account for why people migrate. The aim of the monograph is the introduction of the spatial dimension of population into mathematical demography."

412. Schmidt, Robert E.; Campbell, M. Earl, *Highway Traffic Estimation*, Saugatuck, Conn. (Eno Foundation for Highway Traffic Control), 1956, 247 pp. Paperback. [56-12510]

Introduction; Land Use Generation: Dwelling Unit and Residential Area, Central Business District, Off-Center Commercial Area, Off-Center Employment Area, City As a Generator, Estimating Urban Traffic Patterns; Generation of Terminals: Parking and Storage, Mode Translation; Distribution of Traffic: on Free Roads, Traffic Assignment and Road-User Economic Studies, Traffic Estimates for Toll Facilities; Traffic Growth—Methods of Forecasting: Mechanical, Analytical, Example of Proposed Method, Comparison with Average Factor Method; Critical Hour; Appendix: Traffic Data, Critical Hour Traffic Data.

413. Schnore, Leo F. (Editor), *Social Science and the City—A Survey of Urban Research*, New York (Praeger), 1968, 335 pp. Paperback. [68-26898]

Social Science and the City—Survey of Research Needs (Leo F. Schnore; Eric E. Lampard); The Assimilation of Migrants to Cities—Anthropological and Sociological Contributions (Lyle W. Shannon; Magdaline Shannon); Strategies for Discovering Urban Theory (Anselm L. Strauss); Class and Race in the Changing City—Searching for New Approaches to Old Problems (Eleanor P. Wolf; Charles N. Lebeaux); Toward an Urban Economics (Wilbur R. Thompson); Urban Travel Behavior (John F. Kain); Historical Perspective on Urban Development Schemes (Charles N. Glaab); Urban Geography and City and Metropolitan Planning (Harold M. Mayer); Political Science and the City (Norton E. Long); Comparative Study of Urban Politics (Robert R. Alford); Bibliography, pp. 303-330.

414. Siders, R. A.; Others, *Computer Graphics—A Revolution in Design*, New York (American Management Association), 1966, 160 pp. [66-24180]

Power and Potential of Computer Graphics; Engineering Design Cycle; Evolution of Computer Aids to Design; Role of Passive Computer Graphics; Role of Active Computer Graphics; Who Can Best Use Computer Graphics?; Economics of the

Computer Graphics Decision; Conditions for Success; Impact of Computer Graphics on the Company; Plan of Action; Look at the Future of Computer Graphics; Bibliography, pp. 157-160.

415. Springer, Clifford H.; Herlihy, Robert E.; Mall, Robert T.; Beggs, Robert I., *Statistical Inference,* Volume Three of the Mathematics for Management Series, Homewood, Ill. (Irwin), 1966, 352 pp. [65-8233]

". . . objectives are to increase awareness and understanding of the uses of mathematics and statistics in business decision-making . . . not to develop mathematicians, but rather to develop appreciation of mathematics as a language well-adapted to arriving at logical nonobvious conclusions . . . of practical value in real-life problem situations."

Beyond Arithmetic; Basic Ideas of Probability; Statistical Populations; More Statistical Populations; Principles of Sampling, Statistical Estimation; Drawing Statistical Conclusions; Making Statistical Decisions; Appendices.

416. Teichroew, Daniel, *An Introduction to Management Science: Deterministic Models,* New York (Wiley), 1964, 713 pp. [64-17154]

Formulating Business Problems: Formalizing Business Problems, Problems Involving Progressions and Series, Mathematics of Interest and Periodic Payments; One-Decision Variable Optimization Models: Management Science Models, Optimization of Functions of One Variable, Models Involving Polynomials and Rational Functions, Models Involving Algebraic and Transcendental Functions, Models Involving Integration and Differential Equations; Two-Decision Variable Optimization Models: Optimization of Functions of Two-Decision Variables, Examples of Optimization with Two-Decision Variables, Optimization Subject to Constraints; Linear Systems: Introduction to Matrices and Linear Systems, Matrix Inversion and Linear Systems, Examples of the Use of Matrices in Business Problems, Linear Programming, Examples of Linear Programming, Special Cases of Linear Programming; Multivariate Optimization Models: Optimization Models with Many Variables, Examples of Optimization Models with Many Variables, Quadratic Forms, Dynamic Programming.

417. Thompson, W. W., Jr., *Operations Research Techniques,* Columbus, Ohio (Merrill), 1967, 157 pp. Paperback. [66-29158]

"Proofs and other mathematical developments have been kept to a minimum . . . a selected reading list which accompanies each chapter includes appropriate references to material of a more theoretical nature."

Introduction; Introduction to Linear Programming; Simplex and Transportation Methods of Linear Programming; Basic Inventory Models; Simulation.

418. Tiebout, Charles M., *The Community Economic Base Study,* Supplementary Paper No. 16, New York (Committee for Economic Development), December 1962, 84 pp. [62-22333]

Guide to Economic Base Studies: Importance of Understanding the Community Economic Base, Uses of a Base Study, How to Undertake an Economic Base Study; Economic Base Analysis: Structure of the Local Economy, Measuring the Local Economy, Structural Interrelations in the Local Economy, Forecasting Community Economic Levels; Selected Bibliography, pp. 83-84.

419. Tweraser, Gene C., *Urban Real Estate Research—1965,* Research Monograph 13, Washington, D.C. (Urban Land Institute), 1967, 76 pp. Paper. [59-4179]

Research Activity, 1965—Accomplishments and Work in Process; Inventory of Urban Real Estate Research, 1965: Bibliographies, Central Business District, Community Analysis, Economic Base, Finance, Housing, Industry, Land Planning and Use, Metropolitan Area Studies, Real Estate Business, Real Estate Market, Redevelopment and Renewal, Regional Studies, Shopping Centers and Commercial Areas, Taxation, Textbooks and Reference Works, Transportation, Urban Research, Urbanism.

420. U.S. National Aeronautics and Space Administration, Office of Scientific and Technical Information, *Space, Science, and Urban Life,* Conference Proceedings, NASA SP-37, Washington, D.C. (Superintendent of Documents), 1963, 254 pp. Paper.

Space City, and the Conference Objectives (Wayne E. Thompson); Implications of the Space Effort for Science and Technology (George L. Simpson, Jr.); Insight to the Scope of the National Space Exploration Program (De Marquis D. Wyatt); Goals and Potentials of Scientific Research in Space (Homer E. Newell); Is Our Scientific Research Serving the Needs of the Nation? (George P. Miller); Twenty Years of Economic and Industrial Change (Robert A. Gordon); Impact of Scientific Technology Upon Industry and Society (Jerome B. Wiesner); Some Opportunities and Obstacles in Transferring New Technology Among Various Sectors of the Economy (Charles N. Kimball); Economic, Political, and Sociological Implications of Expanding Space and Scientific Knowledge (J. Herbert Hollomon); Research, Education, and Government—Special Message on Behalf of the State of California (Edmund G. Pat Brown); Perspective and Objectives of Our National Space Program (James E. Webb); What Scientific Developments Will Effect the Transportation, Communication, Power Resources, and Construction Industries in the Years Immediately Ahead? (William O. Baker); Can New Space and Scientific Technology Be Applied to Basic Community Problems of Water Supply, Air Pollution, Public Health and Safety, and Sanitation? (Karl W. Wolf); What Does the Space Program Reveal About Human Engineering and Medical Research? (B. B. McIntosh); Developing and Maintaining Open Channels of Communications Between the Laboratory, Industry, and the Community (George L. Simpson, Jr.); What Specific Implications Does Expanding Technology Have Upon the Problems of Metropolitan Areas? (Martin Meyerson); How Do the Changing Demands for Manpower and Technical Production Affect the Economy of Industry and Community? (Bernard D. Haber); What Immediate Progress Can Be Made to Apply New Space and Scientific Technology to Greater Use in Our Urban and Industrial Communities?

421. Walker, Marshall, *The Nature of Scientific Thought,* Englewood Cliffs, N.J. (Prentice-Hall), 1963, 184 pp. Paperback. [63-9519]

". . . written for the general, educated reader, and no previous knowledge of science or mathematics is assumed."

Scientific Method; Survival Technique; Science and Philosophy; Science and Mathematics; Factors Affecting Measurements; Predictability in Science; Physical

Factors Affecting Models; Biological Factors Affecting Models; Social Factors Affecting Models; Concepts Used in Mathematics; Concepts Used in Physical Science; Models in Biology; Models in Social Science; Science and Ethics; Status of Science Today.

422. Weinberg, Gerald M., *PL/I Programming Primer*, New York (McGraw-Hill), 1966, 278 pp. [66-22793]

Introduction; Basic Program Elements; Program Structuring; Varieties of Elements and Structuring; Subroutine Structure; Manipulation of Multidimensional Arrays; Processing Strings; Keeping Records; Preparing Reports; Programming Mechanics.

Bibliographies

423. Anderson, Stanford, *Planning for Diversity and Choice*, Bibliography: Future-Oriented Writings, pp. 327-334.

424. Batchelor, James H., *Operations Research, An Annotated Bibliography*, St. Louis, Mo. (St. Louis University), Vol. 1, Second Edition, 1959, 865 pp.; Vol. 2, 1962, 628 pp.; Vol. 3, 1963, 384 pp.; Vol. 4, 1964, 474 pp. [59-9440]

"The great variety of publications cited is an imposing indication of the universal nature and wide range of disciplines drawn upon in operations research."

425. Black, Guy, *The Application of Systems Analysis to Government Operations*, Bibliography, pp. 173-181.

426. Chapin, F. Stuart, *Selected References on Urban Planning Methods and Techniques*, Chapel Hill, N.C. (Department of City and Regional Planning, University of North Carolina), 1968, 68 pp.

"Contains nearly 1,000 references on planning methods and techniques."
Urban Economy; Population Studies; Studies of Activity Systems; Studies of Urban Environmental Setting; Land Use Planning; Planning for Industrial and Commercial Areas; Planning for Central Business Districts, Shopping Centers and Other Retail Areas; Other Region-Serving Uses and Special Purpose Areas; Planning for Residential Areas; School Plant Planning; Planning for Open Space and Recreation Areas; Urban Transportation Planning (Thoroughfares & Transit); Thoroughfares and Streets; Parking Analysis; Mass Transportation Planning; Planning for Other Transportation Systems; Planning for Utilities, Wastes, and Related Aspects of Environmental Engineering; The Comprehensive Plan; Plan Effectuation.

427. Coleman, James S., *Introduction to Mathematical Sociology*, References, pp. 531-542.

428. Fite, Harry H., *The Computer Challenge to Urban Planners and State Administrators*, References, pp. 131-140.

429. Gibbs, Jack P. (Editor), *Urban Research Methods*, Bibliography, Compiled in 1959, pp. 581-621.

430. Gilmore, John S.; Ryan, John J.; Gould, William S., *Defense Systems Resources in the Civil Sector: An Evolving Approach, An Uncertain Market*, Bibliography, pp. 165-190.

431. Greenberger, Martin (Editor), *Computers and the World of the Future*, Selected Bibliography, pp. 327-332.

Social Issues and Management Implications; Simulation of Large Systems; Artificial Intelligence; Comparison with Living Organisms; Theory of Computation; Mechanical Translation; Information Storage and Retrieval; Time-Sharing and Man-Machine Interaction; Programming Languages; Processing and Control; Conference Proceedings and Special Collections; Bibliographies.

432. Janke, Rolf, *Architectural Models*, Bibliography, pp. 134-135.

433. Johnson, Richard A.; Kast, Fremont E.; Rosenzweig, James E., *The Theory and Management of Systems*, Bibliography, pp. 330-341.

434. Kilbridge, Maurice D.; O'Block, Robert P.; Teplitz, Paul V., "A Conceptual Framework for Urban Planning Models," Annotated Bibliography of Urban Models . . ., pp. B-262—B-264, General Bibliography of Urban Models, B-264—B-266.

435. Kraemer, Kenneth L.; Lewis, Ralph J., *The Systems Approach in Urban Administration—Planning, Management and Operations*, Bibliography No. 49, Council of Planning Librarians, Monticello, Ill. (Exchange Bibliographies, P.O. Box 229, 61856), 1968, 60 pp.

". . . computer-produced using the KWIC (Key Word In Context) technique."

436. Loomba, N. Paul, *Linear Programming—An Introductory Analysis*, Selected Bibliography, pp. 270-280.

437. Los Angeles, Department of City Planning, *Introduction and Proposed Program, The Mathematical Model Development Program*, Bibliography, pp. 41-48.

438. McLaughlin, James F., *Application of Linear Programming to Urban Planning*, Bibliography No. 45, Council of Planning Librarians, Monticello, Ill. (Exchange Bibliographies, P.O. Box 229, 61856), 1968, 4 pp.

"Thesis abstract and bibliography."

439. Malcolm, D. G., "Bibliography on the Use of Simulation in Management Analysis," *Operations Research*, Vol. 8, No. 2, March-April 1960, pp. 169-177.

440. Martin, Brian V.; Memmott, Frederick W., III; Bone, Alexander J., *Principles and Techniques of Predicting Future Demand for Urban Area Transportation*, Bibliography: Traffic Estimation and Assignment, pp. B1-B8.

441. Miller, Delbert C., *Handbook of Research Design and Social Measurement*, Selected Bibliography on Research Design, p. 51.

442. Murfee, Fuller E.; Hercules, Wendell L., *Operations Research (Supplement): A DDC Report Bibliography*, Document No. AD-426 275, Springfield, Va. (Clearinghouse, U.S. Department of Commerce), January 1964, 1,587 References.

Prepared at Defense Documentation Center. Covers period January 1953-October 1963. Covers special mathematical disciplines and methods developed under operations research and their application to management and engineering problems: allocation, decision, game, and inventory theory; Lanchester equations; linear, dynamic, and nonlinear programming queueing theory; replacement theory; stochastic processes; transportation problem.

443. Norling, A. H. (Project Director), *Future U.S. Transportation Needs*, Bibliography, pp. B-1—B-10.

444. Paterson, Robert W., *Forecasting Techniques for Determining the Potential Demand for Highways*, Bibliography, pp. 117-128.

445. Prest, A. R.; Turvey, R., "Cost-Benefit Analysis: A Survey," Bibliography, pp. 731-735.

446. Schnore, Leo F. (Editor), *Social Science and the City—A Survey of Urban Research*, Bibliography, pp. 303-330.

447. Tummins, Marvin, *Forecasting and Estimating: A Selected Annotated Bibliography With Special Emphasis on Methodology, Supplemented by a Section of Unannotated Entries*, Charlottesville, Va. (Virginia Council of Highway Investigation and Research, Virginia Department of Highways and University of Virginia), September 1961, 96 pp. [61-62777]

General in coverage.

6. METHODOLOGY

Articles, Periodicals

448. Altshuler, Alan, "The Goals of Comprehensive Planning," *Journal of the American Institute of Planners*, Vol. XXXI, No. 3, August 1965, pp. 186-195.

"Comprehensive planning requires of planners that they understand the overall goals of their communities. [Because these are] too general to provide a basis for evaluating concrete alternatives . . . it is difficult to stir political interest in them, and politicians are rarely willing to commit themselves to let general and long-range goal statements guide their considerations of lower-level alternatives. Many planners have themselves abandoned the comprehensive planning ideal in favor of the ideal of middle-range planning [which] has much to recommend it. It provides no basis, however, for planners to claim to understand overall community goals. With it as a guide, therefore, the fundamental distinction between planning and other specialties is likely to become progressively more blurred."

449. Branch, Melville C.; Robinson, Ira M., "Goals and Objectives in Civil Comprehensive Planning," *The Town Planning Review,* Vol. 38, No. 4, January 1968, pp. 261-274.

Planning Objectives: Types, Analytical Formulation, Uncertainty and Comprehensive Planning; Planning Goals: Goals of the Public for City Planning, Staff Planning and the Goals of the Public; Function of City Planning with Respect to Goals and Objectives, Summary.

450. Fitch, Lyle C., "Eight Goals for an Urbanizing America," *Daedalus,* Vol. 97, No. 4, Fall 1968, pp. 1141-1164.

451. Granger, Charles H., "The Hierarchy of Objectives," *Harvard Business Review,* Vol. 42, No. 3, May-June 1964, pp. 63-74.

"There are objectives within objectives, within objectives . . . all require painstaking definition and close analysis if they are to be useful separately and profitably as a whole."

Role and Importance; Tests of Validity; Complete Framework; Steps in Derivation; Practical Uses; Conclusion.

452. Greenhouse, Samuel M., "The Planning-Programming-Budgeting System: Rationale, Language, and Idea-Relationships," *Public Administration Review,* Vol. XXVI, No. 4, December 1966, pp. 271-277.

Basic Concept—Accountability; Objectives; Programs; Program Alternatives; Output; Progress Measurement; Input; Alternative Ways to Do a Given Job; Systems Analysis; Summary.

453. Gross, Bertram M., "What Are Your Organization's Objectives?: A General-Systems Approach to Planning," *Human Relations,* Vol. 18, No. 3, August 1965, pp. 195-216.

Need For a Language of Purposefulness; General-Systems Approach; Performance-Structure Model; Performance Objectives; Structure Objectives; Strategy of Planning.

454. Guttenberg, Albert Z., "The Tactical Plan," in: Webber, Melvin, M.; Others, *Explorations into Urban Structure,* Philadelphia (University of Pennsylvania), 1964, pp. 197-219. [63-15009]

Functional Centrality as an Objective; Tactical Planning; Goal Plan and Tactical Plan, Tactical Plan and Conventional Capital Programming; Tactical Form-Example: Regional Growth and Its Consequences, Tactical Variant; Perspectives: Tactical Variant and Private Interest, Tactical Variant as an Urban Renewal Plan, Tactical Planner and the Public.

455. Haar, Charles M., "The Master Plan: An Inquiry in Dialogue Form," *Journal of the American Institute of Planners,* Vol. XXV, No. 3, pp. 133-142.

". . . an attorney and a city planner discuss some fundamental issues concerned with the role of the city council . . . professional planner . . . master plan . . . more concerned with posing the important questions than with supplying answers."

456. Hill, Morris, "A Goals-Achievement Matrix for Evaluating Alternative Plans," *Journal of the American Institute of Planners,* Vol XXXIV, No. 1, January 1968, pp. 19-29. Brandl, John E., "Comment . . .," *ibid.,* Vol. XXXV, No. 2, March 1969, pp. 139-141; Hill, Morris, "Rejoinder," *ibid.,* pp. 141-142.

Traditional Cost-Benefit Analysis; Balance Sheet of Development; Goals-Achievement Matrix: Procedure, Costs and Benefits in the Goals-Achievement Matrix; Comparison of Quantitative and Qualitative Objectives, Comparison of Goals-Achievement, Limits on the Application of the Goals-Achievement Matrix as a Tool for Plan Evaluation, Preliminary Appraisal of the Goals-Achievement Matrix.

457. Hoover, Robert C., "On Master Plans and Constitutions," *Journal of the American Institute of Planners,* Vol. XXVI, No. 1, February 1960, pp. 5-24.

Some Aspects of Metropolitan Change: Purpose Will Be Emphasized Over Function, Every Job Will Be a Profession, Citizenship Will Be Every Man's Profession; Forward Programming—Major Problem of Planning and Administration: Physical Capital Improvement Programming Is Not Comprehensive, What Is Capital Anyway?; Problem—To Achieve Comprehensive Planning and Preserve Democratic Institutions: Where Are Planning Criteria To Be Found?, Planning As an Advisory Instrument To Policy Making, Concept of Planning As a Fourth Power, Planning As an Executive Function, Concept of an Impermanent Constitution; Proposal for Metropolitan Planning Organization: Metropolitan Direction-Finding Commission's Long-Range Plan, Chief Executive's Plan For Physical Development and Service Operations, Legislature's Policy Statement and Socio-Physical Plan; Summary; Conclusion.

458. Kreditor, Alan, "The Provisional Plan," in: *Proceedings of Seminar on Industrial Development and the Development Plan, 19, 20, 21 April, 1966,* A Project of the Government of Ireland Assisted by the United Nations Special Fund and the United Nations, Dublin, Ireland (Graduate Program in Urban and Regional Planning, University of Southern California), 1967, pp. 27-36.

". . . some specific techniques and problems in the preparation of comprehensive plans in Ireland . . . a discussion of the Provisional Plan which, by Government policy, is the first stage in the planning process . . ."

459. Lamanna, Richard A., "Value Consensus Among Urban Residents," *Journal of the American Institute of Planners,* Vol XXX, No. 4, November 1964, Research Notes, pp. 317-323.

Introduction (Marc Fried); Research Design; Ideal Town; Consensus on Particular Values; Differences in Rank Order of Values; Conclusion.

460. Levy, Ferdinand K.; Thompson, Gerald L.; Wiest, Jerome D., "The

ABCs of the Critical Path Method," *Harvard Business Review*, Vol. 41, No. 5, September-October 1963, pp. 98-107.

What Is the Method?; Example—Building a House; Critical Path Algorithm; Concept of Slack; Handling Data Errors; Cost Calculations; New Developments; Conclusion.

461. Lu, Weiming, "Thoroughfare Planning and Goal Definition," *Traffic Quarterly*, Vol. 17, No. 2, April 1963, pp. 236-248. [50-1781]

Values of Goals; Ideal Thoroughfare Planning Program; Impact Upon Minneapolis; Major Street Planning Goals; Goal Adoption; Conclusions.

462. Meyerson, Martin, "Building the Middle-Range Bridge for Comprehensive Planning," *Journal of the American Institute of Planners*, Vol. XXII, No. 2, Spring 1956, pp. 58-64.

Professionalization; Expanding Functions: Central Intelligence, Pulse-Taking, Policy Clarification, Detailed Development Plan, Feed-Back Review; Implications of These Functions for Planning Agencies.

463. Michael, Donald N., "Urban Policy In the Rationalized Society," *Journal of the American Institute of Planners*, Vol. XXXI, No. 4, November 1965, pp. 283-288.

"During the next two decades much of the customary technical, social and political environment of the planning process will be in a state of rapid flux. . . . Concerned not so much with a new shape that should be given to the planning profession, but with the impact of its environment upon it. . . . The most important thread . . . is that of the *rationalization* of problem solving. . . ."
Future Urban Problem; Computer Technology and the Rationalized Society; Role of Government.

464. Millward, Robert E., "PPBS: Problems of Implementation," *Journal of the American Institute of Planners*, Vol. XXXIV, No. 2, March 1968, pp. 88-94.

". . . PPBS faces severe difficulties before it can become fully operational."
Historical Antecedents of PPBS; Why Has PPBS Been Introduced?; Mechanics of PPBS; Problems of Implementation: Conceptual, Operational; Conclusion.

465. Mitchell, Robert B. (Special Editor), Urban Revival: Goals and Standards, *Annals of the American Academy of Political and Social Science*, Vol. 352, March 1964, pp. 1-151.

Place of Nature in the City of Man (Ian L. McHarg); Physical and Mental Health in the City (Leonard J. Duhl; E. James Leiberman), Urban Social Differentiation and Allocation of Resources (Raymond W. Mack; Dennis C. McElrath); Culture Change and the Planner (Anthony N. B. Garvan); Urban Economic Development (John H. Nixon; Paul H. Gerhardt); Administrative and Fiscal Considerations in Urban Development (Werner Z. Hirsch); Political Side of Urban Development (Scott Greer; David W. Minar); Urban Pattern (Hans Blumenfeld); Public Art of City Building (David A. Crane); City Schools (Patricia Cayo Sexton); Housing and Slum Clearance—Elusive Goals (William G. Grigsby); Social Welfare Planning

(Elizabeth Wood); Recreation and Urban Development—Policy Perspective (Lowden Wingo, Jr.); Urban Transportation Criteria (Henry Fagin).

466. Plumley, H. L., "Long Range Planning: Where Do You Want to Be Five Years from Now?, *Business Management*, Vol. 27, No. 2, February 1965, pp. 36-41 ff.

"How State Mutual Assurance Co. of America sets goals."

467. Robinson, Ira M., "Beyond the Middle-Range Planning Bridge," *Journal of the American Institute of Planners*, Vol. XXXI, No. 4, November 1965, pp. 304-312.

Middle-Range Bridge for Comprehensive Planning: Central Intelligence Function, Pulse-Taking Function, Policy-Clarification Function, Detailed Development Plan Function; Feed-Back Review Function; Community Renewal Program—Concept and Practice: On-Going Data and Information Systems, New Methods for Testing Alternative Renewal Policies, Action-Program for Renewal; Next Step—Community Development Programming: Programming Process, Prospects of CDP.

468. *Saturday Review*, The New Computerized Age, 23 July 1966, pp. 15-36, 67.

The New World Coming (John Diebold); McLuhanism Reconsidered (Eric Barnouw); No Life Untouched (David Sarnoff); Automated Government (John W. Macy, Jr.); Plug-In Instruction (Patrick Suppes); For Each Student a Teacher (Don D. Bushnell); Book-Publisher's Salvation? (John Tebbel); The Town Meeting Reborn (Vernon F. Miller); Whither Personal Privacy (John Lear).

469. Schick, Allen, "The Road to PPB: The Stages of Budget Reform," *Public Administration Review*, Vol. XXVI, No. 4, December 1966, pp. 243-258.

Functions of Budgeting; Stages of Budget Reform; Control Orientation; Management Orientation; Planning Orientation; What Difference Does It Make?; Conclusion.

470. Young, Robert C., "Goals and Goal-Setting," *Journal of the American Institute of Planners*, Vol. XXXII, No. 2, March 1966, pp. 76-85.

Semantic Difficulties; Role of Goals in the Planning Process; Conceptual Nature of Goals—Goals as Ends and Means; Conflicts in Public Goals; Setting the Goals: Establishment of the Perimeter of Concern, Establishment of the Range of Choice, Examination of Relationships of Goals, Relative Evaluation of Goals or Sets of Goals, Establishment of Goals as Policy; Application to Comprehensive Planning.

Books, Reports, Pamphlets

471. Advisory Commission on Intergovernmental Relations, *Urban and Rural America: Policies for Future Growth*, Washington, D.C. (Superintendent of Documents), April 1968, 186 pp. Paper.

Pattern of Urbanization; Economic Growth—Regional, State, and Local Experience;

Impact of Recent Urbanization Trends; New Communities in America and Their Objectives; Guidance and Controls for Large-Scale Urban Development and New Communities; Conclusions and Recommendations; Appendices.

472. Alexander, Robert E., *Long Range Development Plan, University of California, San Diego*, San Diego, Cal. (University of California), October 1963, 53 pp. Paper.

Summary; History; Program; Site; Concept; Long Range Development Plan: Landscape, Grading, Circulation and Parking, Use of Land and Buildings, Utilities and Services, The Community, Stages, Architecture, Features; Appendices.

473. Archibald, Russell D.; Villoria, Richard L., *Network-Based Management Systems (PERT/CPM)*, New York (Wiley), 1967, 508 pp. [66-25216]

Network Planning—What It Is, How It Works; Implementing the System; Case Studies; Pitfalls and Potentials; References, Selected Bibliography, pp. 466-470.

474. Bartholomew, Harland, *A Comprehensive City Plan for Wichita, Kansas*, Wichita, Kan. (City Plan Commission), 1923, 128 pp. Paper.

Introduction; Major Street Plan; Transit; Transportation; Recreation; Housing and Sanitation; Civic Art; Zoning; Legal and Financial Matters; Appendix. Example of an early U.S. master city plan; available only in libraries.

475. Bauer, Raymond A. (Editor), *Social Indicators*, Cambridge, Mass. (M.I.T.), 1966, 357 pp. Paperback. [66-25166]

Preface—Historical Note on Social Indicators (Bertram M. Gross); Detection and Anticipation of Impact—Nature of the Task (Raymond A. Bauer); Social Indicators and Goals (Albert D. Biderman); State of the Nation—Social Systems Accounting (Bertram M. Gross); Anticipatory Studies and Stand-By Research Capabilities (Albert D. Biderman); Problems of Organizational Feedback Processes (Robert A. Rosenthal; Robert S. Weiss).

476. Bolles, John S. and Associates, *Northern Waterfront Plan*, San Francisco, Calif. (14 Gold St., 94133), 1968, 146 pp.

Introduction: Problem and Challenge, Planning Method; Goals and Objectives: General Goal, Specific Goals, Objectives; Background for Planning: History, Existing Conditions, Development Problems and Opportunities; General Development Plan: Elements of the Plan, Land Use, Transportation, Open Space, Urban Design; Sub-Area Plans: I Fisherman's Wharf, II Maritime Parkway, III Ferry Building, IV China Basin; Implementation: Context, Strategy, Public Improvements, Zoning; Special Studies: Embarcadero Freeway Extension, Ferry Terminals, Storm Sewage Retention Basins, Passenger Ship Terminal.

477. Chicago, *The Comprehensive Plan of Chicago*, Chicago, Ill. (Department of Development and Planning), December 1966, 118 pp. Paper.

Quality of Life; Planning Framework; Policies Plan; Improvement Plan.

478. Community Goals Committee, *Tucson—Community Goals*, Tucson, Ariz. (Office of City Manager), May 1966, 96 pp. Paper.

Goals for the Community; Human Needs for the Community: Human Factor, Social Planning and Welfare, Health and Environmental Sanitation, Housing and Living Conditions, Cultural and Recreational Activities, Education; Physical Environment: Land Use, Circulation, Air and Water Conservation, Land Conservation, Urban Aesthetics, Public Safety and Convenience, Identity; Economy: Economic Background, Development Potentials, Mineral Industries, Research and Development, Manufacturing; Government: Background and Need, Implementation, Discussion.

479. Dror, Yehezkel, *Public Policymaking Reexamined*, San Francisco, Cal. (Chandler), 1968, 370 pp. [68-11023]

". . . attempt[s] . . . to advance and apply to public policymaking various approaches developed in the modern social sciences, decisionmaking theories, and systems analysis. . . . The audience . . . includes . . . policy practitioners such as planners, government officials, and contemplative politicians, and other persons interested in policymaking and public affairs."

Mission; Framework for Evaluating . . . ; Diagnostic Evaluation of Contemporary Public Policymaking; An Optimal Model . . . ; Improving . . . ; The Choice— Shaping the Future or Muddling Through; Appendices; Bibliographic Essay, pp. 327-356.

480. Ewald, William R., Jr. (Editor), *Environment and Policy—The Next Fifty Years*, Bloomington, Ind. (Indiana University), 1968, 453 pp. Paperback. [68-27344]

Introduction (William R. Ewald, Jr.); Urgent and Important: Minority Groups— Development of the Individual (Bayard Rustin), Education for a Full Life (Robert M. Hutchins), Health Services in a Land of Plenty (Odin W. Anderson), Health— The Next Fifty Years (William H. Stewart), Problems and Promise of Leisure (Sebastian de Grazia); Contribution of Urban Form, Transportation and Housing to a New Standard of Life: Possible City (Kevin Lynch), Transportation—Equal Opportunity for Access (Max L. Feldman), Housing in the Year 2000 (Charles Abrams), Manpower Needs for Planning for the Next Fifty Years (Jack Meltzer); National Policy for Development: National Development and National Policy (Lyle C. Fitch), Natural Resources—Wise Use of the World's Inheritance (Joseph L. Fisher), Research for Choice (Herbert A. Simon), New Incentives and Controls (Daniel R. Mandelker), New Institutions to Serve the Individual (Alan Altshuler).

481. Graduate Research Center of the Southwest, *Goals for Dallas*, Dallas, Texas (Southern Methodist University, 75222), 1966, 310 pp. Paperback. [66-29022]

Recommended Goals: Government of the City; Design of the City; Health; Welfare; Transportation; Public Safety; Elementary and Secondary Education; Higher Education; Continuing Education; Cultural Activities; Recreation and Entertainment; Economy of Dallas; Essays: [as above]; Days of Decision; Supplementary Reading, p. 310.

482. Hatry, Harry P.; Cotton, John F., *Program Planning for State, County, City*, Washington, D.C. (State-Local Finances Project, George Washington University), January 1967, 72 pp. Paper.

Considerations in Instituting a Planning-Programming-Budgeting System: What Is

PPBS?, Individual PPBS Characteristics, Program Updating Procedures, Some Administrative-Organizational Aspects of PPBS Implementation, Concluding Comments; Application of Analysis in Planning, Programming, Budgeting Systems: Concept of Systems Analysis, Systems Analysis Process, Cost-Effectiveness Analysis of Selected Manpower Programs, Problems of Analysis, Concluding Comments; Bibliography, pp. 70-72.

483. Heikoff, Joseph M., *Planning and Budgeting in Municipal Management,* Washington, D.C. (International City Managers' Association), 1965, 39 pp. Paper.

Concepts of Planning in Municipal Administration; Budgeting as Planning for Resource Allocation; Policies and Programs; Functional Planning in Municipal Departments; Land Use Planning; Executive Planning and Budgeting; Conclusions; Selected Bibliography, pp. 38-39.

484. International City Managers' Association, *Program Development and Administration,* Washington, D.C. (1025 Vermont Ave., N.W., 20005), 1965, 68 pp. Paper. [66-1057]

Progress Process and Its Basic Characteristics; Participants in the Program; Initiation of Programs; Program: Formulation, Implementation, Review; Appendices: Programming a Multiagency Sewerage System—San Diego (Arnold Anderson), Development and Administration of Refuse Removal Program—Albuquerque (Roland E. Lecher; G. B. Robertson), Development Programming in Philadelphia (Robert B. Shawn), Critical Path to Municipal Management (James J. O'Brien), Systematic Program Evaluation—Problem Solving in Municipal Government (Dennis J. Palumbo); Selected Bibliography, pp. 66-68.

485. Jones, J. H., *A Conceptual Plan for the Development of the City of Toronto Waterfront,* Toronto, Ontario, Canada (60 Harbour St.), January 1968, 36 pp. Paper.

Genesis; Objectives; Principle; Port; Airport; Harbour City; Parks; Engineering and Construction; Steps to Implementation.

486. Keeble, Lewis, *Town Planning at the Crossroads,* London, England (Estates Gazette), 1961, 207 pp.

". . . delves into fundamental public planning policies. . . . Provides an excellent short history of town planning in the United Kingdom, then goes on to suggest measures for improvement. Anyone seriously interested in current issues and problems in planning in that nation should consider this book as required reading."

487. Kent, T. J., *Urban General Plan,* San Francisco, Cal. (Chandler), 1964, 213 pp. [64-14294]

"Planner-city councilman describes the kind of policy control instrument he believes an urban general plan should be, how it should be developed, effectuated, and maintained to be of maximum usefulness to the council and the ever-changing city." Municipal Government, City Planning, and the General Plan; Fifty Years of Experience with the General Plan; Legislative Uses of the General Plan; Characteristics of the General Plan; General-Plan Document; Bibliographic Essay on the Urban General Plan (Holway R. Jones).

488. King County Planning Department, *The Comprehensive Plan for King County, Washington,* Part I—Plan Policies, Part II—Plan Map, Seattle, Wash. (402L King County Court House, 98104), 1964, 204 pp. Paper.

Plan Policies: Purpose and Need for Comprehensive Plan, Population and Land Use—Trends and Projections, Development Goals, Development Concept for King County, Development Policy—Plan Policies, Criteria, and Standards, Appendices; Plan Map.

489. King County Planning Department, King County Fire Commissioners Association, *Fire District Planning,* Seattle, Wash. (309 King County Court House, 98104), 134 pp. Paper.

Summary of Study Method and Recommendations; Summary of State Enabling Legislation; Fire Protection Standards; Station Building and Site Plan; Evaluation of the Existing Fire Protection System; Report Maps; Planning the Location; Recommendations.

490. Landsberg, Hans H.; Fischman, Leonard L.; Fisher, Joseph L., *Resources in America's Future*—Patterns of Requirements and Availabilities, 1960-2000, Baltimore, Md. (Johns Hopkins), 1963, 1,056 pp. [63-7233]

"The most comprehensive study ever undertaken, its purpose is to project the nation's physical needs, and to measure the quantities of resource products that must be obtained to satisfy the needs." Important background for certain metropolitan-urban planning analyses and policies.

491. Lecht, Leonard A., *Goals, Priorities, and Dollars—The Next Decade,* New York (Free Press), 1966, 365 pp. Paperback. [66-19798]

Overall View; Consumer Expenditures and Savings; Private Plant and Equipment; Urban Development; Social Welfare; Health; Education; Transportation; National Defense; Housing; Research and Development; Natural Resources; International Aid; Space; Agriculture; Manpower Retraining; Area Redevelopment.

492. Lindblom, Charles E., *The Policy-Making Process,* Englewood Cliffs, N.J. (Prentice-Hall), 1968, 122 pp. Paperback. [68-27390]

Policy Making and Political Science; Analytic Policy Making: Policy Analysis, Limits on Policy Analysis, Making the Most of Analysis; The Play of Power: in Main Outline, Citizen as Policy Maker, Voter and Party Competition, Interest-Group Leaders, Proximate Policy Makers, Organized and Informal Cooperation Among Proximate Policy Makers; Overview: Reconstructing Preferences, Some Clouded Views; Appendix: Summary of Analysis, To Explore Further.

493. Lockyer, K. G., *An Introduction to Critical Path Analysis,* London, England (Pitman), 1964, 111 pp. [65-7130]

Assumes no prior knowledge and no mathematical expertise.
Elements of a Network; Drawing a Network; Analyzing the Network: Isolating the Critical Path, Float or Slack; Reducing the Project Time; Arrow Diagram and Gantt Chart; Loading; Control and Critical Path Analysis; Uncertainty.

494. Los Angeles, Department of City Planning, *Master Plan, City of Los Angeles*, Los Angeles, Cal. (City Hall, 90012), 1968.

Master Plan Elements (folded maps and supplementary text in looseleaf binder): Interim Public Recreation Plan (2.11); Equestrian and Hiking Trails (2.112); Power Line Rights-of-Way (2.113); Civic Center (2.121); Libraries (2.13); Public Schools (2.14); Cultural and Historical Monuments (2.15); Pueblo de Los Angeles (2.151); Fire Protection (2.171); Drainage (2.21); Sewerage (2.221); Water System (2.23); Power System (2.24); Highways and Freeways (2.311).

495. Lyden, Fremont J.; Miller, Ernest G. (Editors), *Planning, Programming, Budgeting: A Systems Approach to Management*, Chicago, Ill. (Markham), 1967, 443 pp. Paper. [67-29429]

PPB in Perspective: PPBS Comes to Washington (Virginia Held), Road to PPB—Stages of Budget Reform (Allen Schick); Budgeting and the Political Process: Budgeting in a Political Framework (Jesse Burkhead), Public Attitudes Toward Fiscal Programs (Eva Mueller); Approaches to Planning and Program Budgeting: Planning Process—Facet Design (Yehezkel Dror), Toward a Theory of Budgeting (Verne B. Lewis), Comprehensive Versus Incremental Budgeting in the Department of Agriculture (Aaron Wildavsky, Arthur Hammann); PPB Approach to Budgeting: Program Budgeting—Applying Economic Analysis to Government Expenditure Decisions (Murray L. Weidenbaum), Role of Cost-Utility Analysis in Program Budgeting (Gene H. Fisher), Costs and Benefits from Different Viewpoints (Roland N. McKean), Benefit-Cost Analysis—Its Relevance to Public Investment Decisions (Arthur Maass), Quality of Government Services (Werner Z. Hirsch); Systems Base of PPB: Systems Analysis and the Navy (Alain Enthoven), Systems Analysis Techniques for Planning-Programming-Budgeting (E. S. Quade), Guaranteed Income Maintenance—A Public Welfare Systems Model (Helen O. Nicol), Cybernetics (Magoroh Maruyama); Application and Critique of PPB: Limitations, Risks, and Problems (Roland N. McKean, Melvin Anshen), Planning-Programming-Budgeting Systems and Project PRIME (Steven Lazarus), Political Economy of Efficiency—Cost-Benefit Analysis, Systems Analysis, and Program Budgeting (Aaron Wildavsky); Appendices.

496. Lynch, Kevin, *Site Planning*, Cambridge, Mass. (M.I.T.), 1962, 248 pp. [62-13231]

Art of Site Planning; Analysis of Site and Purpose; Location of Activities; Systems of Circulation; Visual Form; Light, Noise, and Air; Problems of Control; Process of Site Planning. Housing; Special Types of Site Planning; Design of Streets and Ways; Utility Systems; Soil, Plants, and Climate; Costs; Selected References, pp. 241-243.

497. Marcou, O'Leary and Associates, *Policies for Planning the Richmond Region*, Richmond, Va. (Regional Planning Commission), 1964, 34 pp. Paper.

Approaches to Metropolitan Planning; Four Alternative Development Models; Regional Growth—Prospects for Change; Regional Development Under Four Alternatives; Criteria For Evaluating Alternatives; Evaluation of Alternatives; Conclusion.

498. Marcou, O'Leary and Associates, *The Richmond Regional Development Plan,* Richmond, Va. (Richmond Regional Planning Commission), June 1967, 114 pp. Paper.

Summary of the Regional Development Plan; Methods for Improving the Regional Planning Process; Recommended Development Policies; Framework for Effective Regional Planning; Regional Development Plan; Regional Planning Process.

499. National Capital Planning Commission, *The Proposed Comprehensive Plan for the National Capital,* Washington, D.C.(1111 Twentieth St. NW, 20576), February 1967, 230 pp. (For sale by the Superintendent of Documents). Paper.

Settings for Planning and Design; Summaries of Plan and Design Objectives; Plan Components—Policies and Proposals; City-Sections—Policies and Proposals; Summary of Program Objectives; Regional Implications of the Comprehensive Plan; Appendices.

500. Nemhauser, George, *Introduction to Dynamic Programming,* New York (Wiley), 1966, 251 pp. [66-21046]

Introduction; Concepts; Basic Computations; Computational Refinements; Risk, Uncertainty, and Competition; Non-Serial Systems; Infinite Stage Systems; Conclusions; References.

501. New Orleans, Bureau of Governmental Research, *Plan and Program for the Preservation of the Vieux Carre,* Historic District Demonstration Study, New Orleans, La. (1308 Richards Bldg., 70112), 170 pp. Paper. [68-59463]

Summary; Approach and Method; History and Architecture of the Vieux Carre; The Tout Ensemble and Change; Recommended Plan; Action Program; Appendix.

502. Nolen, John; City Planning and Zoning Commissions, *Comprehensive City Plan, Roanoke, Virginia,* Roanoke, Va. (City Planning Commission), 1928, 76 pp. Paper.

Principal Planning Recommendations; Introduction; Thoroughfares; Parks and Parkways; Recreation; Schools; Public Buildings and Public Areas; Business Districts; Development of Waterfront; Transportation; Industry; Areas for Colored Population; Safeguarding City Development; Zoning; Regional Plan; Making City Planning Effective; Résumé of Recent Progress in Civic Development. Example of an early U.S. master city plan; available only in libraries.

503. Novick, David (Editor), *Program Budgeting—Program Analysis and the Federal Budget,* Cambridge, Mass. (Harvard University), 1965, 382 pp. [66-14451]

Origin and History of Program Budgeting (David Novick); Government Decision-making and the Program Budget: Federal Budget as an Instrument for Management (Melvin Anshen), Conceptual Framework for the Program Budget (Arthur Smithies), Role of Cost-Utility Analysis in Program Budgeting (Gene H. Fisher); Actual and Potential Applications of the Program Budget Idea: Department of Defense (David Novick), Space Program (Milton A. Margolis; Stephen M. Barro),

Transportation in the Program Budget (John R. Meyer), Education in the Program Budget (Werner Z. Hirsch), Federal Health Expenditures in a Program Budget (Marvin Frankel), Program Budget for Natural Resource Activities (Werner Z. Hirsch); Implementation and Operation: Limitations, Risks, and Problems (Roland N. McKean; Melvin Anshen), Problems in Implementing Program Budgeting (George A. Steiner), Program Budget in Operation (Melvin Anshen).

504. Oakland City Planning Commission, *Oakland—Central District Plan,* Oakland, Cal. (City Hall, 94612), 1966, 80 pp. Paper.

Introduction: Central District, Plan Objectives, Plan Summary; Plan Elements: Activities, Circulation, Physical Form; Plan Areas: Core, Old City, Outer Loop, Upper Telegraph, Webster Street, Commercial Service Areas, Waverly, Industrial Areas, Civic Center, Lakeside, Madison Square, Peralta Park, Estuary, Jack London Square; Effectuation: Public Action, Control of Development, Private Action, Future Program; Appendix: Study Outline; Tables.

505. Pereira, William L., & Associates, *A Master Plan for Mountain Park,* Los Angeles, Cal. (5657 Wilshire Blvd.), January 1962, 67 pp. Paper.

Introduction; Master Concept; Regional Influences; Studies of the Land; The Master Plan; Implementation; Phasing; Engineering.

506. Philadelphia City Planning Commission, *Comprehensive Plan: The Physical Development Plan for the City of Philadelphia,* Philadelphia, Penn. (City Hall Annex, 19104), 1960, 103 pp. Paper.

Introduction; City and Its History; Costs and Strategy; The City's People; Economy; General Concepts; Plan for: Industry, Commerce, Recreation and Community Facilities, Residence, Transportation; Map of Comprehensive Plan; Map of Existing Land Use.

507. Philadelphia City Planning Commission, *Northwest Philadelphia District Plan,* Philadelphia, Penn. (City Hall Annex, 19104), September 1966, 113 pp. Paper.

Perspective on the Northwest: Land Form, History of Development, Population, Population Projection, Census Characteristics, Existing Land Use; Development of Objectives: District Assets, Problems and Opportunities, Selection of Goals and Objectives; Physical Image: Historical Influences, Development of Principles, District Structure; Plan Proposals: Transportation, Residence, Commerce, Industry, Education, Recreation, Community Facilities, Land Use—1985; Community Plans: Schuykill River, Ridge Avenue, Germantown Avenue, Stenton-Cheltenham Avenue; Costs and Strategy.

508. Ministry of Housing and Local Government, Planning Advisory Group, *The Future of Development Plans,* New York (British Information Service), 1965, 66 pp.

"The advisory group has recommended that the plans submitted for Ministerial approval deal only with main policies and major proposals for development. The plans are to be simple statements of policy illustrated where necessary with sketch maps and diagrams. Detailed plans, containing fairly precise land use allocations and designed to carry out the major policies, will be prepared and adopted by local authorities. This change in plan-making procedure will not materially alter the present system of development control."

509. Potomac Planning Task Force, *The Potomac—A Report on Its Imperiled Future and a Guide for Its Orderly Development,* Washington, D.C. (Superintendent of Documents), 1967, 103 pp. [67-22805]

Crisis and Challenge; River Landscape; River, Land and Man; Urban Potomac; Resolving the Conflict; Foundation for Action.

510. President's Commission on National Goals, *Goals for Americans,* Programs for Action in the Sixties, Englewood Cliffs, N.J. (Prentice-Hall), 1960, 372 pp. Paperback. [60-53566]

Goals at Home; Goals Abroad; Financial Accounting; Concluding Word; Additional Statements by Individual Members of the Commission.

The Individual (Henry M. Wriston); Democratic Process (Clinton Rossiter); National Goals in Education (John W. Gardner); Great Age for Science (Warren Weaver); Quality of American Culture (August Heckscher); Effective and Democratic Organization of the Economy (Clark Kerr); High Employment and Growth in the American Economy (Herbert Stein; Edward F. Denison); Technological Change (Thomas J. Watson, Jr.); Farm Policy for the Sixties (Lauren K. Soth); Framework for an Urban Society (Catherine Bauer Wurster); Meeting Human Needs (James P. Dixon, Jr.); Federal System (Morton Grodzins); Public Service (Wallace S. Sayre); United States Role in the World (William L. Langer); Foreign Economic Policy and Objectives (John J. McCloy); A Look Further Ahead (William P. Bundy).

511. Schaller, Howard G. (Editor), *Public Expenditure Decisions in the Urban Community,* Washington, D.C. (Resources for the Future), 1963, 198 pp. Paper.

Urban Services—Interactions of Public and Private Decisions (William J. Baumol); Changing Patterns of Local Urban Expenditure (Allen D. Manvel); Intergovernmental Aspects of Local Expenditure Decisions (Selma J. Mushkin); General and Specific Financing of Urban Services (William W. Vickrey); Toward Quantitative Evaluation of Urban Services (Russell L. Ackoff); Benefit-Cost Analysis as a Tool in Urban Government Decision Making (Nathaniel Lichfield; Julius Margolis); Costs and Benefits from Different Viewpoints (Roland N. McKean); Quality of Government Services (Werner Z. Hirsch); Spatial and Locational Aspects of Local Government Expenditures (Seymour Sacks).

512. Schnore, Leo F.; Fagin, Henry (Editors), *Urban Research and Policy Planning,* Beverly Hills, Calif. (Sage), 1967, 638 pp. [67-18420]

Some Research Priorities: Social Science and the City, Survey of Research Needs (Leo F. Schnore; Eric E. Lampard), Assimilation of Migrants to Cities, Anthropological and Sociological Contributions (Lyle W. Shannon; Magdaline Shannon), Strategies for Discovering Urban Theory (Anselm L. Strauss), Class and Race in the Changing City, Searching for New Approaches to Old Problems (Eleanor P. Wolf; Charles N. Lebeaux); Economics and Regional Science: Toward an Urban Economics (Wilbur R. Thompson), Urban Travel Behavior (John F. Kain); History and Geography: Historical Perspective on Urban Development Schemes (Charles

N. Glaab), Urban Geography and City and Metropolitan Planning (Harold M. Mayer); Political Science and Political Sociology: Political Science and the City (Norton E. Long), Comparative Study of Urban Politics (Robert R. Alford); Expanding the Horizon of Planning: Evolving Philosophy of Urban Planning (Henry Fagin), Social Planning in the Urban Cosmos (Llyle C. Fitch); Technology in Transition: The New Technology and Urban Planning (Britton Harris), Transportation Planning for the Metropolis (Melvin M. Webber); The Physical World and Human Environment: Urban Design and Urban Development (Henry Fagin; Carol H. Tarr), Urban Housing and Site Design (Robert D. Katz); Policy and Action: Public Administration and Urban Policy (Robert T. Daland), Moving from Plan to Reality (William L. C. Wheaton), Role of the United Nations in Urban Research and Planning (Ernest Weissmann), Evolving Goals of the Department of Housing and Urban Development (Robert C. Weaver); Bibliography, pp. 603-630.

513. Simpson & Curtin, *Coordinated Transit for the San Francisco Bay Area—Now to 1975,* Final Report of Northern California Transit Demonstration Project, Springfield, Va. (Clearinghouse for Federal Scientific and Technical Information, 22151), October 1967, 222 pp. Paper. (No. PB-175 733)

Report in Brief; Introduction; Present Transit Travel; Present Highway Travel; Existing Transit Service; Present Fare Systems; Present Transit Costs; Operating Results in 1965; Travel Relationships; Population and Economic Growth; Transit Travel in 1975; Future Needs—Vehicles, Plant & Manpower; Fares and Passenger Revenue in 1975; Operating Results in 1975; Fare Collection; Summary of Transit Recommendations for 1975; Promotion of Transit Service; Appendices.

514. Tippetts, Abbett, McCarthy, Stratton, *Airport Master Plan, Dallas Fort Worth Regional Airport,* Dallas, Tex. (Mayor's Office, Municipal Bldg.) and Fort Worth, Tex. (Mayor's Office, City Hall), 1967, 112 pp. Paper.

Basic Planning Considerations: Introduction, Regional Demography, Air Traffic Forecasts, Site Considerations, Air Space Configurations, Airfield Requirements, Air Traffic Simulation, Ground Traffic Simulation, Building Space Requirements; Basic Design Criteria; Airport Master Planning: Regional Ground Transportation, Airport Area Development, Airfield Layouts, Water Supply and Sewage Disposal, Heating and Air-Conditioning System, Power Supply System, Hydrant Fueling System, Airport Master Plan; Passenger Terminal Complex: Linear Terminal Concept, Terminal Complex Growth, Transit System, Automated Baggage System, Passenger Terminal Module—Transverse Sections, Terminal Module Plans, Passenger Terminal Module—Sections and Elevations, Passenger Terminal Complex Elevations, Terminal Complex Perspectives.

515. (City of) Toronto Planning Board, *Proposals for a New Plan for Toronto,* Prepared for Public Discussion, Toronto, Ont. (City Hall, 1), 1966, 64 pp. Paper.

This report is not an official plan. New Plan for Toronto; Plan; Residential Areas; Commerce; Industry; Major Parks and the Waterfront; Education, Research and the Arts; Transportation; Quality of the City; Implementation of the Plan.

516. U.S. Advisory Commission on Intergovernmental Relations, *Urban and Rural America: Policies for Future Growth,* Washington, D.C. (Superintendent of Documents), April 1968, 186 pp.

Pattern of Urbanization: Regional Growth, Urban Mosaic, Dynamics of Mobility and Migration, Urban-Rural Contrasts, Intrametropolitan Differences, Summary Observations; Economics Growth—Regional, State, and Local Experience: Its Meaning and Measurement, Mechanism of Community Economic Growth, Relationship of Population and Public Expenditures, Population Size and the Private Sector, Summary of Findings; Impact of Recent Urbanization Trends: Ghetto Riots and In-Migration, Social and Psychological Effects of Urbanization, Diseconomies of Scale, Business Location Decisions and Jobs, Rural America—What Kind of Remainder?, Burdens of Sprawl, Summary Observations; New Communities in America and Their Objectives: New Towns and New Communities Defined, Development of New Towns and New Communities, European Experience, American Experience, Current New Community Development, New Entrepreneurs, Planning for New Community Location, Financing the Development of New Communities, Government and New Communities, Federal and State Programs Available for New Community Facilities, New Community Development Opportunities, Problems Facing New Community Development, Challenge of New Communities; Guidance and Controls for Large-Scale Urban Development and New Communities: Land-Use Controls, Practices, and Problems—Overview, Land-Use Controls in the Large-Scale Development and New Community, Concluding Observations; Conclusions and Recommendations: Policies Dealing with Urban Growth, Possible Components of Urban Growth Policies.

517. U.S. Congress, Joint Economic Committee, Subcommittee on Urban Affairs, *Urban America: Goals and Problems,* 90th Congress, 1st Session, Washington, D.C., August 1967, 303 pp. Paper.

Values, Goals, Priorities: Hidden Dimension (Edward T. Hall), Goals for Urban Development (Lyle C. Fitch), Community Size—Forces, Implications, and Solutions (Werner Z. Hirsch), Goals and Social Planning (Homer C. Wadsworth), Concept of Community and the Size for a City (Percival Goodman); Functional Problems: Urban Planning and Policy Problems (Donald N. Michael), Effective Research on Urban Problems (Charles Kimball), An Attack on Poverty—Historical Perspective (Roger Starr), Quality of Urban Life—Analysis from the Perspective of Mental Health, Selected Bibliographies (Leo Levy; Harold M. Visotsky), On Urban Goals and Problems (Wilbur R. Thompson); Rules of the Game—Public Sector: Institutional Setting of Urban Affairs (Frederick Gutheim), Poverty and Public Finance in the Older Central Cities (James Heilbrun), Inner City and a New Urban Politics (Harvey S. Perloff; Royce Hanson), Two Essays on the Neighborhood Corporation (Milton Kotler), Urbanization and Federalism in the United States (Daniel J. Elazar); Rules of the Game—Private Sector: Toward a Better Understanding of Urban America (Thomas B. Curtis), Business Welfare and the Public Interest (Charles Abrams), New Perspectives for Urban America (Edgardo Contini), Private Sector and Community Development—Cautious Proposal (Chester W. Hartman), Role of the Private Sector in Urban Problems (Urban America, Inc.), Responsibilities of the Private Sector (Robert B. Choate).

518. Wagner, Pete, *The Scope and Financing of Urban Renewal and*

Development, Statement by the NPA Business Committee, Planning Pamphlet No. 119, Washington, D.C. (National Planning Association), April 1963, 55 pp. Paperback. [63-17295]

Background (H. Christian Sonne); Statement; Scope and Financing of Urban Renewal and Development: Background and Origin, Progress and Shortcomings, Importance of a New Approach, Transportation Problem, Investment Required and Its Impact on Our Resources, Financing; Appendices.

519. Washington Metropolitan Area Transit Authority, *Metro—Adopted Regional Rail Transit Plan and Program,* Washington, D.C. (1634 Eye St., N.W., 20006), 1 March 1968, Revised 7 February 1969, 44 pp. Paper.

Background; Facilities; Schedules; Routes; Equipment; Facilities Design; Timetable for Provision of Facilities; Provision of Facilities; Capital Cost Estimate; Estimated Ridership; Probable Fares; Feeder Bus Service; Estimated: Operating Expenses, Revenue; Financial Plan; Conclusion; Chronology; Consultants, Staff, Officers, Contractors.

Bibliographies

520. Archibald, Russell D.; Villoria, Richard L., *Network-Based Management Systems (PERT/CPM),* Selected Bibliography, pp. 466-470.

521. Brennan, Maribeth, *PERT and CPM, A Selected Bibliography,* Bibliography No. 53, Council of Planning Librarians, Monticello, Ill. (Exchange Bibliographies, P.O. Box 229, 61856), 1968, 11 pp.

522. Farmer, J. (Compiler), *Hatrics—A Bibliography of Critical Path Methods,* Southampton, England (Hampshire Technical Research Industrial Commercial Service), 1966, 72 pp. Paper.

Critical Path Analysis; Planalog; Program Evaluation and Review Technique (PERT); Least Cost Estimating and Scheduling (LESS); Multivariate Analysis and Prediction of Schedules (MAPS); Planned Networks (PLANNET); Resource Allocation and Multi-Project Scheduling (RAMPS); Scheduling and Control by Automated System (SCANS); Scheduling, Planning, Evaluation, Cost and Control (SPECTROL); Scheduling Generally; Appendix.

523. Heikoff, Joseph M., *Planning and Budgeting in Municipal Management,* Selected Bibliography, pp. 38-39.

524. International City Managers' Association, *Program Development and Administration,* Selected Bibliography, pp. 66-68.

525. Schnore, Leo F.; Fabin, Henry (Editors), *Urban Research and Policy Planning,* Bibliography, pp. 603-630.

526. U.S. Congress, Joint Economic Committee, Subcommittee on Urban

Affairs, *Urban America: Goals and Problems,* Selected Bibliographies on Mental Health and Quality of Urban Life, pp. 108-112.

Migration (Mobility) and Mental Health; Sensory Deprivation; Population Density and Mental Health.

7. INSTITUTIONALIZATION

Articles, Periodicals

527. Abruzzi, Adam, "The Production Process: Operating Characteristics," *Management Science,* Vol. 11, No. 6, April 1965, pp. B—98-118.

"It is the thesis of this paper that the productive process has an operating logic whose kernel is found in an organizational plan and program."

Introduction; Production Process—Technical and Operating Aspects: Basic Considerations, Two Types of Organization, Master Production Plan, Operating Logic, Process Classification, Loadings and Schedules, Work-Cycle Development, Operative Activities—Horizontal, Workers' Role, Operating Organizations, Operation Activities—Vertical, Workers' Role, Operating Organizations, Control Activities: Horizontal, Vertical; Production Process—Manpower and Wage-Incentive Reflections: Direct Workers, Indirect Workers, Service Workers; Management Implications.

528. Adams, Frederick J.; Hodge, Gerald, "City Planning Instruction in the United States: The Pioneering Days, 1900-1930," *Journal of the American Institute of Planners,* Vol. XXXI, No. 1, February 1965, pp. 43-51.

Forerunners; Formative Stage, 1909-1920; Portents of the Future.

529. American Institute of Architects, American Institute of Consulting Engineers, American Institute of Planners, American Society of Civil Engineers, American Society of Landscape Architects, Consulting Engineers Council of the United States, National Society of Professional Engineers, "Professional Collaboration in Environmental Design," *AIP Newsletter,* Vol. 1, No. 7, July 1966, pp. 5-8.

Preface; Tenets of the Collaborating Design Professions; Selection and Compensation of Environmental Design Professionals.

Guide approved and adopted by the above organizations (copies may be procured from any of them).

530. Bass, Bernard M., "Business Gaming for Organizational Research," *Management Science,* Vol. 10, No. 3, April 1964, pp. 545-556.

"A non-computer game is described which can be used to test hypotheses about the effects of different organizational structures on material and social psychological outcomes."

Why Haven't Games Been Used for Research?; What Kind of Game Has Research

Used?; Example; Competition; Reducing Interaction Potential by Rule; Tangible Products; Economic Environment; An Industrial Climate; Personnel Compensation; Simple vs. Complex Organization; Line-Staff vs. Overlapping Committees; Three-Way Competition; Conclusions.

531. Beckman, Norman, "The Planner As a Bureaucrat," *Journal of the American Institute of Planners*, Vol. XXX, No. 4, November 1964, pp. 323-327.

Case of Conflicting Identities; Vulnerability of the Planner; Survival in a Bureaucracy; An Illustration—Bureau of the Budget; Planning Is the Art of the Possible; Conclusion.

532. Branch, Melville C., "Comprehensive Planning: A New Field of Study," *Journal of the American Institute of Planners*, Vol. XXV, No. 3, August 1959, pp. 115-120; also Chap. 15 in: Branch, Melville C., *Planning—Aspects and Applications*, New York (Wiley), 1966, pp. 315-328. [66-13523]

"Many professional fields are now involved directly in different applications of planning . . . recent intellectual advances . . . are being applied successfully to planning problems. From a synthesis of these professional and intellectual activities, a theory, principles, and methodology of comprehensive planning are likely to develop. . . ."

533. Craig, David W., "A Plea for the Eventual Abolition of Planning Boards," in: *Planning 1963*, Chicago, Ill. (American Society of Planning Officials), November 1963, pp. 68-81. [39-3313]

Ineffectiveness of the Planning Board at Maturity; Need To Eliminate the Duality; Purposeful Transfer of Planning Activities to the Governing Body; Rebuttal to Objections; How To Do It; Conclusion.

534. Friedmann, John, "Regional Planning As a Field of Study," *Journal of the American Institute of Planners*, Vol. XXIX, No. 3, August 1963, pp. 168-175.

Core of Regional Planning Study; Spectrum of Spatial Planning Activities; Education for Regional Planning.

535. Heikoff, Joseph M., "Planning Is the Responsibility of the Executive," *Public Management*, Vol. XLVII, No. 5, May 1965, pp. 157-163.

"Planning pervades the entire structure of municipal government . . . Yet . . . city planners . . . neglect over-all coordination of the planning specialties. . . . Three kinds of planning—functional, land use, and executive—are outlined, and the part each should play in coordinated planning efforts is described."

536. Karmin, Monroe W., "Fleeing the Hot Seat; More Mayors Bow Out, Concluding Frustrations of Jobs Are Too Great," *The Wall Street Journal*, Pacific Coast Edition, Vol. LXXX, No. 78, Monday, 21 April 1969, pp. 1 & 18.

"Naftalin of Minneapolis Finds Cash and Public Support Lacking for Key Programs. Revenue Sharing Called Vital."

Diminished Prestige; Is It Worth the Frustration?; What Is Needed; "The End of My Rope"; Dissatisfied Blacks; Narrow Interests; Limited Options.

537. Kent, T. J., Jr., "Home Rule and Limited Metropolitan Government," in: American Society of Planning Officials, *Planning 1968*, Selected Papers from the ASPO National Planning Conference, San Francisco, May 4-9, 1968, pp. 206-216. Paperback, [39-3313]

1969—Critical Year in Perspective; Two Basic Assumptions; Policies of the Central City; Is the Bay Area Unique?; Two Issues That Need More Attention.

538. Mahoney, Thomas A., "Managerial Perceptions of Organizational Effectiveness," *Management Science*, Vol. 14, No. 2, October 1967, pp. B-76-91.

Rationale for the Studies; Design of the Studies; Analysis and Findings; Summary and Implications; Appendix.

539. Nash, P. H.; Durden, Dennis, "A Task-Force Approach to Replace the Planning Board," *Journal of the American Institute of Planners*, Vol. XXX, No. 1, February 1964, pp. 10-22.

Need For Re-Evaluation; Original Endorsement Role Obsolete; Evolution of Formal Citizen Participation; Case For Abolition—Efficiency, Political Realism; Case Against Abolition—Inherent Obstacles and Practical Consequences; Channels of Public Planning Participation; Lessons From the Abolitionists; One Alternative to Abolition—Ex Uno Plura (Task Forces of Citizen Experts); Steps Toward the Task Force Approach.

540. Perloff, Harvey S., "Education of City Planners: Past, Present and Future," *Journal of the American Institute of Planners*, Vol. XXII, No. 4, Fall 1956, pp. 186-217.

Introduction; View from the Administrative Tower—Development of Professional Education: "Natural History" of Professional Education, A Look at History and Trends; Development of the City Planning Field: Earlier Phase—Role of the Architect, Landscape Architect, and Engineer, Laying the Foundation for a Broad View of Planning, Role of the Lawyer and Social Scientist, Administration to the Forefront, 1940s and 1950s; Development of Planning Education; Some Trends and Their Implications: More Federal Activity in the Urban Field, Growth of Departmental and Other Agency Planning, Planning as a General Staff Activity, Another Trend—Growth of Urban and Regional Studies and Specializations; Foundations of Education for City Planners: Requirement #1—General Education, #2—A Planning "Core," #3—Specialized Training, Tasks of Leadership; Appendix A— Some Landmarks in the History of City Planning and of Planning Education in the United States; Appendix B—Intellectual and Professional Contributions To, and Influences On, City Planning (Some Examples).

541. Petshek, Kirk R., "A New Role for City Universities—Urban Extension Programs," *Journal of the American Institute of Planners*, Vol. XXX, No. 4, November 1964, pp. 304-316.

Urban Extension as a Unique Institution; What Should University Urban Extension Do?; Urban Extension Activities: Advice to Public Decision-Makers, Special

Groups in the Community, Geographic Areas and Problems of Minority Groups; Values and Clients; Community Demands; Research; University Personnel; Urban Extension and the Professionals.

542. Rabinovitz, Francine F.; Pottinger, J. Stanley, "Organization for Local Planning: The Attitudes of Directors," *Journal of the American Institute of Planners*, Vol. XXXIII, No. 1, January 1967, pp. 27-32.

New Panacea—Executive Department Agency; Conclusions—Time For Flexibility.

543. Tugwell, Rexford G., "Implementing the General Interest," *Public Administration Review*, Vol. 1, No. 1, Autumn 1940, pp. 32-49.

Thoughtful commentary on the role and activities of the New York City Planning Commission several years after the City Charter revision of 1938. The observations and fundamental concepts are completely relevant today.

544. Webber, Melvin M., "Comprehensive Planning and Social Responsibility, Toward an AIP Consensus on the Profession's Roles and Purposes," *Journal of the American Institute of Planners*, Vol. XXIX, No. 4, November 1963, pp. 232-241.

Time for Re-Examination; To Extend Access to Opportunity; To Integrate Larger Wholes; To Expand Freedom in a Pluralistic Society.

Books, Reports, Pamphlets

545. Advisory Commission on Intergovernmental Relations, *Alternative Approaches to Governmental Reorganization in Metropolitan Areas*, Washington, D.C. (726 Jackson Pl., N.W., 20575), June 1962, 88 pp. Paper.

Introduction; Reasons for Reorganization of Local Government in Metropolitan Areas; Criteria for Appraising Different Approaches to Reorganization of Local Government in Metropolitan Areas; Analysis of Alternative Approaches to Reorganization of Local Government in Metropolitan Areas; Concluding Observations.

546. Advisory Commission on Intergovernmental Relations, *Impact of Federal Urban Development Programs on Local Government Organization and Planning*, 88th Congress, 2nd Session, Washington, D. C. (Government Printing Office), 30 May 1964, 198 pp. Paper.

Purpose, Scope, and Coverage of the Report; Criteria for Analysis; Analysis of Federal Programs; Recommendations; Appendices: Suggested Language for a Unified Federal Urban Development Policy, Descriptions and Evaluations of [43] Individual Federal Programs, Methodological Note, Effects of Channeling Federal Aid Through the States.

547. Advisory Commission on Intergovernmental Relations, *Performance of Urban Functions: Local and Areawide*, Washington, D.C. (726 Jackson Pl., N.W., 20575), 1963, 281 pp.

"Outlines a method by which citizens and public officials in large urban areas might analyze urban functions and draw conclusions whether they should be performed on a local, intermediate, or area-wide basis. Also provides useful information on current practices in administering urban services, analyzing and ranking fifteen functions (fire protection, public education, refuse collection and disposal, libraries, police, health, urban renewal, housing, parks and recreation, welfare, hospitals and medical-care facilities, transportation, planning, water supply and sewage disposal, and air pollution control).

548. Argyris, Chris, *Integrating the Individual and the Organization*, New York (Wiley), 1964, 330 pp. [64-13209]

Introduction to Individual and Organizational Effectiveness; Input; Organizational Dilemma; Nature of the System: Lower Levels, Managerial Levels; Organizational Effectiveness and Ineffectiveness; "Mix" Model; Empirical Work Illustrating Aspects of the "Mix" Model; Organizational Structures of the New System; Organizational Staffing and Job Design; Managerial Controls, Rewards and Penalties, and Incentive Systems; Evaluating, Hiring, and Terminating Employees; Organization of the Future: Summary View; Using the Model to Explore Other Studies; Organizational Pseudo-Effectiveness and Individual Pseudo-Health.

549. Banfield, Edward C. (Editor), *Urban Government—A Reader in Administration and Politics*, New York (Free Press), Revised Edition 1969, 718 pp. [69-11169]

Urban Government as a Subject for Study: Aristotle and the Study of Local Government (Norton E. Long), We Need to Shift Focus (James Q. Wilson), Management of Metropolitan Conflict (Edward C. Banfield); Urban Government in the Federal System: Intergovernmental Relations in New York City (Wallace S. Sayre; Herbert Kaufman), Federal and State Impacts (Morton Grodzins), Historical Development of the City's Position (Frank J. Goodnow), Reappraisal of Constitutional Home Rule (Chicago Home Rule Commission), City Government in the State Courts (Harvard Law Review); "One Man, One Vote" Applies to Local Governments (Supreme Court of the United States); Effect of State Reapportionment on the Cities (William J. D. Boyd); Metropolitan Organization: Metropolitan Districts (John C. Bollens), What Is a "Metropolitan Problem"? (Oliver P. Williams; Harold Herman; Charles S. Liebman; Thomas R. Dye), The Desirable and the Possible (Edward C. Banfield; Morton Grodzins), Needed—New Layer of Self-Government (Luther Gulick), Proper Role for the Federal Government (Robert H. Connery; Richard H. Leach); The Machine and Its Reform: Machine at Work (Martin Meyerson; Edward C. Banfield), Rings and Bosses (James Bryce), How to Get a Political Following (George Washington Plunkitt), Attachment of the Immigrant to the Boss (Oscar Handlin), How the Boss Runs the Organization (Frank R. Kent), Nature of Political Obligations (William Foote Whyte), New-Fashioned Machine (Edward N. Costikyan), Latent Functions of the Machine (Robert K. Merton), Mr. Dooley on Why Rayformers Fail (F. P. Dunne), Separation of Powers Necessitates Corruption (Henry Jones Ford), Who, or What, Started the Evil? (Lincoln Steffens), Class Basis of the Reform Ideal (Jane Addams), The Locus of Corruption Has Changed (Edward N. Costikyan); Good Government: Municipal Affairs Are Not Political (Andrew D. White), Absurdity of Partisanship (Brand Whitlock), Unorganized Politics of Minneapolis (Robert L. Morlan), Promotion of the City Manager Plan (Don K. Price), Role of the City Manager (Leonard D.

White), City Manager as Leader (Charles R. Adrian), Resignation of Elgin Crull (Bruce Kovner), Three Fundamental Principles (Harold A. Stone; Don K. Price; Kathryn H. Stone), Group Conflict is Inevitable in a Major City (Charles A. Beard), Effects of the Reform Ideal on Policy (Robert L. Lineberry; Edmund P. Fowler), Philadelphia—"Good Government" Leads to Moral Frustration (James Reichley); Trend of Urban Politics: New Middle Class (Samuel Lubell), Silent Revolution in Patronage (Frank J. Sorauf), Dilemmas of a Metropolitan Machine (Edward C. Banfield), No-Party Politics of Suburbia (Robert C. Wood), New Hurrah (Joseph Lyford), Negro and Philadelphia Politics (John Hadley Strange), "Black Beater" (Lee Sloan); Influence and Leadership: Power Structure of Regional City (Floyd Hunter), Economic Notables of New Haven (Robert A. Dahl), How to Study Community Power (Nelson W. Polsby), Power as Non-Decision-Making (Peter Bachrach; Morton S. Boratz), Local Community as an Ecology of Games (Norton E. Long), "Hard" Reporting on *The New York Times* (Paul H. Weaver), Organized Labor in City Politics (Edward C. Banfield; James Q. Wilson); Problems of Management: Interview with Mayor Lindsay (Nat Hentoff), Recent Concepts in Large City Administration (Charles R. Adrian), Mayor and the Planning Agency (V. O. Key), Comprehensive Planning, An Impossible Ideal (Alan Altshuler), Planner as Advocate (Paul Davidoff), Reflections of an Advocate Planner (Lisa Peattie), Capital Programming in Philadelphia (William H. Brown, Jr.; Charles E. Gilbert), Five Functions for Planning (Martin Meyerson), Coming Revolution in City Planning (Anthony Downs), Three Concepts for Planners (Edward C. Banfield); Formation of Policy: Metro Toronto—Forming a Policy-Formation Process (Harold Kaplan), Autonomy vs. "Political Control" of the Schools (Robert H. Salisbury), Some Alternatives for the Public Library (Edward C. Banfield), Police Administrator as a Policymaker (James Q. Wilson), Welfare for Whom? (Richard A. Cloward; Frances Fox Piven), Milwaukee's National Media Riot (H. R. Wilde); Bibliography, pp. 707-712.

550. Berger, Marjorie S., *Opportunities in City Planning*, New York (Vocational Guidance Manuals), 1961, 100 pp. Paperback.

Introduction; What City Planning Is; What a Planner Does; What Education Is Needed; What the Career Opportunities Are; Supplementary Reading, pp. 97-99.

551. Bullis, Andrew S.; Williams, Lawrence A., *Organizing Municipal Governments for Civil Defense*, Washington, D.C. (American Municipal Association), October 1963, 318 pp. Paper.

Summary of Findings and Recommendations; Civil Defense Responsibility of Municipal Government; Case Studies: Broken Arrow, Oklahoma, Marietta, Georgia, Springfield, Illinois, Hartford, Connecticut, Portland, Oregon, Cleveland, Ohio; Selected Bibliography, pp. 311-318.

552. Calmfors, Hans; Rabinovitz, Francine F.; Alesch, Daniel J., *Urban Government for Greater Stockholm*, New York (Praeger), 1968, 190 pp. [68-23359]

"This descriptive case study . . . traces the evolution of the city's international leadership in planned urban growth, employing comparisons with other cities when appropriate."

553. Cohn, Angelo, *Careers in Public Planning and Administration*, New York (Walck), 1966, 112 pp. [66-10442]

Twentieth-Century Professions; The Past Is Brief; City Managers on the Job; Meeting Specific Problems; What Public Planners Do; Living Examples of Planning; Planners and Consultants in Business; Preparing for Your Career; Getting the First Job; Two Kinds of Reward; What the Future Holds; Reading List, pp. 107-109.

554. Dale, Ernest; Urwick, Lyndall, F., *Staff in Organization*, New York (McGraw-Hill), 1960, 241 pp. [60-11964]

Load on Top Management; Effect of Overload on Executive Health; How Executives Spend Their Time; Methods of Reducing Executive Burdens; Military Use of Staff; Comparison of Military and Business Staffs; United States Presidency and the Staff System; Assistant in Business—Composite Profile; Nice Derangement of Epitaphs; Case Studies of Unsuccessful Assistants-to; Practical Use of General Staff Positions in Business.

555. Etzioni, Amitai, *Modern Organizations*, Foundations of Modern Sociology Series, Englewood Cliffs, N.J. (Prentice-Hall), 1964, 120 pp. Paper. [64-17073]

Rationality and Happiness: Organizational Dilemma; Organizational Goal: Master or Servant?; Classical Approach; From Human Relations to the Structuralists; Bureaucracies: Structure and Legitimation; Organizational Control and Leadership; Organizational Control and Other Correlates; Administrative and Professional Authority; Modern Organization and the Client; Organization and the Social Environment. Selected References, pp. 117-118.

556. Faraci, Piero, *Expenditures, Staff, and Salaries of Local Planning Agencies*, Report No. 232, ASPO Planning Advisory Service, Chicago, Ill. (1313 East 60th St., 60637), March 1968, 52 pp. Paper.

Expenditures: Per Capita, for Consultants; Staff, Professional Planning Positions; Salaries, of Planning Directors; Unfilled Positions; State Planning Agencies.

Tables: Expenditures and Planning Staff by Individual Agency, Jurisdiction, and Population Group; Summary of Expenditures by Jurisdiction and Population Group; Number of Levels of Professional Planners by Jurisdiction and Population Group; Median Salaries of Three Lowest Job Levels; Salaries and Number of Authorized Professional Planners at Position Level by Individual Agency and Number of Levels in Agency; Regional Summary of Salaries at Position Level by Number of Levels in Agency; Summary of Directors' Salaries by Jurisdiction and Population Group; Number and Rate of Unfilled Positions by—Jurisdiction 1966-1968, Jurisdiction and Population Group, Number of Levels in Agency, by Region; Staff Size and Salary by Position Level for State and Provincial Planning Agencies; Sources of Revenue and Expenditures for State and Provincial Planning Agencies.

557. Fulton, Lord (Chairman), *The Civil Service*, Surveys and Investigations, London, England (Her Majesty's Stationery Office), 1968, 478 pp.

". . . including foregoing proposals to set up planning units in all Ministries and to establish a new professional position of 'policy advisors' in all Ministries."

558. Hanson, Royce, *Metropolitan Councils of Governments,* An Information Report, Washington, D.C. (Advisory Commission on Intergovernmental Relations), August 1966, 69 pp. Paper.

Metropolitan Cooperation Movement; Councils of Governments in Perspective; Organization and Structure of Councils of Governments; Councils of Governments and Federal Programs; Conclusions; Selected Bibliography, pp. 68-69.

559. Houle, Cyril O., *The Effective Board,* New York (Young Men's Christian Associations), 1960, 174 pp. [60-6560]

". . . designed to serve both those . . . interested in boards in general and those who would like to improve a specific board."

How to Think About a Board; Human Resources of the Board; Improving the Organization of the Board; The Board, the Executive, and the Staff; Improving the Operation of the Board.

560. Hughes, Charles L., *Goal Setting, Key to Individual and Organizational Effectiveness,* New York (American Management Association), 1965, 157 pp. [65-26864]

Motivation and Management: Conflict—Inevitable or Not?, Needs, Purpose and Organization, Corporate Planning and Individual Achievement, Theories of Motivation and Management, Goal-Oriented People, Increasing Goal-Oriented Action; Goal-Achievement Process: Conditions for Motivation, Motivation Opportunities, Motivation Media; Goal-Setting Systems: Systems Approach to Motivation, General Model for Goal-Setting Systems, Organizational Goal-Setting Systems, Individual Goal-Setting Systems, Individual and Organizational Goal Interaction, Abilities and Performance, Total System for Motivation; Final Thoughts: Role of Maintenance Needs, Personal and Private Goals.

561. Kent, T. J., Jr., *City and Regional Planning for the Metropolitan San Francisco Bay Area,* Berkeley, Calif. (Institute of Governmental Studies, University of California), 1963, 22 pp. Paper.

Introduction; Long-Range Local and Metropolitan Objectives: Two-Level System of Government, Five Metropolitan Functions; Limiting the Metropolitan Government; Proposed Bay Area Regional Planning District: Duties of the . . . District, Organization and Financing, Significant Features of the Proposal; Conclusions.

562. Kupinsky, Mannie (Project Director), *Community Adjustment to Reduced Defense Spending,* U.S. Arms Control and Disarmament Agency Publication 33, Washington, D.C. (Superintendent of Documents), December 1965, 402 pp. Paper.

Purposes, Assumptions, and Organization of Study; National Economic Setting; Subnational Setting; Employment Dislocation in Three Metropolitan Areas Under Several Arms Reduction Assumptions; Review of Selected Past Economic Readjustment Experiences; Policies, Programs, and Administrative Machinery for Meeting Regional Readjustment Problems at the Subnational Level; Community Perspective on Impact Problems and Solutions; Policies and Programs for Meeting Community Dislocation Problems Caused by Defense Cutbacks; General Policy Guidelines for Regional Readjustment to Arms Reduction; Appendices.

563. Landis, James M., *The Administrative Process*, New Haven, Conn. (Yale University), 1938, 160 pp. Paperback. [38-29177]

Place of the Administrative Tribunal; Framing of Policies—Relationship of the Administrative and Legislative; Sanctions to Enforce Policies—Organization of the Administrative; Administrative Policies and the Courts.

564. Leavitt, Harold J., *Managerial Psychology*, An Introduction to Individuals, Pairs, and Groups in Organizations, Chicago ,Ill. (University of Chicago), Revised Edition, 1964, 437 pp. Paperback. [64-16950]

People One at a Time, The Units of Management; People Two at a Time, Problems of Influence and Authority; People in Threes to Twenties, Efficiency and Influence in Groups; People in Hundreds and Thousands, Problems of Organizational Design; Questions and Suggested Readings.

565. Miller, Harold V., *Mr. Planning Commissioner*, Chicago (Public Administration Service), 1954, 81 pp. Paper.

So You're the New Planning Commissioner; An Important Man In Local Government; The General Proposition; Getting Started; Basic Data; First Services; You and Your Staff; Your Major Street Plan; On the Subdivision of Land; Let's Look at Your Schools; For the Recreation of Your People; Planning and Utilities; Parking and Traffic; A Zoning Plan; Other Planning Matters; Capital Budget Programming; How About Your Outskirts?; Presenting the Comprehensive Plan; The Continuing Program.

566. Millett, John D., *Organization for the Public Service*, Princeton, N.J. (Van Nostrand), 1966, 159 pp. Paperback. [66-31839]

Institutional Context of Organization for Public Service; Organization As a: Political Problem, Technical Problem, Problem in Human Relations; Art of Organization.

567. Morris, G. M., *Modernizing Government Budget Administration*, The Application of Technical Cooperation in Improving Budget Administration in the Governments of Developing Countries, Prepared by Public Administration Service, Washington, D.C. (Agency for International Development), June 1962, 104 pp.

The Budget in Modern Government; Budgeting and Economic Development; Essential Elements of an Effective Budget System; Organization and Staffing of the Budget Agency; Techniques of Budget Administration; Technical Assistance in Budget Modernization; Selected Bibliography, pp. 103-104.

". . . intended to aid host country officials and technicians, whether working in their own countries or studying and observing within the United States, to view the problems of budget reform in better perspective."

568. Municipal Manpower Commission, *Governmental Manpower for Tomorrow's Cities*, New York (McGraw-Hill), 1962, 201 pp. [62-20507]

The Problem—The Metropolitan Area; The Environment—Leaders, Institutions and Attitudes; The Facts—Manpower Picture; Recommendations—Agenda for Metropolis; Appendices: Municipal Executives—Statistical Profile, Excerpts from "Careers in the Rebuilding and Management of Cities."

569. Munzer, Martha E., *Unusual Careers,* New York (Knopf), 1962, 138 pp.

One of the Borzoi Books for Young People series. In a chapter called "Man the Designer: Planning for Tomorrow's World," city planning is linked with seven other careers concerned with natural resources. "An excellent addition to the far-too-little literature we have on planning as a career."

570. Parkinson, C. Northcote, *Parkinson's Law,* and Other Studies in Administration, Cambridge, Mass. (Riverside), 1957, 113 pp. [57-9981]

Parkinson's Law, or The Rising Pyramid; The Will of the People, or Annual General Meeting; High Finance, or The Point of Vanishing Interest; Directors and Councils, or Coefficient of Inefficiency; The Short List, or Principles of Selection; Plans and Plants, or The Administration Block; Personality Screen, or The Cocktail Formula; Injelititis, or Palsied Paralysis; Palm Thatch to Packard, or A Formula for Success; Pension Point, or The Age of Retirement.

571. Perloff, Harvey S., *Education for Planning: City, State, & Regional,* Baltimore (Johns Hopkins), 1957, 189 pp. [57-12448]

Education of City Planners: Past, Present, and Future; Education for Regional Planning and Development; Education and Research in Planning: a Review of the University of Chicago Experiment.

572. Pfiffner, John M.; Presthus, Robert, *Public Administration,* New York (Ronald), Fifth Edition, 1967, 567 pp. [67-15468]

Environment of Public Administration; Functions of the Administrator; Organization; Personnel Administration; Financial Administration; Administrative Law and Regulation; Administrative Responsibility; Selected Bibliography, pp. 551-554.

573. United Nations, Department of Economic and Social Affairs, *Training for Town and Country Planning,* United Nations Publication Sales No.: 1957, IV, 11 New York (Room 1059, 10017), 119 pp.

Introduction; Relation of Planning Education to Physical Planning and Economic and Social Development; Training for the Planning Profession; Research and Training Institutions for Planning in Latin America; Appendices.

574. Walker, Robert Averill, *The Planning Function in Urban Government,* Chicago (University of Chicago), Second Edition, 1950, 410 pp. [50-10651]

Development of the Planning Function: Origins of Modern City Planning, Expanding Scope of Urban Planning, Development of the Law Relating to Planning, Nature of the Planning Function; Organization for Planning: Independent Planning Commission, Planning as an Administrative Function, Planning Staff; Case Studies in Urban Planning: Chicago—Planning in Evolution, Studies Illustrating Current Planning Activity; Status of Urban Planning, 1940: Findings and Conclusions; Status of Urban Planning, 1950: Developments During World War II and Its Aftermath, Nature of the Planning Function—Reappraisal.

575. Willmann, John B., *The Department of Housing and Urban Development,* New York (Praeger), 1967, 205 pp. [67-24679]

Before HUD; How HUD: Happened, Is Organized; Mortgage Credit—FHA and FNMA; Renewal and Housing Assistance; Metropolitan Development; Demonstrations and Intergovernmental Relations—"Model Cities" Program; HUD, Congress, and the Home Builders; How HUD Is Doing—Some Pros and Cons; Where HUD Is Heading; Appendices; Bibliography, pp. 197-201.

Bibliographies

576. Banfield, Edward C. (Editor), *Urban Government—A Reader in Administration and Politics,* Bibliography, pp. 707-712.

577. Cohn, Angelo, *Careers in Public Planning and Administration,* Reading List, pp. 107-109.

578. Pfiffner, John M.; Presthus, Robert, *Public Administration,* Selected Bibliography, pp. 551-554.

579. Stoots, Cynthia F., *Metropolitan Organization for Planning,* Bibliography No. 50, Council of Planning Librarians, Monticello, Ill. (Exchange Bibliographies, P.O. Box 229, 61856), May 1968, 5 pp. Paper.

Metropolitan Planning; Metropolitan Goals and Objectives; Managing Our Urban Environment; Regional Development; Periodicals.

580. Willmann, John B., *The Department of Housing and Urban Development,* Bibliography, pp. 197-201.

8. MANAGEMENT, DECISION-MAKING

Articles, Periodicals

581. Beged-Dov, A. G., "Why Only Few Operations Researchers Manage," *Management Science,* Vol. 12, No. 12, August 1966, B-580—B-593.

". . . the single most reliable indicator of how effective is OR is the proportion of operations researchers who eventually attain general management responsibilities . . . certain alarming developments in the practice and education of Operations Research must be arrested."

582. Burck, Gilbert, "Management Will Never Be the Same Again," *Fortune,* Vol. LXX, No. 2, August 1964, pp. 125-126, 199-200, 202, 204.

The Word is Recentralize; "Intuition Can Be Structured;" More Hierarchial Than Ever; Many Dice, Each with a Hundred Sides; Survivors and Thrivers; Tomorrow's Managers.

583. Deardon, John, "Myth of Real-Time Management Information," *Harvard Business Review*, Vol. 44, No. 3, May-June 1966, pp. 123-132.

". . . it would not be practical to operate a real-time *management control system . . .* and such a system would not help solve any of the critical problems even if it could be implemented."

584. Etzioni, Amitai, "Mixed-Scanning: A 'Third' Approach to Decision-Making," *Public Administration Review*, Vol. XXVII, No. 5, December 1967, pp. 385-392.

Rationalistic Approach; Incrementalist Approach: Morphological Assumptions of the Incremental Approach, Critique of the Incremental Approach as a Normative Model, Conceptual and Empirical Critique of Incrementalism; Mixed-Scanning Approach: Can Decisions Be Evaluated?; Morphological Factors.

585. Hammond, John S., III, "Better Decisions with Preference Theory," *Harvard Business Review*, Vol. 45, No. 6, November-December 1967, pp. 123-141.

"Now decision-making technique can be tailored to risk-taking attitudes."
Decisions & Risks; Case of Petro Enterprises; Expected Value Analysis; Fresh Approach; Using Preference Curves; Analysis with Preferences; Commonly Observed Curves; Company vs. Individual; Important Distinction; Summing Up.

586. Holt, Herbert, M.D.; Ferber, Robert C., "The Psychological Transition from Management Scientist to Manager," *Management Science*, Vol. 10, No. 3, April 1964, pp. 409-420.

First Appearance of the Management Scientist; Biography of Bill Gallagher; From Management Consultant to Manager; Conflict; Call for Help; Summary; References, p. 420.

587. Mace, Myles L., "The President and Corporate Planning," *Harvard Business Review*, Vol. 43, No. 1, January-February 1965, pp. 49-62.

Administrative Focus; Realistic Evaluation; Corporate Objectives; Analyzing the Present; Predicting the Future; Reviewing the Forecasts; Evaluating the Program; Creating the Goals; Planning Staff; Planners' Problems; Appendix: Outline of a Five-Year Forecast.

588. Magee, John F., "Decision Trees for Decision Making," *Harvard Business Review*, Vol. 42, No. 4, July-August 1964, pp. 126-128; "How to Use Decision Trees in Capital Investment," *ibid.*, Vol. 42, No. 5, September-October 1964, pp. 79-96.

"The decision tree can clarify the choices, risks, objectives, monetary gains, and information needs involved in long-range planning . . . it allows management to combine analytical techniques with a clear portrayal of the impact of future decision alternatives and events. The interactions between present decision alternatives, uncertain events, and future choices and their results become more visible."

589. Miller, George A., "The Magical Number Seven, Plus or Minus Two: Some Limits on Our Capacity for Processing Information," *Psychological Review*, Vol. 63, No. 2, March 1956, pp. 81-97.

590. Newell, Allen; Simon, Herbert A., "Computer Simulation of Human Thinking," *Science*, Vol. 134, No. 3495, 22 December 1961, pp. 2011-2024.

"A theory of problem solving expressed as a computer program permits simulation of thinking processes."
Behavioral Phenomena; Nonnumerical Computer Program as a Theory; General Problem Solver; Testing the Theory; Conclusion.

591. Poppy, John, "It's OK to Cry in the Office," *Look*, Vol. 32, No. 14, 9 July 1968, pp. 64, 66, 69, 70, 72-76.

"California aerospace contractor is one of hundreds of companies learning that honestly shared emotions help get the work done better. Executives who have long given lip service to 'leveling' now find they can be trained to do it."

592. Ramo, Simon, "Man and Intelligent Machines," in: *McGraw-Hill Yearbook of Science and Technology*, New York (McGraw-Hill), 1962, pp. 53-64. [62-12028]

General Characteristics of Intelligent Machines; Artificial Intelligence; High-Level Machines; Applications of Intellectronics; Future Intellectronic Systems.

593. Ramo, Simon, "Management of Government Programs," *Harvard Business Review*, Vol. 43, No. 4, July-August 1965, pp. 6-8, 10, 12, 163.

"The concept of program management, already well known in the Department of Defense and NASA, is now 'catching fire' in other branches of government. What is industry's stake in better management of government programs? What kinds of managers are needed, how can the supply of them be increased, and how can they be helped to do better?"

594. Reichley, A. James, "A Nightmare for Urban Management," *Fortune*, Vol. LXXIX, No. 3, March 1969, pp. 94-99, 170-172, 174.

The Dragon St. George Couldn't Slay; Labor's Generation Gap; Mules in the Bureaucracy; Learning to Live with Change; No Salt Trucks in Queens; "He May Be Great at Lincoln Center . . ."; Where the Important People Go; A Higher Authority?

595. Seligman, Daniel, "McNamara's Management Revolution," *Fortune*, July 1965, pp. 117-121, 244, 246, 248, 250.

596. Steiner, George A., "The Critical Role of Top Management in Long-Range Planning," Reprint No. 86-1966, Los Angeles, Cal. (Graduate School of Business Administration, Division of Research, University of California, 90024), 1966, 8 pp.; also in *Arizona Review*, April 1966.

Importance of Long-Range Planning; Conceptual Model of Long-Range Planning; Long-Range Planning in Large and Small Business; Top Management's Key Role in Planning; Developing the Plan; Base Decisions on Plans; Planning Takes Time; Resumé.

597. Sumner, Charles E., Jr., "The Future Role of the Corporate Planner," *California Management Review,* Vol. III, No. 2, Winter 1961, pp. 17-31.

"Breakthroughs in technology and the often overpowering complexity of the corporate structure have spawned a new, versatile breed of managers. Known as planners, their business is change and their eyes must always be focussed on the big picture. Here is a profile of the job and the men who must be part seers, part detective-cost accountants, part sociologists to fill it."

Line and Staff; Function of Planning; Growing Importance; Technological Complexity; Three Variables; Interrelated Key Factors; Complexity Fosters Planning Role; What's Going to Happen Next?; Divisionalization Creates Jobs; Mergers Create Planning Jobs; Planning Hierarchy; Positions Via Decentralization; Speeds Integrated Changeovers; Takes Planning to Speed Up; Bosses or Consultants?; Commitment Time; Institutional Commitment; Independent Thinkers Needed; Must Take Long View; Summary; A New Profession; Accounting Skill Necessary; Wanted: Methods Plus Intuition; Decision Matrix Method; Wisdom Through Security.

598. Tilles, Seymour, "The Manager's Job: A System Approach," *Harvard Business Review,* Vol. 41, No. 1, January-February 1963, pp. 73-81.

Systems Approach; Defining the Company; Setting System Goals; Creating Subsystems; Conclusion.

599. Trull, Samuel G., "Some Factors Involved in Determining Total Decision Success," *Management Science,* Vol. 12, No. 6, February 1966, pp. B-270-280.

Decision Success: Compatibility with Existing Operating Constraints, Proximity to Optimum Time for Decision, Proximity to Optimum Amount of Information, The Problem Solver's Influence on the Decision, Avoidance of Conflict of Interests, Reward-Risk Factor, Degree of Understanding; Conclusion.

600. Wheaton, William L. C., "Public and Private Agents of Change in Urban Expansion," in: Webber, Melvin M.; Others, *Explorations into Urban Structure,* Philadelphia, Penn. (University of Pennsylvania), 1963, pp. 154-196. [63-15009]

Investment Mix: Volumes and Types of Investment, Controlling Criteria of Decision, Fact and Value in Market Decisions, Decision Chains and Nonmarket Criteria; Pluralism of Metropolitan Decisions: Public Decision Processes and Institutions, Past and Present Decisions, Private Systems of Power, Decision-Making in the Mixed Economy; To Improve Rationality and Consensus: Guiding the Fact Component, Guiding the Standard Component; Conclusions.

601. Wiest, Jerome D., "Heuristic Programs for Decision Making," *Harvard Business Review,* Vol. 44, No. 5, September-October 1966, pp. 129-143.

Meaning and Significance; Heuristic Programming; Wide-Ranging Uses; Current Usefulness; Effect on Management; Conclusion; Appendix: Scheduling Projects.

Books, Reports, Pamphlets

602. Ackoff, Russell L.; Rivett, Patrick, *A Manager's Guide to Operations Research,* New York (Wiley), 1963, 107 pp. [63-14115]

"It is the purpose of this book to enable the industrial executive to reduce the faith he requires to undertake OR in his organization."
Nature of OR; Form and Content of Problems; Relationship with Other Management Services; Organization and Administration; Further Reading.

603. Barish, Norman R., *Economic Analysis for Engineering and Managerial Decision Making,* New York (McGraw-Hill), 1962, 752 pp. [61-18720]

Engineering and Business Decision-Making; Costs; Methods for Tangible Evaluation of Alternatives; Capital Management; Determinations of Minimum Cost and Maximum Profit; Risk, Uncertainty, and Intangibles; Elements of Economic Measurement, Analysis, and Forecasting.

604. Barnard, Chester I., *The Functions of the Executive,* Cambridge, Mass. (Harvard University), 1958, 334 pp. [39-849]

Classic in management literature, originally published in December 1938.
Preliminary Considerations Concerning Cooperative Systems; Theory and Structure of Formal Organizations; Elements of Formal Organizations; Functions of Organizations in Cooperative Systems.

605. Beer, Stafford, *Decision and Control—The Meaning of Operational Research and Management Cybernetics,* New York (Wiley), 1966, 556 pp. [66-25668]

Nature of Operational Research; Activity of Operational Research; Relevance of Cybernetics; Outcomes.

606. Bross, Irwin D. J., *Design For Decision,* New York (Free Press), 1953, 276 pp. Paperback. [53-12977]

". . . for the nonspecialist . . . in terms the layman can comprehend. Making only minimal use of mathematics."
History of Decision; Nature of Decision; Prediction; Probability; Values; Rules for Action; Operating a Decision-Maker; Sequential Decision; Data; Models; Sampling; Measurement; Statistical Inference; Statistical Techniques; Design for Decision; Further Reading, pp. 267-272.

607. Bursk, Edward C.; Chapman, John F. (Editors), *New Decision-Making Tools for Managers,* Mathematical Programming as an Aid in the Solving of Business Problems, New York (New American Library), May 1965, 412 pp. Paperback. [BNB65-14822]

General: "Operations Research" for Management (Cyril C. Herrmann; John F. Magee), Mathematical Programming—Better Information for Better Decision Making (Alexander Henderson; Robert Schlaifer), How to Plan and Control with PERT (Robert W. Miller), Meaningful Costs for Management Action (Robert Beyer), Econometrics for Management (Edward C. Bennion); Finance: How to Evaluate

New Capital Investments (John G. McLean), Mathematical Models in Capital Budgeting (James C. Hetrick); Marketing: Simulation—Tool for Better Distribution (Harvey N. Shycon; Richard M. Maffei), Marketing Costs and Mathematical Programming (William J. Baumol; Charles H. Sevin), Tests for Test Marketing (Benjamin Lipstein), Less Risk in Inventory Estimates (Robert G. Brown); Product Strategy: Prudent-Manager Forecasting (Gerald A. Busch), Strategies for Diversification (H. Igor Ansoff), Selecting Profitable Products (John T. O'Meara, Jr.); Production: Mathematics for Production Scheduling (Melvin Anshen; Others), The Statistically Designed Experiment (Dorian Shainin), Quality Control (Theodore H. Brown).

608. Clark, Terry N. (Editor), *Community Structure and Decision-Making: Comparative Analysis*, San Francisco, Cal. (Chandler), 1968, 498 pp. [68-12245]

Who Governs, Where, When, and With What Effects?, Social Stratification, Differentiation, and Integration, Concept of Power, Community or Communities?, Community Structure and Decision-Making (Terry N. Clark); Power and Community Structure (Peter H. Rossi), Community Power and Decision-Making—Quantitative Examination of Previous Research (Claire W. Gilbert); Political Ethos and the Structure of City Government (Raymond E. Wolfinger; John Osgood Field), Rancorous Conflict in Community Politics (William A. Gamson), Structure and Values in Local Political Systems—Case of Fluoridation Decisions (Donald B. Rosenthal; Robert L. Crain), Suburban Differentiation and Municipal Policy Choices—Comparative Analysis of Suburban Political Systems (Bryan T. Downes), Functions of Urban Political Systems—Comparative Analysis and the Indian Case (Donald B. Rosenthal); Process of Decision-Making Within the Context of Community Organization (Paul A. Miller), Institutional and Occupational Representations in Eleven Community Influence Systems (William V. D'Antonio; William H. Form; Charles P. Loomis; Eugene C. Erickson); Reputation and Resources in Community Politics (William A. Gamson), On the Structure of Influence (Ronald L. Nuttall; Erwin K. Scheuch; Chad Gordon); Comparative Study of Decision-Making in Rural Communities (Kenneth Kammeyer), Community Power and Urban Renewal Success (Amos L. Hawley), Purposive Community Change in Consensus and Dissensus Situations (Roland L. Warren; Herbert H. Hyman); Life-Style Values and Political Decentralization in Metropolitan Areas (Oliver P. Williams), Differential Patterns of Community Power Structure—Explanation Based on Interdependence; Present and Future Research on Community Decision-Making—Problem of Comparability (Terry N. Clark).

609. Dale, Ernest; Michelon, L. C., *Modern Management Methods*, Cleveland, Ohio (World), 1966, 211 pp. [65-27421]

Job of Management; New Developments in Human Relations and Leadership; Challenge of Organization; Managing and Communicating by Objectives; New Dimensions in Organizational Communication; Long-Range Planning, Financing, and Control; Government-Business Relations; Managerial Decision-Making; Reducing Costs by Value Analysis; Introductory Statistics for Management; Payoff Tables and Decision Trees; Critical Path Analysis—PERT; Management and the Computer; Management and the Future.

610. Feinberg, Sam, *How Do You Manage,* New York (Fairchild), 1966, 288 pp. [64-24707]

Our Managerial Economy; What Makes a Top Executive Run?; Are You an Executive in Deed?; Wanted: Creative Men, Not Just Bodies; Executive Courses in Flux; Education for Leadership; What's Your Executive Development Score?; Executive Recruitment; Firings Masquerading as Resignations; Everyone Is for Ethics; Family vs. Professional Management; Emotional Health; Psychiatry-Psychology in Practice; How Retailers Handle Tensions; The Psychologist in Retailing.

611. Fisk, George (Editor), *The Psychology of Management Decision,* Lund, Sweden (Gleerup), 1967, 309 pp. [67-103360]

Goals of Individuals and Organizations: Structure of Individual Goals—Implications for Organization Theory (Peer Soelberg), Human Group Model of Organization (Mason Haire), Statistical Decision Theory and Benefit-Cost Analysis for Preferred-ness of Choice Among Alternative Projects (Morris Hamburg), Aspiration Levels and Utility Theory (William H. McWhinney); Organization for Communicating Performance Information: Perceived and "Real" Organization Behavior (Ezra S. Krendel; Joel N. Bloom), Computer Aided Planning, Command and Control (Andrew Vazsonyi), Associative Memory in Heuristic Problem Solving for Man-Machine Decisions (Noah S. Prywes); Individual Behavior Patterns—Attitudes Toward Risk, Learning, and Personality Influence in Decision Making: Behavioral Experiment in the Economics of Information (Paul E. Green), On the Two-Armed Bandit Problem (Donald F. Morrison), Computer Simulation of Learning and Decision Processes in Poker (Leonard Garrett; Robert McClure; Ron Schnitzer), Strategic Thinking as a Function of Social Attitudes—An Experiment with Prisoner's Dilemma (Lars Dencik; Hakan Wiberg), Interaction Effect of Personality Variables in a Prisoner's Dilemma Management Game (J. M. Keenan; H. A. Hoverland), Interaction of the Manager's Personality with His Environment—Model for Solving the Adaptation Problem (Martin Pfaff); Future Directions: Significance of Buyer Perception in Studies of Market Behavior (Alfred R. Oxenfeldt; F. E. Brown).

612. Gilbert, Charles E., *Governing the Suburbs,* Bloomington, Ind. (Indiana University), 1967, 364 pp. [67-13024]

Society; Politics; Government; Policy and Performance; Municipal Government; Conclusions; Appendices: Theory, Method.

613. Goldman, Thomas A. (Editor), *Cost-Effectiveness Analysis—New Approaches in Decision-Making,* New York (Praeger), 1967, 231 pp. [66-26561]

Introduction and Overview (Edward S. Quade); Measures of Effectiveness (William A. Niskanen); Choice of Analytic Techniques (Alfred Blumstein); Use of Cost Estimates (Harry P. Hatry); Estimating Systems Costs (James D. McCullough); Armed Forces' Use of Cost-Effectiveness Analysis (R. S. Berg); Defense Contractor Use of Cost-Effectiveness Analysis (Eugene R. Brussell); Analysis of Tactical Air Systems (Murray Kamrass; Joseph A. Navarro); Cost-Effectiveness Analysis for Government Domestic Programs (William M. Capron); Cost-Effectiveness Analysis for the "War on Poverty" (Stanley M. Besen; Alan E.

Fechter; Anthony C. Fisher); Analysis of Metropolitan Transportation Systems (John F. Kain); Structure of Incentive Contracts (Collin W. Scarborough); Structure of Military Industrial Funds (Martin J. Bailey).

614. Gore, William J.; Dyson, J. W. (Editors), *The Making of Decisions: A Reader in Administrative Behavior,* Glencoe, Ill. (Free Press), 1964, 440 pp. [64-23084]

Perspectives Toward the Decision-Making Process: Administrative Decisions—Scheme for Analysis (Craig C. Lundberg), Approaches to the Study of Decision-Making Relevant to the Firm (Martin Shubik), Rational Behavior and Economic Behavior (George Katona), Socioeconomic Decisions (Paul Diesing), Culture, Science, and Politics (Vincent Ostrom), Meaning of "Political" in Political Decisions (Bruno Leoni); Decision-Making Strategies: Behavioral Model of Rational Choice (Herbert A. Simon), Criteria for Choice Among Risky Ventures (Henry Allen Latané), On the Rationale of Group Decision Making (Duncan Black), Science of "Muddling Through" (Charles E. Lindblom), Towards a Decision-Making Model in Foreign Policy (Joseph Frankel), Basic Frameworks for Decisions (Charles Wilson; Marcus Alexis); Organizational Variables Influencing the Decision-Making Process: Social Behavior as Exchange (George C. Homans), Normative Regulation of Authoritative Behavior (Jay Jackson), Norms, a Feature of Symbolic Culture—A Major Linkage between the Individual, Small Group, and Administrative Organization (Fremont Shull; Andre Del Beque), Notes on the Concept of Commitment (Howard S. Becker), Role of Expectations in Business Decision-Making (R. M. Cyert; W. R. Dill; J. G. March), Decision Making in the House Rules Committee (James A. Robinson), Contradictory Functional Requirements and Social Systems (Gideon Sjoberg), Subjective Probability and Decision Under Uncertainty (N. T. Feather); Functions of the Decision-Making Process: Notes on a Theory of Political Actualization—Paradigm of the Political Process (Robert S. Cahill; Marshall N. Goldstein), Interpersonal Decision Making—Resolution of Dyadic Conflict (John M. Atthowe, Jr.), Three Modes of Conflict (Anatol Rapoport), Termination of Conflict (Lewis A. Coser).

615. Guzzardi, Walter, Jr., *The Young Executives—How and Why Successful Managers Get Ahead,* New York (New American Library), 1966, 191 pp. Paperback.

The Man and—Myth, His Attitudes, the House He Does Not Enter, the Corporation, His Decisions, Style, Secret Sharer, Computer, Sea Around Him; The Cocoon That Surrounded Him; Toward Journey's End; Selected Bibliography, pp. 190-191.

616. Hitch, Charles J., *Decision-Making for Defense,* Berkeley, Calif. (University of California), 1965, 90 pp. [65-27885]

Four lectures, 1789-1960 ("administrative development of the U.S. defense establishment from its earliest beginnings to the present day"); Planning-Programming-Budgeting; Cost-Effectiveness; Retrospect and Prospect.

617. Hunter, Floyd, *Community Power Structure—A Study of Decision Makers,* Chapel Hill, N.C. (University of North Carolina), 1953, 297 pp. [53-10042]

Introduction; Location of Power in Regional City; What Some Leaders Are Like;

Structure of Power in Regional City; Power Structure in a Sub-Community; Regional City in Political Perspective; More Private Aspects of Power; Projects, Issues, and Policy; Organized Community and the Individual; Bibliography, pp. 275-289.

618. Kozmetsky, George; Kircher, Paul *Electronic Computers and Management Control*, New York (McGraw-Hill), 1956, 296 pp. Paper. [55-12105]

Introduction: Why Be Interested in Computers?; What Is an Electronic Computer?; Survey of Electronic Methods of Data Processing; Studies and Applications of Electronic Systems; Administrative Problems Experienced in Introducing Computer Systems; Management and the Scientific Approach; Management Planning and Control; Programming, Scheduling, and Feedback; Integrated Business Systems; Automation and Scientific Computation; Role of the Executive in Selection of an Electronic System; Challenge to the Executive. Bibliography, pp. 245-250.

619. Leavitt, Harold J.; Pondy, Louis R., *Readings in Managerial Psychology*, Chicago, Ill. (University of Chicago), 1964, 642 pp. Paper. [64-15811]

Inside the Individual, Personality Theory and Assessment: Emotional Side of Man, Problem-Solving Side of Man, Assessing the Whole Individual; Between Individuals, Interpersonal Influence: Communication and Influence, Leadership and Influence, Power and Influence, Motivation and Influence; Collections of Individuals, Group Behavior: Group Pressures on the Individual, Group Effectiveness, Groups in Conflict; People in Complex Systems, Formal Organizations: Decision Making in, Structure of, Human Aspects of, Technological Aspects of.

620. Levin, Richard I.; Kirkpatrick, C. A., *Quantitative Approaches to Management*, New York (McGraw-Hill), 1965, 379 pp. [64-24603]

"an approach to quantitative methods in management that is clear and understandable . . . intended to . . . give college and business people without mathematical background an understanding of some of the quantitative methods that can be used in management."

Subjects include: scientific method, breakeven analysis, probability, statistics, inventory, vectors, matrix algebra, linear programming, game theory, Markov analysis, and queueing.

621. Likert, Rensis, *New Patterns of Management*, New York (McGraw-Hill), 1961, 279 pp. [61-13167]

Introduction; Leadership and Organizational Performance; Group Processes and Organizational Performance; Communication, Influence, and Organizational Performance; Effect of Measurements on Management Practices; Some General Trends; Effective Supervision—An Adaptive and Relative Process; An Integrating Principle and an Overview; Some Empirical Tests of the Newer Theory; Voluntary Organizations; Nature of Highly Effective Groups; Interaction-Influence System; Function of Measurements; Comparative View of Organizations; Looking to the Future.

622. Lindsay, Franklin A., *New Techniques for Management Decision Making*, New York (McGraw-Hill), 1958, 129 pp. Paper. [58-11183]

Introduction to Mathematical Analysis; Techniques of Mathematical Analysis; Application of Mathematical Analysis to Management Problems; Conclusion.

623. (Air Vice-Marshall) McCloughry, E. J. Kingston, *The Direction of War*, A Critique of the Political Direction and High Command in War, New York (Praeger), 1955, 261 pp. [55-11306]

". . . Chief Operations Planner at H.Q. Allied Expeditionary Air Force . . . setting down the historical lessons about High Command and then deducing principles and the various organizations and requirements necessary to meet different sets of circumstances . . . Political Direction and High Command are concerned so essentially with personalities and those antinomies of human nature in war or peace that all other considerations are secondary."

624. Maynard, H. B. (Editor-in-Chief), *Handbook of Business Administration*, New York (McGraw-Hill), 1967, 1916 pp. [66-20719] 177 contributing authors.

Introduction to Business Administration; Organization; General Management; Common Concerns of All Managers; Management: Research and Development, Materials, Manufacturing, Marketing, Financial; Accounting and Control; Management of Human Resources; Managing External Relations; Secretarial and Legal Activities; Office Administration; Systems and Data Processing; Management of International Operations; Tools and Techniques of Management Decision Making and Control.

625. Myers, Charles A. (Editor), *The Impact of Computers on Management*, Cambridge, Mass. (M.I.T.), 1967, 310 pp. Paperback. [67-14097]

Introduction (Charles A. Myers); Impact of Information Technology on Organizational Control (Thomas L. Whistler); Computers and Organization Structure in Life-Insurance Firms—External and Internal Economic Environment (George E. Belehanty); Tasks, Organization Structures, and Computer Programs (David Klahr; Harold J. Leavitt); Implications of On-Line, Real-Time Systems for Managerial Decision-Making (Donald C. Carroll); Computers and Profit Centers (John Dearden); Total-Systems Concept—Its Implications for Management (John A. Beckett); Changes in Management Environment and Their Effect Upon Values (Charles R. DeCarlo); Comments on the Conference Discussion (Jay W. Forrester); Appendix—Impact of Computerized Programs on Managers and Organizations—Case Study of an Integrated Manufacturing Company.

626. Quade, Edward S. (Editor), *Analysis for Military Decisions*, Chicago, Ill. (Rand McNally), 1964, 390 pp. [65-3032]

Orientation; Elements and Methods; Special Aspects; Summary; Appendixes. "The real impact of the book . . . is the guidepost that it establishes for the development of systems analyses in other areas of national decision. . . . As a nontechnical primer . . . this book is a milestone in clearly explaining the systems-analysis art to the layman."

627. Schoderbek, Peter P. (Editor), *Management Systems*, New York (Wiley), 1967, 483 pp. [67-17350]

Systems Concept: General Systems Theory—The Skeleton of Science (Kenneth E. Boulding), Utility of System Models and Developmental Models for Practitioners (Warren Bennis; Others), Organization Theory—Overview and Appraisal (William G. Scott); Information Technology: Application of Information Technology (John Diebold), Management Information Crisis (D. Ronald Daniel), Thrust of Information Technology on Management (Oliver W. Tuthill); Information Technology and the Organization: Management in the 1980's (Harold J. Leavitt; Thomas L. Whistler), Information Technology and Decentralization (John F. Burlingame), Impact of Information Technology on Organization (James C. Emery); Breaking the Chain of Command (S. C. Blumenthal); Design of Management Systems: Designing Management Systems (Richard A. Johnson; Others), Development and Installation of a Total Management System (R. L. Martino), Basic Concepts for Designing a Fundamental Information System (A. F. Moravec), Designing a Behavioral System (Stanley Young); Total Management Systems: Is the Total System Concept Practical? (A. T. Spaulding, Jr.), Systems Can Too Be Practical (Allen Harvey), The Total Systems Myth (W. M. A. Brooker), Management Information Systems (J. W. Konvalinka; H. G. Trentin); Human Problems of Systems: Designing the Participative Decision-Making Systems (Bruce De Spelder), Human Side of a Systems Change (Lawrence K. Williams), Resistance to Change—Its Analysis and Prevention (Alvin Zander), Psychology for the Systems Analyst (Robert E. Schlosser); Management Control Systems: Approach to Computer Based Management Control Systems (Donald G. Malcolm; Alan J. Rowe), Computer Management Control Systems Through the Looking Glass (George F. Weinwurm), Research Problems in Management Control (Alan J. Rowe), Industrial Dynamics and the Design of Management Control Systems (Edward B. Roberts); Cybernetics: Development of Cybernetics (Charles R. Dechert), Perceptive Feedback (Joseph B. Bonney, Jr.), What Has Cybernetics to Do With Operational Research (Stafford Beer); Electronic Computers and Management Organization (George J. Brabb; Earl B. Hutchins), Management Concept in Electronic Systems (Virgil F. Blank), Electronic Power Grab (Robert L. McFarland); Models and Simulation: Models (Irwin D. J. Bross), Evaluation of Models (Karl W. Deutsch), Management Control Simulation (Joel M. Kibbee); Measurement: Fundamentals of Measurement (Paul Kircher), Why Measure? (C. West Churchman), Appraisal of Some of the Problems of Measurement in Operational Research (R. W. Shepard); PERT-PERT/COST: Program Evaluation Review Technique (David G. Boulanger), PERT/Cost—The Challenge (Don T. DeCoster), Its Values and Limitations (Peter P. Schoderbek), Sociological Problems of PERT (Peter P. Schoderbek); Real Time Systems: Management Control in Real Time Is the Objective (H. C. Hartmann), Management in Real Time (Sherman C. Blumenthal), On-Line Management Information (Norman J. Ream), The SABRE System (R. W. Parker), Myth of Real-Time Management Information (John Dearden); Information Retrieval: Information Systems for Management Planning (John T. Jackson), Growing Data Volume—Can It Be Mastered? (Walter F. Williams), Information Retrieval—1966 (Benjamin F. Cheydleur), Comments on Information Retrieval (Gary R. Martins); Prologue to the Future.

Brief treatment of each topic, but inclusive coverage.

628. Simon, Herbert A., *Administrative Behavior,* New York (Free Press), Second Edition, 1957, 259 pp. Paperback. [57-5627]

Decision-Making and Administrative Organization; Some Problems of Administrative Theory; Fact and Value in Decision-Making; Rationality in Administrative Behavior; Psychology of Administrative Decisions; Equilibrium of the Organization; Role of Authority; Communication; Criterion of Efficiency; Loyalties and Organizational Identification; Anatomy of Organization; What Is an Administrative Science?

629. Simon, Herbert A., *The New Science of Management Decision*, New York (Harper & Row), 1960, 50 pp. [60-15199]

Executive as Decision Maker; Traditional Decision-Making Methods; New Techniques for Programmed Decision Making; Heuristic Problem Solving; Organizational Design: Man-Machine Systems for Decision Making.

630. Steiner, George A., *Top Management Planning*, New York (Macmillan), 1969, 800 pp. [69-17783]

Nature and Concept of Business Planning: Nature of Business Planning and Plans, Conceptual and Operational Model of Corporate Planning, Importance of Comprehensive Planning, Top Management's Role in Planning; Process of Developing Plans: Organizing for Corporate Planning, Network of Corporate Aims I & II, Appraising the Future Environment for Planning, Nature and Development of Business Strategies, Business Policies and Procedures, From Strategic Planning to Current Action; Tools for More Rational Planning: Rationality in Planning, Older Tools for Making More Rational Planning Decisions, Systems Approach to Decision-Making, Newer Quantitative Techniques for Rational Decisions, Management Information Systems, Computers and Management Information Systems; Planning in Selected Major Functional Areas: Marketing, Product, Financial, Diversification, Research and Development, Planning in Other Functional Areas; Concluding Observations: Current State-of-the-Art and the Future of Comprehensive Corporate Planning; Bibliography.

631. Tucker, Samuel A. (Editor), *A Modern Design for Defense Decision— A McNamara-Hitch-Enthoven Anthology*, National Security Management, Washington, D.C. (Industrial College of the Armed Forces), 1966, 259 pp.

Introduction (Alain C. Enthoven); The Budget as a Management Tool for Integrating National and Defense Policy (Robert S. McNamara): Managing the Department of Defense, Decision-Making in the Department of Defense, Formulation of Political Objectives and Their Impact on the Budget, Foundation for Defense Planning and Budgeting, Department of Defense Budget and the National Economy (Charles J. Hitch); Organizational Framework and Agency Responsibilities: Evolution of the Department of Defense (Charles J. Hitch), Agency Responsibilities for Planning and Budgeting (Robert S. McNamara), Joint Chiefs of Staff and the Defense Budget (Robert S. McNamara; General Maxwell D. Taylor); Programming System (Charles J. Hitch): Development and Salient Features of the Programming System, Programming and Budgeting in the Department of Defense (Alain C. Enthoven), Program Packages, Retrospect and Prospect; Systems Analysis (Alain C. Enthoven): Cost Effectiveness (Charles J. Hitch), Choosing Strategies and Selecting Weapons Systems, Operations Research at the National Policy Level, Systems Analysis and the Navy, Cost-Effectiveness Analysis of Army Divisions; U.S. Defense Policy for the 1960's (Alain C. Enthoven); Appendices: Introduction of New Government-Wide Planning and Budgeting System (President Lyndon B. Johnson), Illustrative Example of Systems Analysis (William P. Snyder); Suggestions for Further Reading, pp. 249-251.

Bibliographies

632. Bross, Irwin D. J., *Design For Decision*, Further Reading, pp. 267-272.

No references later than 1952.
Statistical Decision; Probability; General Texts; Sampling; Design of Experiments; Special Applications; Historical Landmarks; Related Topics.

633. Fry, B. L., "Selected References on PERT and Related Techniques," *IEEE Transactions on Engineering Management*, Vol. EM-10, No. 3, September 1963, pp. 150-153.

"Literature of network-type management control systems is organized by author, title, author's organization, chronology, and a file number . . . two listings—one indexed by author and one by title."

634. Holt, Herbert, M.D.; Ferber, Robert C., "The Psychological Transition from Management Scientist to Manager," *Management Science*, References, p. 420.

635. Olive, Betsy Ann, *Management—A Subject Listing of Recommended Books, Pamphlets and Journals*, Ithaca, N.Y. (Graduate School of Business and Public Administration, Cornell University), 1965, 222 pp. Paperback.

Books and Pamphlets: Administration and Management, Personnel Administration, Industrial and Labor Relations, Marketing, Quantitative Methods, Economics, International Economics, Business and Trade, Business Economics, Politics, Government and Public Administration, Law and Government Regulation, Problems of Developing Countries, Education, General Reference Sources; Periodicals; Author Index.

636. Steiner, George A., *Top Management Planning*, References, pp. 733-770.

637. Wasserman, Paul; Silander, Fred S., *Decision-Making—An Annotated Bibliography Supplement, 1958-1963*, Ithaca, N.Y. (Graduate School of Business and Public Administration, Cornell University), 1964, 178 pp. Paperback. [58-4160]

Decision Making—General and Theoretical Material; Leadership As a Factor in Decision Making; Behavioral Decision Theory: General, Learning, Game Theory, Utility, Personal Characteristics, Cognitive Dissonance, Simulation; Decision Making in Small Groups: General and Theoretical Material, Conflict, Conformity and Competition, Quality of Decisions—Individual vs. Group, and Group Size, Group Characteristics and Operating Conditions; Community Decision Making; Communications and Information Handling; Techniques and Methods; Cases, Illustrations, and Applications.

9. EFFECTUATION

Articles, Periodicals

638. Babcock, Richard F.; Bosselman, Fred P., "Suburban Zoning and the Apartment Boom," *University of Pennsylvania Law Review*, Vol. III, No. 8, June 1963, pp. 1040-1091.

Early Regulation of Multiple-Family Dwellings; Trends in Residential Building; Economics of the Apartment Boom; Suburban Reaction to Multiple-Family Dwellings; Shouted Reasons; Whispered Reasons; Current Judicial Approach; Two Case Studies; Prospects for the Future; Conclusion.

639. Bair, Frederick H., Jr., "Improving Zoning—Some New Approaches," *Land Use Controls—A Quarterly Review*, Vol. 2, No. 4, Fall 1968, pp. 1-16.

Planned Developments Generally; Planned Development—Housing (PD-H) Districts; Guides and Standards for Planned Developments—Housing; Applications in Conventional Zoning; Micro-Zoning; Performance Standards; Coordinated Regulatory System for Use, Development, Construction, and Occupancy; Plan of Attack; Expected Results.

640. Bair, Frederick H., Jr., "Opportunities for Community Relations," in *Planning 1963*, Chicago, Ill. (American Society of Planning Officials), November 1963, pp. 251-259. Paperback. [39-3313]

Preparing the Proposal; Public Consent and Public Opinion; What the Public Hearing Is For; After the Hearing; In Summary.

641. Burke, Edmund M., "Citizen Participation Strategies," *Journal of the American Institute of Planners*, Vol. XXXIV, No. 5, September 1968, pp. 287-302.

Dilemmas of Citizen Participation; Staff Supplement Strategy; Cooptation; Community Power Strategies; Conclusions.

642. Chapin, F. Stuart, Jr., "Taking Stock of Techniques for Shaping Urban Growth," *Journal of the American Institute of Planners*, Vol. XXIX, No. 2, May 1963, pp. 76-87.

Background; General Plan for the Metropolitan Region; Urban Development Policies Instrument; Metropolitan Area Public Works Program; Urban Development Code; Informed Metropolitan Community; Co-Ordinated Use of Techniques—A Guidance System.

643. Clavel, Pierre, "Planners and Citizen Boards: Some Applications of Social Theory to the Problem of Plan Implementation," *Journal of the American Institute of Planners*, Vol. XXXIV, No. 3, May 1968, pp. 130-139.

". . . reports on a study of planning as expert advice to nonpartisan citizen boards—and the means by which this advice is implemented or blocked in a semirural county."

Exchange, Status, and Inequality; Planners and Citizen Boards; Correlates of Status Inequalities; Theoretical Implications of the Analysis; Practical Applications of the Analysis.

644. Coughlin, Robert E., "The Capital Programming Problem," *Journal of the American Institute of Planners,* Vol XXVI, No. 1, February 1960, pp. 39-48.

"Capital programming is done everywhere, but virtually nowhere is theory to be found on how to do it. Here an analytic framework is set forth for formulating a capital program, given a comprehensive plan. . . ."
Levels of Decision; Parts of the Problem: Catalogue of Proposed Projects, Program Period, Budget, Problem in Brief, Objective; Six-Year Program As a Whole—Choices Among Functional Groups: Comprehensive Plan Profile; Six-Year Program As a Whole—Choices Among Projects: Sequence Importance, Profile Allocation Chart in Practice, Each Year of the Six-Year Program, Conclusion.

645. Davidoff, Paul, "Advocacy and Pluralism in Planning," *Journal of the American Institute of Planners,* Vol. XXXI, No. 4, 1965, pp. 331-338.

"City planning is a means of determining policy. Appropriate policy in a democracy is determined through political debate. The right course of action is always a matter of choice, never of fact. Planners should engage in the political process as advocates of the interests of government and other groups. Intelligent choice about public policy would be aided if different political, social, and economic interests produced city plans. . . ."

646. Ewing, David W., "Corporate Planning at the Crossroads," *Harvard Business Review,* Vol. 45, No. 4, July-August 1967, pp. 77-86.

"New directions are needed to avoid the danger that this function will take the low road while top management travels the high road."
Functional Purpose? Human Side?; Market Forecasts?; Conclusion.

647. Godschalk, David R.; Mills, William E., "A Collaborative Approach to Planning Through Urban Activities," *Journal of the American Institute of Planners,* Vol. XXXII, No. 2, March 1966, pp. 86-95.

"Urban activities are workable bases for continuing planner-citizen dialogues. Seeking both to inform and to involve citizens, these exchanges also provide the planner with an avenue of collaboration with his client community. Findings from the surveys may be maintained in an activities base, which includes both quantitative and qualitative data on activities. Policy and planning decisions benefit from activities base data on the concerns and potentialities of subcommunities. A pilot study of household activities demonstrates the usefulness and feasibility of the collaborative approach [which] seems particularly appropriate in light of the wider definition of the planning realm now being recognized."

648. Grebler, Leo, "Land Assembly and Relocation in Urban Renewal: A Study of European Methods," *Appraisal Journal,* January 1963, pp. 13-21; Reprint No. 16, Real Estate Research Program, Graduate School of Business Administration, University of California, Los Angeles, 1963.

Land Assembly Through: Condemnation, Other Means; Contrast with the United States; Methods of Land Disposition; Relocation; Summary.

649. Hill, Carroll V., "Planning, Property Taxation, and Zoning," in: American Society of Planning Officials, *Planning 1967*, Selected Papers from the ASPO National Planning Conference, Houston, Texas, April 1-6, 1967, Chicago, Ill. (1313 East 60th St., 60637), pp. 282-290. Paperback. [39-3313]

Leadership Role for the Planner; Zoning and Urbanization; Property Taxation Reform; Need for Improved Assessment Procedures; Coordination of Land-Use Planning, Assessments, and Property Taxation; Research Needs for Comprehensive Planning; A Trilogy in Urban Development.

650. Hyman, Herbert H., "Planning With Citizens: Two Styles," *Journal of the American Institute of Planners*, Vol. XXXV, No. 2, March 1969, pp. 105-112.

Planning Area; Planning Orientation; Involvement of Citizen Groups; Planner and the Central Office; Planner's Use of Influence; Importance of Basic Approach; Recent Developments; Conclusions.

651. Kaplan, Marshall, "Advocacy and the Urban Poor," *Journal of the American Institute of Planners*, Vol. XXXV, No. 2, March 1969, pp. 96-101.

Advocacy Planning Models; Hunters Point—Directed Advocacy; Oakland—Non-directed Advocacy; Conclusions—Technical Assistance Versus Ideology, Determining Local Priorities, White Professional in the Black Ghetto, Citizen Participation.

652. Levine, Aaron, "Opportunities for Community Relations," in: *Planning 1963*, Chicago, Ill. (American Society of Planning Officials), November 1963, pp. 259-266. Paperback. [39-3313]

Definition of Citizen; Women's Groups; Top Business Organizations; Newspaper Support; Planning Publications; Town Meetings; Planning Exhibits; Community Participation in Highway Planning.

653. Lovelace, Eldridge H., "Planner's Notebook: Control of Urban Expansion, The Lincoln, Nebraska, Experience," *Journal of the American Institute of Planners*, Vol. XXXI, No. 4, November 1965, pp. 348-352.

Physical Setting; Governmental Setting; The Issue—The Stevens Creek Matter; Lessons Learned.

654. Mainer, Robert, "The Case of the Stymied Strategist," *Harvard Business Review*, Vol. 46, No. 3, May-June 1968, pp. 36-38, 40-41, 45-46, 48, 172, 174, 178, 180.

"A new company president wants his executives to do strategic planning—but does the company really need it? What happens when a management group turns its attention to strategic planning? Is such planning worth the unusual demands it places on management? Why does it perplex some executive teams which are otherwise effective in running their businesses?"

655. Makielski, S. J., Jr., "Zoning: Legal Theory and Political Practice," *University of Detroit Journal of Urban Law*, Vol. 45, Fall 1967, pp. 1-22.

Introduction; Historical Background; Legal Theory of Zoning; Political Practice of Zoning; An Evaluation.

656. Mays, Arnold H., "Zoning for Mobile Homes—A Legal Analysis," *Journal of the American Institute of Planners*, Vol. XXVII, No. 3, August 1961, pp. 204-211.

Background: Who Lives in Mobile Homes, What Mobile Homes Are Like, Mobile-Home Parks; Zoning Power and Mobile Homes: Use of Zoning to Exclude, Conditional Zoning—The "Special Exception" Device, Controlling Location of Individual Mobile Homes, Controlling Location of Mobile Home Parks, Use of Zoning to Control Size, Nonconforming Use Doctrine; Special Restrictive Legislation: Time Limitations, Applied to Individual Mobile Homes to Mobile-Home Parks, As a Total Exclusionary Device.

657. Myers, Sumner, "How to Sell New Ideas to the Cities," *Harvard Business Review*, Vol. 46, No. 4, July-August 1968, pp. 111-118.

"The cities are in crisis and anxious to buy innovations from industry to help them solve their many problems."

Channeling Business Effort; Problems in Perspective; Budging the Bureaucracy; Concentration of Effort; Conclusion.

658. Olson, William A., "City Participation in the Enforcement of Private Deed Restrictions," in: American Society of Planning Officials, *Planning 1967*, Selected Papers from the ASPO National Planning Conference, Houston, Texas, April 1-6, 1967, Chicago, Ill. (1313 East 60th St., 60637), pp. 266-270. Paperback. [39-3313]

"In the absence of zoning . . . to regulate the use of land and to maintain some degree of integrity of the residential neighborhood . . . [Houston] now participates in the enforcement of private deed restrictions."

659. Peattie, Lisa R., "Reflections on Advocacy Planning," *Journal of the American Society of Planning Officials*, Vol. XXXIV, No. 2, March 1968, pp. 80-88.

Press of Bureaucratic Management; Advocacy Planning; Clients; Approaches to Planning; Conclusions.

660. Schaffer, Robert H., "Putting Action into Planning," *Harvard Business Review*, Vol. 45, No. 6, November-December 1967, pp. 158-166.

"The Secret of success begins with concentrating on goals that are immediate rather than remote from today's needs."

Traditional Starting Points: Unanticipated Hazards, Placing the Blame; More Workable Approach: Key Principles, Extending the Scope; Framework for Progress.

661. Stanford Research Institute, "The Tribulations of Hawkeye—A Study in Planning," *Stanford Research Institute Journal*, Vol. 5, Fourth Quarter 1961, pp. 133-169.

The Company and Its Background; Planning Committee in Action; Long-Range Planning; Research and Development; Marketing and Distribution; Diversification; Production and Inventory; International.

662. Volpert, Richard S., "Creation and Maintenance of Open Spaces in Subdivisions: Another Approach," *UCLA Law Review,* Vol. 12, No. 3, March 1965, pp. 830-855.

663. Yearwood, Richard M., "Accepted Controls of Land Subdivision," *University of Detroit Journal of Urban Law,* Vol. 45, Winter 1967, pp. 217-257.

Introduction; Plat Registration; Utility Services; Performance Bonding; Requiring Unnecessary Improvements; Cost the Developer Should Bear for Required Improvements; Conclusion.

Books, Reports, Pamphlets

664. American Society of Planning Officials, *Problems of Zoning and Land-Use Regulation,* Research Report No. 2, Chicago, Ill. (1313 East 60th St., 60637), 1968, 80 pp.

Future Urban Land Needs; Problems of Land-Use Controls; Fiscal Policy and Land Use; Federal Involvement.

665. American Society of Planning Officials, *The Text of A Model Zoning Ordinance,* Chicago, Ill. (1313 East 60th St., 60637), Third Edition, 1966, 99 pp. Paperback. [60-14648]

With commentary by Fred H. Bair, Jr. and Ernest R. Bartley.

666. Anderson, Desmond L. (Editor), *Municipal Public Relations,* Chicago, Ill. (International City Managers' Association), 1966, 273 pp. [66-20874]

Perspective: Public Relations in Society (Desmond L. Anderson), Public Relations and the Administrative Process (Desmond L. Anderson), Research and the Public Relations Process (Kent M. Lloyd); Program Involvement: City Council—Focal Point of Influences (Robert J. Huntley), Multitudinous Publics (Garth N. Jones), Serving the Public (Robert B. Callahan), Employee-Citizen Team (Harvey W. Wertz), Community Group Relationships (Garth N. Jones), Municipal Police and Public Relations (G. Douglas Gourley); Informational Reporting: Roles of Reporters and Mass Media (Robert F. Wilcox), Reporting through the Media (Robert F. Wilcox), Municipal Reports and Events (Robert M. Christofferson), Reporting in Person (David Mars; Gale L. Richards), Printing Arts and Publications Distribution (Arthur A. Atkisson, Jr.); Integrating the Whole: Organizing for Public Relations (Robert M. Christofferson), Employee Relations and Training (Robert B. Callahan); Appendices; Selected Bibliography, pp. 253-258.

667. Anderson, Robert M., *American Law of Zoning—Zoning, Planning, Subdivision Control,* Rochester, N.Y. (Lawyers Co-operative Publish-

ing Company), San Francisco (Bancroft-Whitney Company), 4 vols., 1968, c. 2600 pp. [68-28408]

". . . materials are arranged and finely subdivided in a manner intended to serve the attorney or planner . . . seeking an answer or lead, rather than . . . undertaking to read a chapter or more. The main focus . . . is the current state of planning law. . . ."

668.   Babcock, Richard F., *The Zoning Game—Municipal Practices and Policies,* Madison, Wis. (University of Wisconsin), 1966, 202 pp. [66-22864]

Players: The Stage—Historical and Current, The Layman as Public Decision-Maker, as Private Decision-Maker, Planner, Lawyer, Judge; Rules: Purpose of Zoning, Principles of Zoning, Interested Parties, Bases for Decision-Making.

669.   Bair, Frederick H., *Local Regulation of Mobile Home Parks, Travel Trailer Parks and Related Facilities,* Chicago, Ill. (Mobile Homes Manufacturer's Association), 1965, 94 pp.

"The latest work on mobile home regulation. Suggests a number of solutions to the vexing problems of developing sensible standards and regulations."

670.   Beuscher, J. H., *Land Use Controls—Cases and Materials,* Madison, Wis. (College Printing & Typing), Fourth Edition, 1966, 582 pp.

Early Legislative Control or Land Use—Ten Statutes Spanning Six Centuries; Judicial Control of Land Use Through Waste and Related Doctrines; Private Law Devices To Assure Land Development Plans; Master Plan and Official Map; Regulation of Land Subdivision; Regulation of Land Use Through Zoning; Some Legal Tools for Urban Renewal; Other Public Actions To Control Land Use.

671.   Birch, David L., *The Businessman and the City,* Boston, Mass. (Graduate School of Business Administration, Harvard University), 1967, 219 pp. [67-30740]

Three Keynote Addresses; President Johnson's Address; Housing and Urban Renewal; Urban Transportation; Education; Social Problems; Air and Water Pollution; Conclusions.

672.   Bourne, Larry S., *Private Redevelopment of the Central City—Spatial Processes of Structural Change in the City of Toronto,* Toronto, Ont. (Department of Geography, University of Chicago), 1967, 199 pp. Paperback. [66-30638]

Introduction; Conceptual Background and Formulation; Redevelopment as a Spatial Process; Empirical Background and Formulation—Toronto; Location Factors in Structural Change; Structural Change in Toronto; Area and Site Correlates of Structural Change; Conclusions; Appendices; Selected Bibliography, pp. 190-199.

673.   Burke, Gerald L., *Greenheart Metropolis—Planning the Western Netherlands,* New York (St. Martin's), 1966, 172 pp. [66-71195]

Introduction—Problem of Population Pressure and Approaches to Its Solution; Economic Background—Bases of the Dutch Economy; Reclamation of the Zuiderzee —Scheme As a Whole, Description and Criticism of the North-East Polder; Delta

Plan—The Undertaking and Its Effect Upon the Western Region; Randstad Holland: I. Analysis of the Problem and Report of the Committee for the Western Region—II. Problems of the Southern Circumference, South Holland Waterway Region, Hague Agglomeration—III. Problems of the Northern Circumference, The Ijmond District (Haarlem and South Kennemerland, Amsterdam Agglomeration, the Gooi District, Remainder of the North Holland Province)—IV. Eastern Circumference, Utrecht, City and Province, Utrechtse Heuvelrug; Structure Plan for the Southern IJsselmeer Polders; Comparisons and Conclusions; Appendix—Organization of Planning in the Netherlands; Bibliography, pp. 163-166.

674. Cleland, David I., King, William R., *Systems Analysis and Project Management,* New York (McGraw-Hill), 1968, 315 pp. [68-17179]

Basic Systems Concepts: Systems Concept in Management; Systems Analysis for Strategic Decisions: Systems Analysis, Conceptual Framework for Systems Analysis, Methods of Systems Analysis, Planning, Planning-Programming—Budgeting and Systems Analysis; Project Management in Executing Decisions: Project Environment, Organizational Concepts of Project Management, Organizational Chart—Systems Viewpoint, Project Authority, Project Control, Project Planning and Control; Appendices.

675. Delafons, John, *Land-Use Controls in the United States,* Cambridge, Mass. (Joint Center for Urban Studies of the Massachusetts Institute of Technology and Harvard University), 1962, 100 pp. and Appendices. Paper.

Context; History; Objectives; Methods; Utility; Conclusion; Appendix.

676. Fiser, Webb S., *Mastery of the Metropolis,* Englewood Cliffs, N.J. (Prentice-Hall), 1962, 168 pp. Paperback. [62-7452]

Introduction: Problems and Purposes; Future of the Metropolis; Context of National Forces; Intermingling of Public and Private; Case of Urban Renewal; Governmental Reorganization; Citizen Action; Mastery of the Metropolis.

677. Gellhorn, Walter, *Ombudsmen and Others: Citizens' Protectors in Nine Countries,* Cambridge, Mass. (Harvard University), 1966, 488 pp. [66-23465]

". . . describes and analyzes how the concept of the ombudsman—a high official who investigates complaints against bureaucratic injustice and neglect—works in the nine countries where it is practiced."

678. Green, Philip P., Jr., *Cases and Materials on Planning Law and Administration,* Chapel Hill, N.C. (Institute of Government, University of North Carolina), 1962, 455 pp.

". . . designed for use in University training of students preparing for professional careers as city planners. . . . Intended to be supplemented with a text such as [*Principles and Practice of Urban Planning*] (International City Managers' Association, Fourth Edition, 1968)."

Federal, State, and Local Governments; Governmental Organization for Planning; Master Plan, Comprehensive Plan, General Plan, Development Plan—or Planning?; Administrative Devices for Controlling Public Projects; Acquisition of Property for

Public Purposes; Problems Relating to Particular Facilities; Law of Nuisance; Restrictive Covenants; Building Regulations and Similar Measures; Subdivision Regulations; Zoning; Urban Renewal; Some Special Problems; Local Governmental Measures to Assist Economic Development.

679. Hackney, John W., *Control and Management of Capital Projects*, Dynamic Estimating, Control, and Management by Owner Corporations of the Cost, Time, and Value of Engineering-Construction Projects, New York (Wiley), 1965, 305 pp. [65-26846]

Elements of the System, Nature of Engineering-Construction Projects and Their Control Requirements; Capital-Cost Control Cycle; Cost Coding, A Common Language for the Project; Predesign Estimating; Adjustments to Expected Site, Price, and Wage Conditions; Labor Productivity, Its Prediction and Management; Indirect Project Costs, Design Engineering, Procurement Cost, and Field Construction Expense; Accuracy and Contingencies, How Good Is the Crystal Ball?; Estimating Technique, A Typical Project; Elements of Time Control: Planning, Scheduling, and Reporting Progress; Network Diagramming: Pert, CPM, and Their Offspring; Value-Control Cycle; Forecasting Income, Critical Element in Project Value; Determining Operating Costs; Estimating Total Investment, Overall Commitment of Corporate Assets; Computing Return on Investment, Interest-Rate-of-Return Concept; Appraising Value, Project Risk and Its Relationship to Acceptable Return; Dynamic Value Relationships, Relationships Between Income, Operating Costs, Investment, Time, Risk, and Return; Procedures for Presenting and Approving Projects, Why, Who, When, and How; Records, Reports, and Corrective Action, with Comment on PERT/COST, PROMOCOM, and ON-LINE Data Processing; Post-Project Operations: Project Closing Property Records, Performance Reports, and Accuracy Analysis—How Can We Do It Better?; Manning of Projects; Selection and Management of Contracts and Contractors; Project Management and Managers, Essential Ingredients of Successful Projects; Appendices: Check List for Detailed Definition Rating, Analysis of Estimating Accuracy, General Bibliography, pp. 299-300.

680. Hanke, Byron R.; Krasnowiecki, Jan; Loring, William C.; Tweraser, Gene C.; Cornish, Mary Jo., *The Homes Association Handbook*, Technical Bulletin 50, Washington, D.C. (Urban Land Institute), October 1964, 406 pp. Paper. [64-8521]

Basic Data; Getting Down To Cases; Action Guidelines; Legal Analysis; Appendices.

681. Herring, Frances W. (Editor), *Open Space and the Law*, Berkeley, Cal. (Institute of Governmental Studies, University of California), 1965, 160 pp. Paperback.

Introduction (E. Stanley Weissburg); Open Space and the Police Power (I. Michael Heyman); Legal Alternatives to Police Power—Condemnation, Purchase, Development Rights, Gifts (E. Stanley Weissburg); Influence of Taxation and Assessment Policies on Open Space (Franklin C. Latcham; Roger W. Findley); Financing Park and Open Space Projects (Harold E. Rogers, Jr.); Better Use of Old Tools—Are New Tools Necessary? (Frances W. Herring); Selected Recent References, pp. 151-153.

682. Hodgell, Murlin R., Zoning—A Guide to the Preparation, Revision or Interpretation of Zoning Ordinances in Kansas, *Kansas State College Bulletin*, Vol. 42, No. 6, April 1, 1958, 137 pp.

Descriptive Introduction; Title of Ordinance; Ordaining or Enacting Clause; Short Title; Definitions; Establishment of Districts; Application of Regulations; Districts: Agricultural, Residential, Business, Industrial; Supplementary Regulations; Nonconforming Uses and Buildings; Administration and Enforcement; Interpretation of Ordinance; Board of Appeals; Amendments; Separability Clause; Violations and Penalties; Effective Date of Ordinance; Appendices.

683. Howes, Rev. Robert G., *The Church and the Change*, Boston, Mass. (Daughters of St. Paul), 1961, 184 pp.

"Written by one of the very few priests in the United States who has had formal training in city planning, this book focuses on the position of the Roman Catholic Church in the American community. . . . Calls for a new direction of Catholic involvement, and suggests courses of action that should be pursued at the diocesan and at the national levels."

684. Lefcoe, George, *Land Development Law, Cases and Materials,* New York (Bobbs-Merrill), 1966, 1681 pp. [66-18858]

". . . a collection of materials on the legal arrangements characterizing contemporary land development . . . particularly designed to illumine the role of lawyers in the processes of land development."

Government as Vendor and Purchaser; Vendors and Purchasers in the Private Land Markets: Predevelopment Stage; Land Development and Public Regulation; Post Development Period: Land Sale Contracts and Deeds; Land Finance; Recording; Title Insurance; Private Governments in Housing: Land-Use Controls by Contract; Planning and Zoning; Real Property Taxation.

685. Los Angeles County Regional Planning Commission, *Suggested Uniform Zoning Ordinance,* Los Angeles, Cal. (320 W. Temple St., 90012), Second Draft, July 1965, 321 pp. and Index. Paper.

General Provisions; Establishment of Zones; Agricultural Zones; Residential Zones; Commercial Zones; Manufacturing and Industrial Zones; Special Purpose Zones; General Standards of Development; Special Regulations; Administration of Zoning Ordinance.

686. McKeever, J. Ross (Editor), *The Community Builders Handbook,* Washington, D.C. (Urban Land Institute), 1968, 526 pp. [67-24963]

Residential Communities: Preliminary Steps in Community Development, Planning the Development, Protecting the Future of the Development; Planning for Special Types of Land Development: Campus Planning and College Housing, Civic Centers, Community Clubs, Golf Course Subdivisions, Resort Type Subdivisions, Mobile Home Parks, Motels, Office Parks, Medical Buildings, Housing for the Elderly, New Towns, Urban Renewal; Planning and Operation of Shopping Centers: Introduction, Planning Preliminaries, Planning the Shopping Center Site, Architectural and Structural Design . . ., Management, Maintenance, and Operation . . .; Industrial Parks; Appendices.

687. Malo, Paul, *The Binghamton Commission on Architecture and Urban Design, The First Three Years: 1964-1967*, Binghamton, N.Y. (Valley Development Foundation, 212 Security Mutual Bldg., 13901), 1968, 84 pp. Paper. [68-64328]

The Commission: What It Is and What It Does; How It Came About; Review of the Ordinance and the Supplementary Rules and Procedures of the Commission; First Issues—Private Property: Signs, Landmarks; Other Private Property; Private and Public Properties- Urban Renewal; Other Public Property; Relations With Others; Conclusion; Legal Background and Legal Problems of the Binghamton Ordinance (Robert M. Anderson).

688. Mandelker, Daniel R., *Controlling Planned Residential Developments*, Chicago, Ill. (American Society of Planning Officials), 1966, 66 pp. Paper.

New Techniques for Old, The Basis of Change; Procedures for Approving Planned Developments; Substantive Requirements and Standards for Project Approval; Conclusion, Suggestions for Innovation.

689. Mandelker, Daniel R., *Green Belts and Urban Growth*, Madison, Wis. (University of Wisconsin), 1962, 176 pp. [62-9262]

"An account of the procedural experience in administering the British green belt program."

690. Martin, Roscoe C., *The Cities and the Federal System*, New York (Atherton), 1965, 200 pp. [65-24497]

Nation of Cities; American System—The Many and the One; Unequal Partners—Case of the Reluctant State; Emergence of an Urban Partner; Expanded Partnership—Nature; Three Views of the Expanded Partnership; Expanded Partnership—Appraisal.

691. Metzenbaum, James, *Law of Zoning*, Mt. Kisco, N.Y. (Baker, Voorhis), Revised Edition, Three Volumes, 1955, Amended through 1967.

"An extraordinarily complete yet convenient collection of all important Federal and state court opinions on zoning. . . .

"Enhancing the great usefulness of these volumes, [there are] also printed the texts of the zoning ordinances in our larger cities, and . . . a manual of procedures applicable at all levels, from the humblest village board up through the courts of appeal. . . . Cases are segregated both by state and by subject. . . . Indexes are a model of completeness. Thousands of cross-references escort the reader to what he wants."

692. Moak, Lennox L.; Killian, Kathryn W., *Capital Programming and Capital Budgeting*, Chicago, Ill. (Municipal Finance Officers Association), 1964, 152 pp. [64-18473]

"An informative manual identifying the principal elements, essential steps, and objectives involved in the preparation of capital programs and budgets. . . . Describes current practice in 17 large cities in the United States and Canada and suggests sound programming and budgeting practice."

693. New York State, Office of Planning Coordination, *Local Planning and Zoning*—A Manual of Powers and Procedures for Citizens & Governmental Officials, Albany, N.Y. (488 Broadway, 12207), 128 pp. Paper.

Planning by Cities, Towns and Villages; County and Regional Planning; Planning Associations or Federations; Zoning by Cities, Towns and Villages; Building Regulations; Airports and Their Approaches; Appendices.

694. Philadelphia City Planning Commission, *Capital Program 1968-1973*, Philadelphia, Pa. (City Hall Annex, 19104), 14 December 1967, 289 pp. Paper.

Summary of Projects Completed or Under Way 1962-1967; Capital Program and the Comprehensive Plan; Financing the Program; Project Descriptions: City Commissioners, Department of Commerce, Director of Finance, Fire Department, Free Library, Department of Public Health, Police Department, Department of Public Property, Department of Recreation, Redevelopment Authority, Department of Streets, Water Department, Department of Public Welfare, Youth Study Centers.

695. Rathkopf, Arden H., *The Law of Zoning and Planning*, New York (Boardman; Mathew Bender), Annual Cumulative Supplements, 3 Vols.

696. Rody, Martin J.; Smith, Herbert H., *Zoning Primer*, Trenton, N.J. (Chandler-Davis), 1960, 48 pp. Paper. [60-9612]

Principles of Effective Zoning; Constitutional Aspects of Zoning; Purpose of Zoning; Relationship of Zoning to the Master Plan; Is Zoning Important to Rural Communities?; What Are the Methods and Procedures Used in Developing a Zoning Ordinance?; Techniques of the Zoning Ordinance; Variances; Other Problems Found in Present Day Zoning; Trends in Zoning; Summary.

697. San Francisco, Department of City Planning, *San Francisco Downtown Zoning Study*—C-3 and Adjacent Districts, Final Report, San Francisco, Cal. (100 Larkin St., 94102), December 1966, 48 pp. Paper.

Introduction; Guiding Principles; Downtown Zoning Districts; Building Bulk and Intensity of Development; Special Areas; Off-Street Parking and Loading; Coordination with Downtown Public Programs; Summary Tables and Zoning Maps.

698. Scheidt, Melvin E. (Director of Study), Public Works Committee, National Resources Planning Board, *Long-Range Programming of Municipal Public Works*, Washington, D.C. (Superintendent of Documents), 1941, 72 pp. Paper.

An important early study and publication on municipal capital project programming. Programming—Its Objectives and Procedure; Financial Analysis; Listing Proposed Improvements; Preparation and Adoption of the Program; Programming Organization; Appendices; Bibliography, p. 72.

699. Shore, William B.; Keith, John P. (Editor), *Public Participation in Regional Planning*, New York (Regional Plan Association), October 1967, 72 pp. Paper.

Foreword; Public's Role in Regional Planning: Search for Planning Goals, Place of Public Participation in Planning; Publics to Be Consulted, Conclusions; Consulting the Public on the Second Regional Plan: Goals Project Process and Participants, What We Learned from the Goals Project, Detailed Replies of Goals Project Participants, Failures and Hopes; Appendix: 1. Organizations invited to participate in the Goals for the Region Project, 2. Committee on The Second Regional Plan.

700.   Smith, Herbert H., *The Citizen's Guide to Zoning*, West Trenton, N.J. (Chandler-Davis), 1965, 182 pp. Paper. [64-10782]

Introduction; Purpose and Importance of Zoning; Technical Elements of Zoning; Steps and Roles to Get Started; Zoning Administration; Changing the Zoning Ordinance; The Apartment Question; Special Zoning Problems; Caveats and Oppositions.

701.   Spiegel, Hans B. C. (Editor), *Citizen Participation in Urban Development*, Washington, D.C. (NTL Institute for Applied Behavioral Science, National Education Association), 1968, 291 pp. Paperback, [68-8752]

The Many Faces of Citizen Participation—Bibliographic Overview (Hans B. C. Spiegel; Stephen D. Mittenthal), Bibliography, pp. 14-17; Federal Regulations and Advice (Department of Housing and Urban Development); Arguments For and Against Citizen Participation in Urban Renewal (James Q. Wilson); Dilemma of Citizen Participation (Robert C. Seaver); Selection of Indigenous Leadership (Louis A. Zurcher); Participation of Residents in Neighborhood Community-Action Programs (Frances F. Piven); Voice of the People (Peter Marris; Martin Rein); Native Leadership (Saul Alinsky); Saul Alinsky and His Critics; Church and Neighborhood Community Organization (Thomas D. Sherrard; Richard C. Murray); Citizen Participation (Edgar S. and Jean Camper Cahn); Are the Poor Capable of Planning for Themselves? (Harold C. Edelston; Ferne K. Kolodner); Community Status As a Dimension of Local Decision Making (Robert L. Crain; Donald B. Rosenthal); How Much Neighborhood Control? (Subcommittee on Urban Affairs, Joint Economic Committee, U.S. Congress).

702.   Technical Planning Associates, *Subdivisions: Design and Review*, Hartford, Conn. (Connecticut Federation of Planning & Zoning Agencies), 1968, 119 pp. Paper.

" '. . . general guide . . . to refer to professional engineers and planners for advice on the individual problems that arise in the subdivision process . . .' covers problems and solutions on sites—in developed areas, undeveloped areas, on slopes, on flat lands, with different soils, different percolation characteristics . . . water tables . . . vegetative cover . . . access. . . . Multiple-family as well as single-family subdivisions . . . cluster development for both."

703.   U.S. Federal Housing Administration, *Planned-Unit Development with a Homes Association*, Land Planning Bulletin No. 6, Washington, D.C. (Superintendent of Documents), 1963, 64 pp. Paper.

". . . a summary of recent thoughts on and experiences with the planned-unit concept of housing, as well as a guide to FHA standards in this area [of its activity]."

Some ABC's on Planned Units; Will They Sell and Resell?; Is It a Planned Unit or Isn't It?; Land Development Program; The Home's Place in a Planned-Unit; Development Plan; Legal Absolutes—The Foundation; Creating the Association and Its Facilities; How To Get FHA-Insured Financing; In the End, What Do You Have?

704. Whyte, William H., Jr., *Securing Open Space for Urban America: Conservation Easements,* Technical Bulletin 36, Washington, D.C. (Urban Land Institute), December 1959, 67 pp. Paper.

Precedents; Public Purpose; Limits of Zoning; Just Compensation; Gifts; Tax Question; Costs to the Public; Deed; Financing; Agencies; Legislation; Appendices.

705. Williams, Norman, Jr., *The Structure of Urban Zoning,* and Its Dynamics in Urban Planning and Development, New York (Buttenheim), 1966, 351 pp.

Introduction, The Comprehensive Plan; Land Use Control: Pre-Zoning Techniques, Zoning—District Structure, Other Controls, Mapping, Administration; Community Layout, Design and Redesign: Newly-Developed Areas, Reorganization of Blighted Areas; Some Curiosities; Index of Cases Cited, Discussed, Not Discussed; Index of Topics; Articles; Books.

706. Yokley, E. C., *The Law of Subdivisions,* Charlottesville, Va. (Michie), 1963, 492 pp. [63-4315]

Scope, Purposes, and Definitions; General Power To Control Subdivision Development; Community Planning; Planning Commission; Need For Open Space; Streets; Dedication; Effect of Zoning on Subdivision Development; Maps and Plats; Approval of Subdivision Plans; Subdividers and Lot Owners—Rights and Liabilities; Municipal Powers and Obligations in Specific Cases; Procedure, Appeals, Judicial Construction; Statutory Authorization For Subdivision Control; Forms.

707. Yokley, E. C., *Zoning Law and Practice,* Charlottesville, Va. (Michie), Third Edition, 1967. 4 Volumes; Cumulative Supplements.

Vol. 1. Origin, Early Growth and Definitions; Police Power; Comprehensive Plan; Context of the Ordinance; Enactment; Interim Ordinances; Amendment; Spot Zoning; Permits; Enforcement.

Vol. 2. Planning Commission; Subdivision Control and Regulation; Procedure Before the Board of Appeals; Board of Appeals—General Powers; Exceptions and Variances; Nonconforming Uses; Area and Height Requirements; Appellate Jurisdiction and Procedure; Judicial Construction; Restrictive Covenants; Jurisdictional Conflicts.

Vol. 3. Injunction; Mandamus; Urban Redevelopment and Urban Renewal; Building Codes and Ordinances; Airport Zoning; Off-Street Parking; Special Subjects of Zoning Legislation; Forms.
Vol. 4. Tables of Cases; Index.

708. Zoning Procedures Study Committee, *Planning and Zoning for Fairfax County, Virginia—A Proposal,* County Executive's Office (Fairfax County Courthouse, Fairfax, Va.), September 1967, 73 pp.

Introduction; Viable Planning Program for Fairfax County; Organizational Structure for Planning and Zoning; Rules of Conduct for County Officials; Rules of Administrative Procedure; New Development Powers for Fairfax County; Summary of Recommendations.

Bibliographies

709. Betebenner, Lyle, *City Planning and Zoning in American Legal Periodicals,* Bibliography No. 28, Council of Planning Librarians, Monticello, Ill. (Exchange Bibliographies, P.O. Box 229, 61856), January 1965, 137 pp.

Access; Administrative: Law, Procedure, Remedies Exhausted; Air; Pollution, Ports, Space; Amendments and Changes; Antenna; Billboards; Boards; Building: Height, Lines; Cemeteries; Churches; Church-School; Comprehensive; Condemnation; Consent; Constitutionality; Co-op Apartments and Condominiums; Court Decisions; Covenants; Dedication; Delegation; Discrimination; Due Process; Easements; Eminent Domain (and Condemnation); Esthetics; Highways; Housing; Injunction; Marketability; Minimum Area; Non-Conforming Use: Abandon— Non-Use, Amortize and Eliminate, Building Under Construction—Substantial Expenditure Required, Change and Extension, General; Nuisance; Off-Street Parking; Planning—General; Police Power; Redevelopment—Slum; Schools; Segregation; Subdivision; Standards; Trailers and Trailer Camps; Validity; Variance— Exception—Etc.; Vested Rights; Water; Zoning—General; List of Periodicals Cited.

710. Goodman, William I.; Brown, William, *Planning Legislation and Administration, An Annotated Bibliography,* Bibliography No. 57, Council of Planning Librarians, Monticello, Ill. (Exchange Bibliographies, P.O. Box 229, 61856), 1968, 13 pp.

Government, Law, and Administration, Legal Process; Participants in Community Planning and Their Contributions; Activities of the Planning Agency, Organization of the Working Staff; Capital Programming; Place of the Planning Agency; Social Institutions Upon Which Planning Legislation and Practice Are Based; Police Power, Eminent Domain, Early Statutes; Subdivision Regulations—Development, Authorization, Typical Standards and Requirements; Urban Renewal and Housing; Role of Government in Implementing Planning Programs.

711. Hackney, John W., *Control and Management of Capital Projects,* General Bibliography, pp. 299-300.

712. Herring, Francis (Editor), *Open Space and the Law,* Recent References, pp. 151-153.

713. Public Administration Service, *Automation in the Public Service: An Annotated Bibliography,* Chicago, Ill. (Public Automated Systems Service, 1313 East 60th St., 60637), 1966, 70 pp. Paper.

". . . first major bibliography on computer operations in government includes a basic classification system which not only orders the material but also serves as a guide for establishing libraries of automation materials. It lists over 700 items and codes them according to the new system."

10. SYSTEM ELEMENTS

Geography, Geology, Seismism

Articles, Periodicals

714. Bascom, Willard, "Beaches," *Scientific American*, Vol. 203, No. 2, August 1960, pp. 80-94.

"Where the land meets the sea, the waves usually lay down a strip of sand or other uniform material. The constant shifting of this material has created a conservation problem for beach-loving man."

715. Branch, Melville, C., "Rome and Richmond—A Case Study in Topographic Determinism," *Journal of the American Institute of Planners,* Vol. XXVIII, No. 1, February 1962, pp. 1-9; also Chap. 5 in: Branch, Melville C., *Planning—Aspects and Applications*, New York (Wiley), 1966, pp. 109-120. [66-13523]

Situation, Site, and Settlement; Growth; Mature Development.

716. Kerr, Paul F., "Quick Clay," *Scientific American*, Vol. 209, No. 5, November 1963, pp. 122-126, 138, 140, 142.

"It is a water-soaked glacial deposit that sometimes changes suddenly from a solid to a rapidly flowing liquid, causing disastrous landslides. . . ."

717. Leopold, Luna B.; Langbein, W. B., "River Meanders," *Scientific American*, Vol. 214, No. 6, June 1966, pp. 60-70.

"The striking geometric regularity of a winding river is no accident. Meanders appear to be the form in which a river does the least work in turning; hence they are the most probable form a river can take."
Regular Forms from Random Processes; Sine-Generated Curves; Curve of Minimum Total Work; Shaping Mechanism; Riffles and Pools; Obtaining Meander Profiles; Conclusions.

718. Marsden, S. S., Jr.; Davis, S. N., "Geological Subsidence," *Scientific American*, Vol. 216, No. 6, June 1967, pp. 93-100.

"In many parts of the world the pumping of oil, gas or water out of the ground has caused the land to sink. Where oil or gas are involved the subsidence can be forestalled by pumping in water."

Books, Reports, Pamphlets

719. Beaujeu-Garnier, J.; Chabot, G., *Urban Geography*, New York (Wiley), 1967, 470 pp. [68-2591]

Urban Phenomenon—Urban Concentration; Definition of the Town; Cartographic Representation of Towns; The World's Towns: European, U.S.S.R., Australasia and the Americas, North Africa and Non-Soviet Asia, Africa South of the Sahara; Urban Functions: Origin of Towns, General Principles, Military, Commercial, Industrial, Cultural, Town As a Resort, Administrative and Political Function, Variety and Plurality of Functions; Plan and Extent of Towns: Part Played by the Site, Town Plan, Expansion of Towns, Suburbs, Town Clusters and Conurbations, Satellite Towns, Agglomerations; Life in Towns: Concept of Urban Concentration, Problems of Space, Urban Population; Town and Its Region: Nature of Urban Influence, Spheres of Influence, Conclusion; Bibliography, pp. 453-466.

720. Berry, Brian J. L., *Geography of Market Centers and Retail Distribution*, Foundations of Economic Geography Series, Englewood Cliffs, N.J. (Prentice-Hall), 1967, 146 pp. Paperback. [67-13355]

Definitions and Examples: Systems of Central Places in Complex Economies, Systematic Variations of the Hierarchy; Approaches to a Theory: Classical Central-Place Theory, Modern Theoretical Departures; Perspectives of Time and Space: Away from the Complex—Cross-Cultural Patterns, Paths to the Present—Theory and Fact; Mechanics of Application: Marketing Geography, Planning Uses.

721. Dickinson, Robert E., *City and Region—A Geographical Interpretation*, London, England (Routledge & Kegan), 1964, 588 pp. [BNB 64-23449]

Urban Settlement As Regional Centre: Region As a Social Unit; Nature of the City; City As a Regional Centre; Town-Country Relations; Structure of the City: Structure of the City, City As a Whole, Regions Within the City—the Natural Area; City-Region: Regional Relations of the City, Regional Relations of the City—Case Studies, City-Region—in the United States, I and II, Western Europe, Britain; Regionalism and the City-Region: Case for the Region, Regionalism in—France, Britain, United States, Germany, International Aspects of Regionalism—Conclusion.

722. Freeman, T. W., *Geography and Planning*, London, England (Hutchinson), Third Edition, 1967, 192 pp. Paperback.

Planner and Geographer; Physical Landscape; Climate and Weather; Rural Land Use; Aspects of Town Geography; Some Problems of Industrial Location; National Parks; Changing Scene; Notes and References, pp. 178-185.

723. Gottmann, Jean; Harper, Robert A. (Editors), *Metropolis on the Move: Geographers Look at Urban Sprawl*, New York (Wiley), 1967, 203 pp. Paperback. [66-27895]

Urban Sprawl and Its Ramifications (Jean Gottmann); Forces, Pressures and the Form of Urban Sprawl: Pull of Land and Space (Harold Moyer), What Sprawl Has Done to Central-Place Theory (Edwin Thomas), Pressures Brought by Urban Renewal (Peter Nash), Agricultural Land on the Urban Fringe (Edward Higbee); Sprawl and the Functioning City: Journey-to-Work (Robert Dickinson), City As a Place To Live (Robert C. Ledermann), Trading Function (Bart J. Epstein), Manufacturing and Sprawl (James Kenyon); Skyscraper Amid the Sprawl (Jean Gottmann); Urban Sprawl and the Future: Sprawl and Planning (Henry Fagin), Challenge of the New Urbanization to Education (Robert McNee); Bibliography, pp. 193-195.

724. Hodgson, John H., *Earthquakes and Earth Structure*, Englewood Cliffs, N.J. (Prentice-Hall), 1964, 166 pp. Paperback. [64-13244]

What Are Earthquakes Like?; How Do Seismologists Study Earthquakes?; Earthquakes—Where and Why?; What Can We Do About Earthquakes?

725. Ingle, James C., Jr., *The Movement of Beach Sand*, An Analysis Using Fluorescent Grains, Developments in Sedimentology 5, New York (Elsevier), 1966, 221 pp. [65-13238]

Introduction; Field and Laboratory Procedures; General Patterns of Foreshore-Inshore Tracer Transport; Sand Movement Seaward of the Breaker Zone; Sand Movement Around Man-Made Structures; Analysis of Tracer Dispersion; Summary; Appendices; References, pp. 201-209.

726. Johnson, James H., *Urban Geography—An Introductory Analysis*, Commonwealth and International Library of Science, Technology Engineering and Liberal Studies, New York (Pergamon), 1967, 188 pp. [67-21274]

Factors in Urban Growth; Urban Society and Urban Form; Demographic Characteristics of Urban Populations; Occupational Characteristics of Urban Populations; Location, Spacing, and Size of Urban Settlements; City Centre; Residential Suburbs; Manufacturing Areas in Cities; Theories of Urban Structure.

727. Jones, Emrys, *Human Geography—An Introduction to Man and His World*, New York (Praeger), 1964, 240 pp. [66-14506]

Population; Divisions of Mankind; Movements of Mankind; Obtaining Food; Farms and Villages; Mining and Manufacturing; Towns and Cities; Communications.

728. Jones, Emrys, *Towns & Cities*, Opus 13, London, England (Oxford), 1966, 152 pp. Paperback. [66-8567]

What Is a Town?; Process of Urbanization; Pre-Industrial Cities; Western City; Size and Classification of Cities; City and Region; Man and City; Bibliography, pp. 143-145.

729. Mayer, Harold M.; Kohn, C. F. (Editors), *Readings in Urban Geography*, Chicago, Ill. (University of Chicago), 1959, 625 pp. [59-11973]

"Comprehensive survey of current concepts and knowledge in urban geography; selected writings of more than 40 contemporary urban geographers and authorities in planning and related fields."

730. Murphy, Raymond E., *The American City, An Urban Geography*, New York (McGraw-Hill), 1966, 464 pp. [65-24894]

The Urban Inquiry; Basic Concepts and Definitions; Suburban Agglomerations and the Rural-Urban Fringe; The City's Spheres of Influence; Location of Central Places; Urban Hierarchy; Urban Economic Base; Functional Classifications of Cities; Distribution Patterns and Characteristics of Functional Classes; People in the City; Urban Land-Use Maps and Patterns; Theoretical Explanations of City Structure; Transportation in Relation to the City; Commercial Activities and the City; Central Business District; Manufacturing and the City; Manufacturing

Expansion; Urban Residential Patterns; Other Urban Patterns and Associated Land-Use Problems; The Political Factor and Growth of the City.

731. Perpillou, Aime Vincent, *Human Geography,* Geographics for Advanced Study, New York (Wiley), 1966, 513 pp. [66-2077]

Man and the Factors of Human Evolution: Man and the Natural Environment, Man and Civilization, Human Societies; Forms of Adaptation to the Environment—

in Cold Regions, Temperate Regions, Tropics, Dry Regions, Mountain and Coastal Regions; Technical Factors and Stages in Human Emancipation—Development of Techniques, Rise of the Industrial Mode of Life, Evolution and Distribution of the Industrial Economy, Development and Permanence of the Key Industries, Trade and Its Routes; Human Settlement: Growth and Distributions of Population, Human Migration and Overpopulation, Settlement, Towns, States and Nations; Bibliography.

732. Putnam, William C., *Geology,* New York (Oxford), 1964, 480 pp. [64-11237]

Earth; Rock-Forming Minerals; Igneous Rocks and Volcanism; Sedimentary Rocks; Metamorphic Rocks; Structural Geology; Petroleum Geology; Ground Water; Earthquakes and Earth's Interior; Weathering and Mass Movement; Stream Transportation and Erosion; Deserts; Glaciation; Sea; Mountains; Geologic Time and Life of the Past.

733. Ray, Richard G., *Aerial Photographs in Geologic Interpretation and Mapping,* Geological Survey Professional Paper 373, Washington, D.C. (Superintendent of Documents), 1960, 230 pp. Paper.

Introduction; Interpretation; Instrumentation; Source and Identifying Data of Aerial Photographs; References Cited, pp. 224-227.

734. Stamp, L. Dudley, *Applied Geography,* A Pelican Original, Baltimore, Md. (Penguin), 1960, 218 pp.

Meaning and Scope of Applied Geography; Human Environment; Land and People; Geographical Study of Population; Land; Interpretation of the Population and Land-Use Patterns; Land Planning; Classification of Land; Evolution of Town and Country Planning in Britain; Photography; Problems of Climatology; Aspects of Rural Land-Use Planning; Aspects of Urban Geography; Aspects of Industrial Geography; Geographical Aspects of Trade; Conclusions.

735. Stevens, Benjamin H., *Statistical Analysis for Areal Distributions,* Monograph Series Number Two, Philadelphia, Penn. (Regional Science Research Institute), 1966, 172 pp. Paperback.

Relationship Between Statistics and Geography; Areal Distributions; Moments; Measures of Average Position; Measures of Dispersion; Model Surfaces; Relative Dispersion, Spacing Measures, Skewness, and Kurtosis; Areal Association; Areal Inference and Estimation; Final Remarks; Appendices.

736. Taaffe, Edward J.; Garner, Barry J.; Yeates, Maurice H., *The Peripheral Journey to Work—A Geographic Consideration,* Evanston, Ill.

(Transportation Center, Northwestern University), 1963, 125 pp. [63-13481]

Comparison of Peripheral and Central Business District Commuting Patterns; Spatial Distribution of Peripheral Commuters; Application of Probability Models to the Peripheral Commuter Pattern; Other Factors Affecting the Peripheral Pattern; Conclusions and Future Development.

737. Tazieff, Haroun, *When the Earth Trembles,* New York (Harcourt, Brace, World), 1962, 245 pp. [64-11542]

Great Chilean Earthquake; Geography of Earthquakes; Instruments and What They Show; Appendix; Bibliography, pp. 232-235.

738. Vance, James E., Jr., *Geography and Urban Evolution in the San Francisco Bay Area,* Berkeley, Calif. (Institute of Governmental Studies, University of California), 1964, 89 pp.

Preface; Introduction; San Francisco: The Warehouse; Boatman's City; City Around the Bay; City of Parts: Large-Scale Venice of the West; Steam Comes Ashore; Railroad Towns; Trolley Wire Leads into the Country; City of Automobiles; New City; Urban Realms of the Bay Area; Conclusions; Maps.

Bibliographies

739. Beaujeu-Garnier, J.; Chabot, G., *Urban Geography,* Bibliography, pp. 453-466.

740. Gottmann, Jean; Harper, Robert A. (Editors), *Metropolis on the Move: Geographers Look at Urban Sprawl,* Bibliography, pp. 193-195.

741. Howard, William A., *Geographic Aspects of Urban Planning, A Selected Bibliography,* Bibliography No. 54, Council of Planning Librarians, Monticello, Ill. (Exchange Bibliographies, P.O. Box 229, 61856) 1968, 8 pp.

742. Jones, Emrys, *Towns & Cities,* Bibliography, pp. 143-145.

743. Mayer, Harold M., "A Survey of Urban Geography," in: Hauser, Philip M.; Schnore, Leo F. (Editors), *The Study of Urbanization,* pp. 81-113. [65-24223]

Urban Functions; Urban Sites and Situations; Cities as Central Places, Urban Internal Structure and Pattern; Transportation Facilities and Land Uses; Conclusions; Notes.

744. Siddell, William R., *Transportation Geography—A Bibliography,* Manhattan, Kan. (Kansas State University), Revised Edition, 1967, 57 pp. Paper.

General: General, Interaction Models, Cities and Transportation; Transportation Facilities: Ocean Shipping, Seaports, Inland Waterways, Railroads, Highways,

Pipelines, Air Transportation, Other; Regional Studies: Anglo-American, Latin America, Europe, the Mediterranean, and the Atlantic, Asia and Oceania, Africa.

745. Tazieff, Haroun, *When the Earth Trembles*, Bibliography, pp. 232-235.

746. Zelinsky, Wilbur, *A Bibliographic Guide to Population Geography*, Research Paper No. 80, Chicago (Department of Geography, University of Chicago), 1962, 257 pp.

"First comprehensive bibliography devoted exclusively to population geography, [it] contains a finding list of all significant writings—over 2,500 items—known to have been published anywhere since the beginning of specialized research in the subject."

People, Demography, Sociology, Psychology

Articles, Periodicals

747. Durand, John D. (Special Editor), World Population, *The Annals of the American Academy of Political and Social Science*, Vol 369, January 1967, 254 pp. [67-15899]

Long-Range View of World Population Growth (John D. Durand); Population Projections for the World, Developed and Developing Regions—1965-2000 (M. A. El-Badry); Control of Mortality (T. E. Smith); Character of Modern Fertility (Norman B. Ryder); Reproductive Performance and Reproductive Capacity in Less Industrialized Societies (G. W. Roberts); Prospects for Reducing Natality in the Underdeveloped World (Dudley Kirk); Estimating China's Population (John S. Aird); Population and Food Supply (Conrad Taeuber); Population, Natural Resources, and Technology (Edward A. Ackerman); Effects of Population Growth on the Economic Development of Developing Countries (Richard A. Easterlin); Population Growth and Educational Development (B. Alfred Liu); Labor Supply and Employment in Less Developed Countries (Jan L. Sadie); Demographic Transitions and Population Problems in the United States (Irene B. Taeuber); Supplements: Recent Trends—in Ethnology (Robert T. Anderson), Deviant Behavior and Social Control (David J. Bordua).

748. *Fortune*, "The City and the Negro", Vol. 65, No. 3, March 1962, pp. 88-91, 139-140, 144, 146, 151-152, 154.

Negro Migration from South to North and from Country to City. The Changing Face of Chicago. Crucial Difference; Other Side of Jordan; Need for Excellence; "More Powerful Than Apathy"; "They're Paying Attention"; Twenty-Point Drop; Success Story; Underdeveloped Country; Housing Dilemma; Cost of Delay.

749. Gans, Herbert J., "The Balanced Community—Homogeneity or Heterogeneity in Residential Areas?," *Journal of the American Institute of Planners*, Vol. XXVII, No. 3, August 1961, pp. 176-184.

Heterogeneity—and Social Relations, and Democracy, and the Children, and Exposure to Alternatives; Implications for Planning; Appraisal of Present Conditions; Toward a Reformulation of the Issue; Appendix—Heterogeneity for Aesthetic Values.

750. Gans, Herbert J., "Planning and Social Life—Friendship and Neighbor Relations in Suburban Communities," *Journal of the American Institute of Planners,* Vol. XXVII, No. 2, May 1961, pp. 134-143.

". . . while physical propinquity does affect some visiting patterns, positive relationships with neighbors and the more intensive forms of social interaction, such as friendship, require homogeneity of background, or of interests, or of values."

Propinquity, Homogeneity—and Friendship, and Neighbor Relations; Meaning of Homogeneity; Variations in Homogeneity; Role of Propinquity; Limitations of These Observations; Conclusions; Implications for Planning Practice.

751. Greenwood, Ernest, "Relationship of Science to the Practice Professions," *Journal of the American Institute of Planners,* Vol. XXIV, No. 4, 1958, pp. 223-232.

Editor's Note: ". . . a treatise on the relationship between the social work profession and the social sciences. . . . If the reader will substitute 'city planning' where he reads 'social work,' he may interpret Professor Greenwood to be saying that social science and city planning are distinctly separate kinds of activities and that city planning is dependent upon the sciences. . . ."

Nature of the Social Sciences; Nature of Social Work; Social Science—Social Work Relationship. Bibliography, pp. 231-232.

752. Mangin, William, "Squatter Settlements," *Scientific American,* Vol. 217, No. 4, October 1967, pp. 21-29.

"The shantytowns that have sprung up in developing areas are widely regarded as being sinks of social disorganization. A study of such communities in Peru shows that here, at least, the opposite is true."

753. Perloff, Harvey S., "New Directions In Social Planning," *Journal of the American Institute of Planners,* Vol. XXXI, No. 4, 1965, pp. 297-304.

Existing Strengths and New Requirements; Organizational Considerations; Substantive Considerations; Informational and Planning Tools: Households—Focus on Objectives, Regional Economy, Social Structure, Physical-Locational Patterns; Process and Implementation; Conclusion.

754. Revelle, Roger, "The Problem of People," *Harvard Today,* Autumn 1965, pp. 2-9.

"Can man domesticate himself?" Brief, lucid description of population problems and the new Harvard Center for Population Studies.

755. Rosow, Irving, "The Social Effects of the Physical Environment," *Journal of the American Institute of Planners,* Vol. XXVII, No. 2, May 1961, pp. 127-133.

"The assumption of housers that planned manipulation of the physical environment can change social patterns in determinate ways seems to be only selectively true."

Social Pathology; Livability; Community Integration; Aesthetics; References.

756. Schmitt, Robert C., "Interpretations: Density, Health, and Social Disorganization," *Journal of the American Institute of Planners,* Vol. XXXII, No. 1, January 1966, pp. 38-40.

Presents data "broadly indicative but far from conclusive . . . [which] obviously cast doubt on the validity of Jane Jacob's thesis regarding the relative importance of population densities and overcrowding to health and social disorganization."

757. Silberman, Charles E., "The City of the Negro," *Fortune,* Vol. LXV, No. 3, March 1962, pp. 88-91, 139-140, 144, 146, 151, 152, 154.

Filling the Vacuum, Crucial Difference, Other Side of Jordan, Need for Excellence, "More Powerful than Apathy," "They're Paying Attention," Twenty-Point Drop, Success Story, Underdeveloped Country, Housing Dilemma, Cost of Delay.

758. Thoma, Lucy; Lindeman, Erich, "'Newcomers' Problems in a Sub-urban Community," *Journal of the American Institute of Planners,* Vol. XXVII, No. 3, August 1961, pp. 185-193.

Problems of Mobility; Population Characteristics; Screening the Newcomers; Newcomer's Profile; Sectional Differences; Emotional Problems; Initiation Rituals; Scapegoating and Mental Health.

759. Wolf, Eleanor P., "The Tipping-Point in Racially Changing Neighbor-hoods," *Journal of the American Institute of Planners,* Vol. XXIX, No. 3, August 1963, pp. 217-222.

Tipping-Point Concept; Tipping-Point in Russell Woods; Housing Market Factors in Racial Transition.

Books, Reports, Pamphlets

760. Adams, Bert N., *Kinship in an Urban Setting,* Chicago, Ill. (Mark-ham), 1968, 228 pp. [68-20258]

Introduction to Urban Kinship; Dimensions and Importance of Urban Kin Relations; Young Adults and Their Parents; Adult Siblings—Interest and Comparison; Best-Known Cousin and Secondary Kin; Urban Kin Relations—Summary and Conclusions; Bibliography, pp. 213-220.

761. Back, Kurt W., *Slums, Projects, and People—Social Psychological Problems of Relocation in Puerto Rico,* Durham, N.C. (Duke University), 1962, 123 pp. [62-15369]

Introduction; Method of the Study; Living Conditions; Housing Aspirations; Attitudes Toward a Move to Public Housing; Change and Personality; Perception of Housing Projects; Process of Relocation.

762. Bogue, Donald J., *Skid Row in American Cities,* Chicago (Community and Family Study Center, University of Chicago), 1963, 521 pp. [62-22329]

". . . presents the research techniques used to identify and achieve an understanding of this way of life. The skid row sections of 45 cities are mapped. Detailed research identifies the sociological and psychological factors, the characteristics of the skid row inhabitants and the physical conditions of the area in which they settle, that enable skid rows to develop. . . . Outlines a program for the elimination of skid rows."

763. Bracey, H. E., *Neighbours—Subdivision Life in England and the United States,* Baton Rouge, La. (Louisiana State University), 1964, 208 pp. [64-15877]

People; Families on the Move; Choice of Neighbourhood; Neighbourhood Appraisal After Occupation; Neighbouring; Partying; Friends Old and New; Children and Neighbouring; Children and Teenagers—Their Schooling and Social Activities; Church; Adult Social Organizations; Summing Up—Facts, Factors and the Future; Appendices.

764. Burgess, Ernest W.; Bogue, Donald J. (Editors), *Urban Sociology,* Chicago, Ill. (University of Chicago), 1964, 325 pp. Paperback (abridged edition of *Contributions to Urban Sociology* [63-21309]).

Research in Urban Society—Long View (Ernest W. Burgess; Donald J. Bogue); Urban Ecology and Demography: Variables in Urban Morphology (Beverly Duncan), Trends in Differential Fertility and Mortality in a Metropolis—Chicago (Evelyn M. Kittagawa; Philip M. Hauser); City Size As a Sociological Variable (William Fielding Ogburn; Otis Dudley Duncan), Analysis of Variance Procedures in the Study of Ecological Phenomena (Nathan Keyfitz), Cityward Migration, Urban Ecology, and Social Theory (Ronald Freedman); Urban Social Organization and Mass Phenomena: Function of Voluntary Associations in an Ethnic Community —"Polonia" (Helena Znanieki Lopata), Some Factors Affecting Participation in Voluntary Associations (Herbert Goldhammer), Urbanization and the Organization of Welfare Activities in the Metropolitan Community in Chicago (Arthur Hillman), "Street Corner Society" (William Foote Whyte), Police—Sociological Study of Law, Custom, and Morality (William A. Westley); Ethnic and Racial Groups in Urban Society: Approach to the Measurement of Interracial Tension (Shirley A. Star), Social Change and Prejudice (Morris Janowitz), Negro Family in Chicago (E. Franklin Frazier), Occupational Mobility of Negro Professional Workers (G. Franklin Edwards); Urban Social Problems: Catholic Family Disorganization (John L. Thomas), Organized Crime in Chicago (John Landesco), Delinquency Research of Clifford B. Shaw and Henry D. McKay and Associates (Editors).

765. California, Department of Finance, Revenue and Management Agency, *California Population Projections, 1965-2000,* Sacramento, Cal. (State Capitol, Rm. 1145, 95814), March 1966, 26 pp.

U.S. Population Projections, For 1980 and 2000, By Age and Sex; Age-Specific Birth Rates, 1960-2000; California's Total Population by Age and Sex, 1965, 1980, 2000.

766. Charlesworth, James C. (Editor), *Mathematics and the Social Sciences—The Utility and Inutility of Mathematics in the Study of Economics, Political Science, and Sociology,* Philadelphia, Penn. (American Academy of Political and Social Science), June 1963, 121 pp.

Foreword (James C. Charlesworth); Mathematics in Economics—Language and Instrument (Leonid Hurwicz), Limits to the Uses of Mathematics in Economics (Oskar Morgenstern); Use of Mathematics in the Study of Political Science (Oliver Benson); Mathematics and Political Science (Andrew Hacker); Uses of Mathematics in Sociology (Harrison White); Limits to the Uses of Mathematics in the Study of Sociology (Don Martindale).

767. Chicago, Tenants Relocation Bureau, *The Homeless Man on Skid Row*, Chicago, Ill. (320 N. Clark St.), September 1961, 109 pp. Paper.

Approach; Historical Background; Functions of Skid Row and The Type of Men Who Live There; Geography of Skid Row; Characteristics of the Residents of Chicago's Skid Rows; Problem Drinking and Alcoholic Dereliction; Handicapped and Ill; Workingman on Skid Row; Old Man and Pensioner; Migration and Mobility; Death on Skid Row; Selected Attitudes of Skid Row Residents Toward Employment, Housing and Each Other; Chicago's Present Activity; Summary of Problem; Program; Selected Bibliography—Skid Row and Skid Row Problems, pp. 104-107.

768. Clark, Colin, *Population Growth and Land Use*, New York (St. Martin's), 1968, 406 pp. [67-15941]

Reproductive Capacity of the Human Race; Survival and Growth, History of Population Growth; Population and Food; Measurement of Fertility; Sociology of Reproduction; Economics and Politics of Population Growth; Location of Industries and Population; Land Use in Urban Areas.

769. Clark, S. D., *The Suburban Society*, Toronto, Ont. (University of Toronto), 1966, 233 pp. [66-1140]

Process of Suburban Development; Creation of the Suburban Community; Choice of a Suburban Home; Suburban Population; Deprivations of Suburban Living; Repudiation of the Urban Society; Building the New Society; New Society.

770. Connery, Robert H. (Editor), *Urban Riots: Violence and Social Change*, New York (Academy of Political Science, Columbia University), 1968, 190 pp. Paper. [68-27832]

Thoughts on Violence and Democracy (Barrington Moore, Jr.); Urban Violence and American Social Movements (St. Clair Drake); Violence As Protest (Robert M. Fogelson); Ghetto Rebellions and Urban Class Conflict (Herbert J. Gans); Law, Justice, and the Poor (Curtis J. Berger); Community Control of Education (Marilyn Gittell); Economic Program for the Ghetto (James Heilbrun; Stanislaw Wellisz); Leadership of the Poor in Poverty Programs (Stephen M. David); Problem of Residential Segregation (Karl E. Taeuber); Politics of Protest—How Effective Is Violence? (Bruce L. R. Smith); Social Impact of the Urban Riots (Frank J. Macchiarola); Ethics of Violent Dissent (Robert J. McNamara, S.J.); Official Interpretations of Racial Riots (Allan A. Silver); Urban Crisis and the Consolidation of National Power (Richard A. Cloward; Frances Fox Piven); Intergovernmental Response to Urban Riots (David B. Walker); Bibliography on Violence and Social Change (Stanley E. Gunterman), pp. 184-190.

771. Cox, Harvey, *The Secular City*, New York (Macmillan), 1965, 276 pp. [65-16713]

". . . wide ranging commentary on the 'two main hallmarks of our era'—urbanization and secularization. . . . If Mr. Cox is correct in his analysis of our emerging urban culture, then the plans we are preparing today are hopelessly out of step with future needs. While our plans are discussing the need for building stable sub-communities, the author is arguing that anonymity and high rates of residential and employment mobility are not only necessary but they are desirable features of urban living."

772. Dean, Lois R., *Five Towns, A Comparative Community Study*, New York (Random House), 1967, 173 pp. Paperback. [67-13101]

Final Hypothesis; Procedures; Variables and Hypotheses; Community Portraits: Minersville, River City, Hometown, Factoryville, Newtown; The Communities Compared; Prognostications; In Summary.

773. Faris, Robert E. L.; Dunham, H. Warren, *Mental Disorders in Urban Areas: An Ecological Study of Schizophrenia and Other Psychoses*, Chicago (University of Chicago), 1965, 260 pp. Paperback. [65-16168]

"This . . . book bids fair to set a new standard for research—and the presentation of results—in the field of sociological-psychiatric correlations."

Natural Areas of the City; Urban Distribution of Insanity Rates; Typical Pattern in the Distribution of Schizophrenia; Random Pattern in the Distribution of the Manic-Depressive Psychoses; Differential Distribution of the Types of Schizophrenia; Concentration of the Alcoholic Psychoses and Drug Addicts in the Zone of Transition; Association of General Paralysis with Vice Areas; Correlation of Old Age Psychoses with Areas of Tenancy; Insanity Distributions in a Smaller City, Providence, Rhode Island; Mind and Society; Hypotheses and Interpretations of the Distributions; Appendices.

774. Festinger, Leon; Schachter, Stanley; Back, Kurt, *Social Pressures in Informal Groups: A Study of Human Factors in Housing*, New York (Harper), 1950, 240 pp. [50-13371]

775. Freedman, Ronald (Editor), *Population: The Vital Revolution*, Garden City, N.Y. (Doubleday), 1964, 274 pp. Paperback. [64-19296]

". . . 19 essays . . . by experts analyzing important world population trends in non-technical language . . . to acquaint an intelligent world audience of non-specialists with some of the best current scientific knowledge and opinion about population trends."

776. Frieden, Bernard J.; Morris, Robert (Editors), *Urban Planning and Social Policy*, New York (Basic Books), 1968, 459 pp. Paperback. [68-16874]

Approaches to Social Planning: Comprehensive Planning and Social Responsibility (Melvin M. Webber), Emerging Patterns in Community Planning (Robert Morris; Martin Rein), Social and Physical Planning for the Elimination of Poverty (Herbert J. Gans); Housing and Urban Renewal: Some Social Functions of the Urban Slum (Marc Fried; Joan Levin), Fear and the House-as-Haven in the Lower Class (Lee Rainwater), Human Dimension in Public Housing (Preston David), National Community and Housing Policy (Alvin L. Schoor); Racial Bias and Segregation:

Equality and Beyond, Housing Segregation in the Great Society (Eunice and George Grier), Tipping-Point in Racially Changing Neighborhoods (Eleanor P. Wolf), Prospect for Stable Interracial Neighborhoods (Chester Rapkin; William Grigsby), Barriers to Northern School Desegregation (Robert A. Dentler); Citizen Organization and Participation (Charles E. Silberman), Church and Neighborhood Community Organization (Thomas D. Sherrard; Richard C. Murray), Planning and Politics—Citizen Participation in Urban Renewal (James Q. Wilson), Analysis of Influence in Local Communities (Robert A. Dahl); Urban Poverty: American Lower Classes—Typological Approach (S. M. Miller), Technology and Unemployment (Robert M. Solow), Challenge of the Future (Edgar M. Hoover), Where Welfare Falls Short (Eviline M. Burns); Guidelines for Social Policy: Toward Equality of Urban Opportunity (Bernard J. Frieden), From Protest to Politics—Future of the Civil Rights Movement (Bayard Rustin), Common Goals and the Linking of Physical and Social Planning (Harvey S. Perloff), New Public Law—Relation of Indigents to State Administration (Edward V. Sparer), Client Analysis and the Planning of Public Programs (Janet S. Reiner; Everett Reimer; Thomas A. Reiner), New Metropolis and New University (Robert C. Wood), Manpower Planning and the Restructuring of Education (Avner Hovne), Poverty and the Community Planner's Mandate (Martin Rein; Peter Morris), Community Action Program—New Function for Local Government (David A. Grossman).

777. Gans, Herbert J., *The Levittowners, Ways of Life and Politics in a New Suburban Community*, New York (Pantheon), 1967, 474 pp. [66-17359]

Origin of a Community; Quality of Suburban Life; Democracy of Politics; Appendix: Methods of the Study.

778. Gans, Herbert J., *People and Plans—Essays on Urban Problems and Solutions*, New York (Basic Books), 1968, 395 pp. [68-54134]

Environment and Behavior; City Planning and Goal-Oriented Planning; Planning for the Suburbs and New Towns; Planning Against Urban Poverty and Segregation; Racial Crisis; Another Approach to Sociological Analysis and Planning.

779. Gans, Herbert J., *The Urban Villagers, Group and Class in the Life of Italian-Americans*, New York (Free Press), 1962, 361 pp. Paperback. [62-15362]

Introduction; Peer Group Society; West Enders and American Society; Epilogue; Appendix—Methods Used in This Study; Bibliography, pp. 351-358.

780. Gilbert, Ben W., *Ten Blocks from the White House—Anatomy of the Washington Riots of 1968*, New York (Praeger), 1968, 245 pp. Paperback. [68-8903]

Prologue to April—Riotproof City: Thursday Night—First Sparks of Anger, Thursday Night and Friday Morning—Police Problem, Friday—The Thin Line Vanishes, Midday Friday—Hot Words, Friday Afternoon—Washington Burns, Friday Evening to Saturday—Troops Arrive, Occupation of Washington, Fair Trial—Rioters in the Courts, Looters, "All You Need Is a Match, Man," Merchants, Resurrection City—May-June; Epilogue in August—City's Voices; Appendices.

781. Glazer, Nathan; Moynihan, Daniel Patrick, *Beyond the Melting Pot,* Cambridge, Mass. (Harvard University), 1963, 360 pp. [63-18005]

". . . a superb job of questioning the validity of the melting-pot concept in American life. 'The point . . . is that it did not happen' . . . at least not in New York and those cities that in some ways resemble New York. With fascinating detail they describe the various ethnic and racial groups—their backgrounds; economic, political and social status; accomplishments; hindrances to advancement; role in the life of the city; relationships among and between them; habits of life; neighborhoods; and their future."

782. Handlin, Oscar, *The Newcomers,* Cambridge, Mass. (Harvard University), 1959, 171 pp. [59-14737]

"Results of a three-year study of New York's newest immigrants, the Negro and the Puerto Rican, in a changing metropolis . . .. will interest those readers . . . concerned with the problem of our minorities."

783. Hauser, Philip M., *Population Perspectives,* New Brunswick, N.J. (Rutgers University), 1960, 183 pp. [61-7090]

World Population Explosion; United States Population Explosion, Facts, Consequences and Implications; Metropolitan Area Explosion, Facts, Consequences and Implications; Overview and Conclusions.

784. Hauser, Philip M.; Duncan, Otis D. (Editors), *The Study of Population, An Inventory and Appraisal,* Chicago (University of Chicago), 1958, 864 pp: [58-11949]

Overview and Conclusions (Philip M. Hauser; Otis Dudley Duncan); Demography as a Science: Nature of Demography, Data and Methods, Demography as a Body of Knowledge, as a Profession (Philip M. Hauser; Otis Dudley Duncan); Development and Current Status of Demography: Introduction, Development of Demography (Frank Lorimer), Development and Perspectives of Demographic Research in France (Alfred Sauvy), Development of Demography in Great Britain (E. Grebenik), Demography in Germany (Hermann Schubnell), Contributions of Italy to Demography (Alessandro Costanzo), Demographic Studies in Brazil (Giorgio Mortara), Survey of the Status of Demography in India (C. Chandrasekaran), Demographic Research in the Pacific Area (Irene B. Taeuber), Development and Status of American Demography (Rupert B. Vance); Elements of Demography: Introduction, World Demographic Data (Forrest E. Linder), Population Composition (Amos H. Hawley), Population Distribution (Donald J. Bogue), Fertility (N. B. Ryder), Mortality (Harold F. Dorn), Population Growth and Replacement (Hannes Hyrenius), Internal Migration (Donald J. Bogue), International Migration (Brinley Thomas), Population Estimates and Projections (John V. Grauman), Family Statistics (Paul C. Glick), Working Force (A. J. Jaffe), Population and Natural Resources (Edward A. Ackerman); Population Studies in Various Disciplines: Introduction, Ecology and Demography (Peter W. Frank), Human Ecology and Population Studies (Otis Dudley Duncan), Geography and Demography (Edward A. Ackerman), Physical Anthropology and Demography (J. N. Spuhler), Genetics and Demography (Franz J. Kallmann, M.D.; John D. Rainer, M.D.), Economics and Demography (Joseph J. Spengler), Sociology and Demography (Wilbert E. Moore).

785. Hawkins, Brett W., *Nashville Metro—The Politics of City-County Consolidation*, Nashville, Tenn. (Vanderbilt University), 1966, 162 pp. [66-20048]

Introduction; Political Science and "The Metropolitan Problem"; Nashville—the Setting; Propositions About Opposition and Support for Metropolitan Integration; Formative Years—1951-1958; Annexation and City-County Dissension—1958-1962; 1962 Charter Commission; 1962 Campaign—Issues, Interests, and Activities; Voter Opposition and Support; General Conclusions; Selected Bibliography, pp. 153-155.

786. Inkeles, Alex, *What Is Sociology?: An Introduction to the Discipline and Profession*, Foundations of Modern Sociology Series, Englewood Cliffs, N.J. (Prentice-Hall), Fourth Printing, June 1965, 120 pp. Paper. [64-13092]

The Subject Matter of Sociology; The Sociological Perspective; Models of Society in Sociological Analysis; Conceptions of Man in Sociological Analysis; Basic Elements of Social Life; Fundamental Social Processses; Modes of Inquiry in Sociology; Sociology As a Profession. Selected References, p. 118.

787. Jacobs, Paul, *Prelude to Riot—A View of Urban America from the Bottom*, New York (Random House), 1966, 298 pp. [66-21487]

Police; "Welfare Bureau"; Employment; Housing; Health and Medical Care; Schools; McCone Commission; Sources.

788. Judson, Arnold S., *A Manager's Guide to Making Changes*, New York (Wiley), 1966, 186 pp. [65-28641]

Defining Change and Its Causes; How People Are Affected by Changes; Factors that Influence an Individual's Attitudes Toward a Change; How People React to Changes; Predicting the Extent of Resistance; Minimizing Resistance to Changes: Concepts, Methods; Differences in the Perception of Changes; A Systematic Approach to Making Changes; Implications for Managerial Competence.

789. Lambert, William W. and Wallace E., *Social Psychology*, Foundations of Modern Psychology Series, Englewood Cliffs, N.J. (Prentice-Hall), Second Printing, August 1964, 120 pp. (paper) [64-10868]

Social Psychology, Its Major Concerns and Approaches; Socialization; Perceiving and Judging Social Events; Social Significance of Attitudes; Social Interaction; The Individual in Group Settings; Culture and Social Psychology. Selected Readings, p. 117.

790. Lyford, Joseph P., *The Airtight Cage, A Study of New York's West Side*, New York (Harper & Row), 1966, 356 pp. [65-14687]

". . . describes how and why this once agreeable part of the city became a place of torment for the poor and a zone of fear for so many."
Introduction; The View; Of the People; By The People; A Question of Spirit; For the People; The Airtight Cage.

791. Mahood, H. R. (Editor), *Pressure Groups in American Politics*, New York (Scribners), 1967, 305 pp. Paperback. [67-10457]

Introduction (Phillip R. Monypenny); Theory of Groups: Foreword, Group Basis of Politics—Notes for a Theory (Earl Latham), . . . Notes on Analysis and Development (Robert T. Golembiewski); Pressure Groups and Society: Foreword, Businessmen in Politics (Andrew Hacker; Joel D. Aberback), Seven Fallacies of Business in Politics (Michael D. Reagan), Organized Labor in Electoral Politics—Some Questions for the Discipline (Harry M. Scoble), Organized Labor Bureaucracy as a Basis of Support for the Democratic Party (Nicholas A. Masters), Changing Political Role of the Farmer (Gilbert Fite), National Farm Organizations and the Reshaping of Agricultural Policy in 1932 (William R. Johnson), Other Interests, Congress of Racial Equality and Its Strategy (Marvin Rich), John Birch Society (Alan F. Westlin); Pressure Groups and Government: Foreword, Pressure Groups in Congress (Emanuel Cellar), Pressure Politics and Resources Administration (Robert J. Morgan), Interest Groups, Judicial Review, and Local Government (Clement Vose); Pressure Groups and Regulation: Pressure Groups—Threat to Democracy (H. R. Mahood).

792. Niebanck, Paul L., *The Elderly in Older Urban Areas—Problems of Adaptation and the Effects of Relocation*, Philadelphia, Penn. (Institute for Environmental Studies, University of Pennsylvania), 1965, 174 pp. [66-13]

General Focus of the Study; Overview of the National Relocation Population; Income Needs and Resources; Housing Needs and Resources; Social and Psychological Needs and Resources; Renewal Areas and Relocation; Relocation—Positive Tool.

793. Park, Robert E.; Burgess, Ernest W.; McKensie, Roderick D., *The City*, Chicago, Ill. (University of Chicago), 1967, 239 pp. [66-23694]

"From the period 1915-1940, the writings of the Chicago school of urban sociology were extensive and their impact diverse."

The City—Suggestions for the Investigation of Human Behavior in the Urban Environment; Growth of the City; Ecological Approach; Natural History of the Newspaper; Community Organization and Juvenile Delinquency; Community Organization and the Romantic Temper; Magic, Mentality, and City Life; Can Neighborhood Work Have a Scientific Basis?; Mind of the Hobo—Reflections upon the Relation Between Mentality and Locomotion; Bibliography of the Urban Community (Louis Wirth), pp. 161-228.

794. Schlivek, Louis B., *Man in Metropolis*, New York (Doubleday), 1965, 432 pp. [65-16555]

"Through pictures and narrative, the author traces the lives of some 15 people and in so doing, depicts the problems and prospects of the New York metropolitan region. . . . Urban renewal, rehabilitation, mass transportation, suburbia, and exubia are among the topics discussed."

795. Seeley, John R.; Sim, R. Alexander; Loosley, Elizabeth W., *Crestwood Heights: A Study of the Culture of Suburban Life*, New York (Wiley), 1963, 505 pp. Paper.

"For five years the authors observed and interviewed the residents, and they describe the social life, with special reference to the child-rearing process and its implications for mental health, of an upper-middle-class, family- and child-centered community—in a sense, the realization of the American dream . . . of interest not only as a detailed description of a modern community, but also as an account of social science research and the difficulties it has encountered."

796. Sherrard, Thomas D. (Editor), *Social Welfare and Urban Problems,* New York (Columbia University), 1968, 210 pp. [68-19758]

Foreword (Whitney M. Young, Jr.); Abstracts; Introduction (Thomas D. Sherrard); Neighborhood Organizations in the Delivery of Services and Self-Help (Bertram M. Beck); Immobile Poor (Joseph W. Eaton); Slums and Ethnicity (Nathan Glazer); Ombudsman—Quasi-Legal and Legal Representation in Public Assistance Administration (Geoffrey C. Hazard, Jr.); Variability of Ghetto Organization (Richard J. Hill; Calvin J. Larson); Social and Physical Planning for the Urban Slum (Jack Meltzer; Joyce Whitley); Public Welfare in an Urbanizing America (Edward E. Schwartz).

797. Stockwell, Edward G., *Population and People,* Chicago, Ill. (Quadrangle Books), 1968, 307 pp. [67-10246]

Introduction; Mortality; Fertility; Migration and Mobility; Population Size and Growth; Population Composition; Population Distribution; Conclusions.

798. Taeuber, Karl and Alma, *Negroes in Cities,* Chicago (Aldine), 1965, 284 pp. [65-12459]

"It documents in cold, sober statistics the well-known fact that Negroes have been sharply restricted to few areas of the city. . . ."

Pattern of Negro Residential Segregation; Development of an Urban Negro Population; Negro Residential Segregation in United States Cities; Inter-City Variation in Residential Segregation; Process of Neighborhood Change; Prevalence of Residential Succession; Changing Character of Negro Migration; Concomitants of Residential Succession; Race and Residential Differentiation.

799. Warner, W. Lloyd and Associates, *Democracy in Jonesville—A Study in Quality and Inequality,* New York (Harper & Row), 1949, 311 pp. Paperback. [49-10212]

Course of Human Events; Status of the Democracy of Jonesville; Facts of Life; Social Mobility—Rise and Fall of Families; Democracy of Childhood; Room at the Top; the Mill—Its Economy and Moral Structure; the Joiners—Male and Female; Status Aspirations and the Social Club; Sacred and Profane Worlds of Jonesville; The Norwegians—Sect and Ethnic Group; Status in the High School; Party Politics— Unequal Contest; Town and Country—Structure of Rural Life; Jonesville Goes to War; We Hold These Things to Be True—Social Logics of Jonesville.

800. Wiesner, Jerome B.; The Life Sciences Panel, President's Science Advisory Committee, *Strengthening the Behavioral Sciences,* Washington, D.C. (Superintendent of Documents), 20 April 1962, 19 pp. Paper.

Introduction; Development and Present State of Behavioral Science: Study of Communication, Mechanisms of Personality Development, Motives and the Brain, Study of Cultures and Societies, Studies of Thinking Processes, Closing Remarks;

Recommendations: General Education in Behavioral Sciences, Specific Training of Behavioral Scientists, Systematic Collection of Basic Behavioral Data for the United States, Collection and Processing of Data on Other Societies and Cultures, Need for Larger Units of Support for Basic Research, Providing Advice to Government, Research and Development in Agencies with Action Missions, Research Relevant to Education, International Relations.

801. Winslow, Hall, *Artists in Metropolis*, Brooklyn, N.Y. (Planning Program, Pratt Institute), 1964, 165 pp.

"This unique study [thesis] of New York's art colony in the context of its urban setting is aimed primarily at the need to recognize and plan for living lofts for artists, especially in areas where urban renewal is wiping out their quarters. . . . Presents material on the profession, incomes, related shops and galleries, space needs, environment, and importance to the city."

802. Wood, Elizabeth, *Social Planning—A Primer for Urbanists*, Brooklyn, N.Y. (Planning Department, Pratt Institute), 1965, 91 pp. Paper.

". . . considers the ineffectiveness of planners and administrators in [social planning] and proposes a body of principles and policies which can be used as readily as knowledge about the organization of land and buildings . . ."

Goal of Social Planning; Upward Mobility—Natural Process; Making the Natural Process Work Better; Social Welfare Services and Social Planning; Urban Renewal and Social Planning—Summary and Case Study; Appendices.

Bibliographies

803. Adams, Bert N., *Kinship in an Urban Setting*, Bibliography, pp. 213-220.

804. Bolan, Lewis, *The Role of Urban Planning in the Residential Integration of Middle Class Negroes and Whites*, Bibliography No. 41, Council of Planning Librarians, Monticello, Ill. (Exchange Bibliographies, P.O. Box 229, 61856), 1968, 6 pp.

Thesis abstract and bibliography.

805. Bonjean, Charles M.; Hill, Richard J.; McLemore, S. Dale, *Sociological Measurement—An Inventory of Scales and Indices*, San Francisco, Cal. (Chandler), 1967, 580 pp. [67-24968]

". . . intended to be an aid to those engaged in sociological research . . . especially helpful during the initial stages of research—in reviewing the literature and selecting measures of the phenomena with which an investigator is concerned."

806. Chicago, Tenants Relocation Bureau, *The Homeless Man on Skid Row*, Selected Bibliography—Skid Row and Skid Row Problems, pp. 104-107.

807. Gutman, Robert, *Urban Sociology: A Bibliography*, New Brunswick, N.J. (Urban Studies Center, Rutgers University), 1963, 44 pp. Paper.

"The relevance of sociology to the policy problems of contemporary urban life was a basic criterion for selection in this bibliography of the major published works of urban sociology since 1945. Titles are arranged under nine major headings, which include 'Space and Land Use in Urban Society, the Culture of Urban Settlements, and Sociology and Urban Planning.' Important works from other disciplines supplement those from urban sociology to make a valuable record of titles for the planner . . . aware of the far-reaching consequences of decisions involving housing, transportation and spatial relationships."

808. Hawkins, Brett W., *Nashville Metro—The Politics of City-County Consolidation*, Selected Bibliography, pp. 153-155.

809. Jones, Dorothy M. (Compiler), *Aging in the Modern World—An Annotated Bibliography*, Washington, D.C. (Superintendent of Documents), 1963, 194 pp. Paperback.

Economic, Social, and Physical Aspects of Aging; Listing of Periodicals; Author Index.

810. Park, Robert E.; Burgess, Ernest W.; McKensie, Roderick D., *The City*, Bibliography of the Urban Community (Louis Wirth), pp. 161-228.

811. Schneidermayer, Melvin, *The Metropolitan Social Inventory: Procedures for Measuring Human Well-Being in Urban Areas*, Bibliography No. 39, Council of Planning Librarians, Monticello, Ill. (Exchange Bibliographies, P.O. Box 229, 61856), 1968, 8 pp.

Thesis abstract and bibliography.

Economics, Finance, Taxation, Real Estate

Articles, Periodicals

812. Altschuler, Alan A., "Transit Subsidies: By Whom, For Whom?," *Journal of the American Institute of Planners*, Vol. XXXV, No. 2, March 1969, pp. 84-89.

"Urban transit subsidies are needed to enhance the mobility of the poor and the physically handicapped whose relative mobility has been steadily decreasing. Analysis of the overall transit situation suggests that such subsidies should be specifically tailored to needy individuals rather than to transit companies. Transportation user charges, including highway user tax payments, offer one logical source of subsidies."

813. Blumenfeld, Hans, "The Economic Base of the Metropolis," *Journal of the American Institute of Planners*, Vol. XXI, No. 4, Fall 1955, pp. 114-132.

Summary; Concept of the Economic Base; Identification of "Basic" with "Export" Activities; Limitations of the "Basic-Nonbasic" Concept in Time and Space;

Mercantilistic and Physiocratic Overtones of the "Basic-Nonbasic" Concept; Development of Techniques of Measurement: Manufacturing versus Services, Proportional Apportionment; Specific Problems: Replacement of Imports by Local Production, An Extreme Case of the Effect of the Establishment of a New "Basic" Industry on the Balance of Payments, Indirect Primary Activity, Inter-Urban Transportation, Public Employees, Students, etc., Discrepancy Between "Basic" Employment and Outside Earnings, "Basic-Nonbasic" Ratio, "Multiplier"; Application of the "Basic-Nonbasic" Method to the Metropolis: "Multiplier," Promotion of "Basic" Industries, Real Economic Base of the Metropolis.

814. Brazell, E. C., "Comparative Costs For Open Space Communities: Rancho Bernardo Case Study," *Land Use Controls—A Quarterly Review*, Vol. 1, No. 2, April 1967, pp. 35-40.

"It is not until the density of the [Planned Unit Development] reaches more than 12 units per acre that [its] costs begin to compare with the standard subdivision development of three to four units per acre."

815. Carter, Anne P., "The Economics of Technological Change," *Scientific American*, Vol. 214, No. 4, April 1966, pp. 25-31.

"The effects of such change are brought out by a comparison of input-output tables listing the transactions among all sectors of industry in the U.S. for the years 1947 and 1958."

816. Hayes, Samuel L.; Harlan, Leonard M., "Real Estate as a Corporate Investment," *Harvard Business Review*, Vol. 45, No. 4, July-August 1967, pp. 144-152, 155-160.

"Equity participation in real estate developments affords unrecognized potential." Demand Factors; Supply Sources; ABC's of Profitability in Real Estate; Opportunity for Investors; Early Entries; Role in Real Estate; Conclusion.

817. Herzog, John P., "The Property Tax vs. Land Value Tax as a Policy Instrument," *Western City*, Vol. 42, No. 7, July 1966, pp. 22, 24, 29.

Why the property tax has such staying power and how we might constructively cope with it. Some criteria which a good tax should meet are listed; how the property tax measures up is discussed; ways are suggested of improving the real property tax.

818. Meyers, Harold B., "Tax-Exempt Property: Another Crushing Burden for the Cities," *Fortune*, Vol. LXXIX, No. 5, 1 May 1969, pp. 76-79, 112, 114.

"Erosion of the local tax base has reached scandalous proportions all across the nation. Now desperate public officials are counteracting in imaginative new ways." $12 Billion in Lost Revenue; Break for God, Grandpa, and Dead Pets; Time to Share the U.N.; World's Tallest Exception; Making Suburbanites Pay Their Way; Where a University Should Go; Worldly Approach to Earthly Estates; Abusing the Privilege.

819. Reiner, Thomas A., "Review Article: Economic Planning," *Journal of the American Institute of Planners*, Vol. XXXII, No. 2, March 1966, pp. 115-119.

National Economic Planning; Urban Planning and Economic Plans; Conclusion. References, pp. 118-119.

820.   *Saturday Review,* Taxes—The Collection and Distribution of Your Money, A Special Issue Prepared In Cooperation With The Research Institute of America, 22 March 1969, pp. 22-35, 73-74.

Tax Reform—The Time Is Now (Joseph W. Barr); Should the Government Share Its Tax Take? (Walter W. Heller); Can Taxes Do More Than Raise Revenue? (Haig Babian); Prospects For Lower Taxes (Bert A. Gottfried).

821.   Shenkel, William M., "The Economic Consequences of Industrial Zoning," *Land Economics,* Vol. XL, No. 3, August 1964, pp. 255-265.

Demand for Industrial Land; Development of Districts Zoned for Industry; Industrial Zoning Practices; Relative Elasticity of Demand for Industrial Space; Conclusion.

822.   Sternlieb, George, "The Future of Retailing in the Downtown Core," *Journal of the American Institute of Planners,* Vol. XXIX, No. 2, May 1963, pp. 102-112.

"The decline of retailing downtown is a concomitant of deep-seated changes in residential and transportation patterns that are not likely to be reversed by attempts to revitalize the central business district."

823.   Tanner, James C., "One-Company Town: Bartlesville Prospers Under Long Economic Domination by Phillips," *The Wall Street Journal,* Pacific Coast Edition, Vol. LXXV, No. 24, 4 August 1966, pp. 1, 9.

Oil Firm's Bright Young Men Home Management Skills in Civic Work; Few Spats; Nice Lawns but No Nightlife. Some Company Towns Left; Leaning Over Backwards; Engineers' Advice; Busy, Busy; Cocktails & Sports; A Friend of LBJ; Moving Up the Clock.

Books, Reports, Pamphlets

824.   Bank of America, *Focus on Los Angeles-Long Beach Metropolitan Area,* Los Angeles, Cal. (660 S. Spring St., 90012), June 1966, 57 pp. Paper.

Introduction; Population; Employment; Personal Income; Industry: Manufacturing, Agriculture, Mineral Extraction, Retail Trade, Wholesale Trade, Services, Finance, Insurance, Real Estate, Construction, Government, Transportation, Communication, Utilities; Los Angeles Basin: Growth, County as a Share of the Basin, Effect of Basin Growth on Los Angeles County; Summary and Outlook; Statistical Appendix: Data on Los Angeles County Subareas, Data on Selected Cities.

825.   Bird, Frederick L., *The General Property Tax: Findings of the 1957 Census of Governments,* Chicago, Ill. (Public Administration Service), 1960, 77 pp. [60-9554]

". . . how the findings of the government census relate to the financial position of

American local governments . . . how the tax load is distributed among the major use categories of property. . . ."

826. Blecke, Curtis J., *Financial Analysis for Decision Making*, Englewood Cliffs, N.J. (Prentice-Hall), 1966, 200 pp. [66-12900]

Monthly Financial Status Report; Balance Sheet Management; Capital Investment Decision; Source and Application of Funds; Distribution Cost Control; Direct Versus Absorption Costing; Profitability Review by Products; Financial Analysis for Cost Reduction; Financial Analysis for Improved Asset Utilization; Evaluation of Research and Development Projects; Evaluation of Mergers and Acquisitions; Annual Performance Rating with Competition; Long-Range Profit Planning; Operations Research and Analysis.

827. Bowman, Mary Jean; Haynes, W. Warren, *Resources and People in East Kentucky: Problems and Potentials of a Lagging Economy*, Baltimore, Md. (Johns Hopkins), 1963, 448 pp.

"After a thorough case study of a depressed rural area . . . the authors conclude that East Kentucky is not under- but rather overdeveloped. Because of its inherent limitations, its economic potential will continue to shrink. The underdeveloped resource is people, and any aid program should focus on them rather than on land and physical improvements. . . . Presents a challenge to the typical area development programs prescribed for depressed areas . . . and urges a reassessment of both goals and methods."

828. Brown, Robert K., *Real Estate Economics—An Introduction to Urban Land Use*, Boston (Houghton Mifflin), 1965, 388 pp. [65-3251]

Evolution of Urban Land Use Pattern: Problems and Prospects, Private Privileges —Public Responsibilities, Structural Development of the Urban Form, Changing Pattern of Urban Development; Real Estate Economics and Urban Development: Real Estate Market, Classic Concept of Value, Real Estate Contracts—Contract of Sale and Deed, Mortgage and the Lease, Structure of the Mortgage Market, Forecasting—Functional Predictor; Patterns of Urban Land Use: Commercial Development of Patterns, Location of Industry, Residential Development Patterns, Air Rights in Urban Development, Depreciation and Its Impact on the Urban Property Inventory; Emerging Patterns in Urban Land Use: Structure of Metropolitan Finance, Real Estate Taxation, Mass Transit—Problems and Prospects, Urban Planning, Federal Urban Renewal, Public Housing, Negro and the City; Appendix I —Developing a Small Industrial Park, Appendix II—Development of Rosedale Heights Annex, A Residential Subdivision.

829. Chinitz, Benjamin (Editor), *City and Suburb, The Economics of Metropolitan Growth*, Englewood Cliffs, N.J. (Prentice-Hall), 1964, 181 pp. Paperback. [64-23569]

City and Suburb (Benjamin Chinitz); Economic Structure and Growth: Pittsburgh Takes Stock of Itself (Edgar M. Hoover), Labor Market Perspectives of the New City (Arnold R. Weber); Urban Transportation Problems: Knocking Down the Straw Men (John R. Meyer); Planning in the Metropolitan Area: Myth and Reality of Our Urban Problems (Raymond Vernon); Metropolitan Organization and Finance: Metropolitan Financial Problems (Lyle C. Fitch), Some Fiscal Implications of Metropolitanism (Harvey E. Brazer), Political Economy of the Future (Robert C. Wood); Suggested Readings, pp. 179-181.

830. Committee on Economic Development, *Community Economic Development Efforts—Five Case Studies,* Supplementary Paper No. 18, New York (711 Fifth Ave., 10022), December 1964, 349 pp. Paperback. [64-66259]

Community Adjustment—Lessons of Experience (John H. Nixon); Economic Redevelopment of the Burlington, Vermont Area (Donald R. Gilmore); Economic Redevelopment Efforts in the Utica-Rome, New York Area (V. C. Crisafulli); Chronic Unemployment in Altoona, Pennsylvania (Jacob J. Kaufman; Halsey R. Jones, Jr.); Economic Redevelopment for Evansville, Indiana—Case Study of a Depressed City (J. W. Milliman; W. G. Pinnell); Helena-West Helena, Arkansas— Case Study in Economic Readjustment (W. Paul Brann).

831. Downs, Anthony, *Who Are the Urban Poor?,* New York (Committee for Economic Development), 1968, 57 pp. Paper [68-58543]

Summary of Factual Findings; What Is Poverty?; Extent of Urban Poverty; Specific Types of Urban Poverty; How Social Institutions Reinforce Poverty; Population Changes and the Urban Poor; Policy Issues.

832. Eckstein, Otto, *Public Finance,* Foundations of Modern Economics Series, Englewood Cliffs, N.J. (Prentice-Hall), 1964, 120 pp. Paperback. [64-13088]

Scope of Government Activity; Efficiency in Government Expenditures; Public Finances of State and Local Governments; Economics of Metropolitan Areas; Taxation: Principles and Issues of Fairness; Taxes, Efficiency, and Growth; Budget Policy for Economic Stability.

833. Finney, H. A.; Miller, Herbert E., *Principles of Accounting—Intermediate,* Englewood Cliffs, N.J. (Prentice-Hall), Sixth Edition, 1965, 840 pp. [65-18498]

Accounting Procedures Reviewed; Working Papers—Closing Procedures; Financial Statements; Net Income Concepts and Corrections of Prior Years' Earnings; Capital Stock; Surplus and Dividends; Miscellaneous Topics Relating to Stockholders' Equity; Generally Accepted Accounting Principles; Cash; Receivables; Inventories, I, II, III; Investments, I, II; Tangible Fixed Assets, I, II, III; Intangible Fixed Assets; Liabilities and Reserves; Interpretation of Accounting Statements; Analysis of Working Capital; Analysis of Operations; Statements from Incomplete Records; Quasi-Reorganizations, Business Combinations, Divisive Reorganizations; Income Tax Allocation; Price-Level Impact on Financial Statements.

834. Fishman, Leo (Editor), *Poverty Amid Affluence,* New Haven, Conn. (Yale University), 1966, 246 pp. Paperback. [66-12495]

Definition and Measurement of Poverty: Poverty from the Civil War to World War II (Oscar Handlin), Population Change and Poverty Reduction, 1947-1975 (Robert J. Lampman); Social Attitudes, Social Organization, and Poverty: Poverty and Social Organization (Harold A. Gibbard), Poverty and the Individual (I. Thomas Stone; Dorothea C. Leighton; Alexander H. Leighton); Special Cases of Poverty: Poverty and the Negro (Herman P. Miller), Poverty in Appalachia (Donald A. Crane; Benjamin Chinitz), Poverty and Resources Utilization (Joseph L. Fisher); Approaches to the Elimination of Poverty: Public Approaches to Minimize Poverty (Theodore W. Schultz), Unemployment and Poverty (Harry G. Johnson), Strategies in the War Against Poverty (Otto Eckstein), Ends and Means in the War Against Poverty (Robert J. Lampman).

835. Gillies, James (Editor), *Essays in Urban Land Economics*, Los Angeles, Calif. (Real Estate Research Program, University of California), 1966, 351 pp. [66-65516]

Housing Economics: Housing As Social Overhead Capital (Leland S. Burns), Twenty Years of Rent Control in New York City (Ernest M. Fisher), Some Theoretical Considerations for the Structure of the Housing Market (Leo H. Klaassen), Rates of Ownership, Mobility and Purchase (Sherman J. Maisel), Residential Rehabilitation (M. Carter McFarland), Income Level of New Housing Demand—Some Comments (Arthur M. Weimer), Land Values and the Dynamics of Residential Location (Paul F. Wendt; William Goldner); Housing Finance: Some Economic Aspects of Income Producing Real Estate (Robert M. Fisher), FHA-FNMA Policy and Mortgage Interest Rate (Robert C. Weaver), Land Development and Local Public Finance (C. E. Elias, Jr.), Place, Prosperity vs. People Prosperity —Welfare Considerations in the Geographical Redistribution of Economic Activity (Louis Winnick); Housing Industry: Some Preliminary Observations on the Publicly Held Company and Management in the Light Construction Industry (James Gillies); City: Enduring City (Fred E. Case), Entrepreneurial Influences in Shaping the American City (Frank G. Mittelbach), Price Discrimination Against Negroes in the Rental Housing Market (Chester Rapkin).

836. Gillies, James; Berger, Jay S., *Financing Homeownership: The Borrowers, the Lenders, and the Homes*, Part 4, Profile of the Los Angeles Metropolis: Its People and Its Homes, Los Angeles, Cal. (Real Estate Research Program, Graduate School of Business Administration, University of California, 90024), 1965, 80 pp. Paper.

". . . adds a further dimension to the profile of Los Angeles by shedding light on how people finance their housing."

837. Graduate School of Public Administration, New York University, *Financing Government in New York City*, New York (New York University), June 1966, 717 pp. [66-25151]

Summary of Findings and Recommendations: Setting and Problem, Outlook for City Expenditures, Improved Fiscal Leadership, Intergovernmental Fiscal Relations, City's Revenue System; Staff Papers: Appraisal of the City's Fiscal Position (6), Financing Selected Services (2), Additional Fiscal Resources for New York City (2), Reforming the City's Tax Structure (6).

838. Heilbroner, Robert L.; Bernstein, Peter L., *A Primer on Government Spending*, V-233, New York (Vintage), 1963, 120 pp. [63-16856]

Preface; Wealth and Waste; Who Buys the Nation's Output?; Who Can Buy More?; Spending and Borrowing; Government Debts; Inflation; Burden of the Debt; Has

It Worked in the Past?; Foreign Complications; Is There an Alternative?; End of the First Lesson; Appendix—Test of Understanding.

839. Heilbrun, James, *Real Estate Taxes and Urban Housing*, New York (Columbia University), 1966, 195 pp. [66-20489]

Introduction; House-Operating Firm; Supply and Demand; Rental-Housing Industry; Digression Concerning Slums; Varieties of Local Real Estate Tax; Effect of Local Real Estate Taxes on: Operating Decisions, Investment Decisions; Digression Concerning Evidence; Weighting Alternatives; Appendices.

840. Henry, S. G. B., *Elementary Mathematical Economics*, London (Routledge), 1969, 111 pp. Paperback.

". . . aimed at the most elementary level . . ."
Functions and Graphs; Calculus; Functions of Several Variables; Constrained Maxima and Minima; Guide to Further Reading, pp. 106-108.

841. Hirsch, Werner, Z. (Editor), *Regional Accounts for Policy Decisions*, Baltimore, Md. (Johns Hopkins), 1966, 248 pp. [66-230-00]

Introduction; Urban Renewal: Objectives, Analysis, and Information Systems (Lowden Wingo, Jr.); Regional Accounts for Public School Decisions (Werner Z. Hirsch); Urban Economic Development (Wilbur R. Thompson); State Planning and Development in a Federal System (Robert S. Herman); Influence of National Decisions on Regional Economies (Marvin Hoffenburg; Eugene J. Devine); Criteria for the Evaluation of Regional Development Programs (George H. Borts).

842. Hoover, Edgar M.; Vernon, Raymond, *Anatomy of a Metropolis, The Changing Distribution of People and Jobs Within the New York Metropolitan Region*, Cambridge, Mass. (Harvard University), 1959, 345 pp. [59-12971]

Cities and Suburbs; Jobs; People; Jobs, People, and the Future; Appendices.

843. Ketchum, Marshall D. (Editor), *Public Finance and Fiscal Policy*, Selected Readings, New York (Houghton Mifflin), 1966, 590 pp. Paperback.

Introduction to Public Finance. Fiscal Role of the Federal Government—Principles of Budget Determination (Richard A. Musgrave), Institutional Background to American Fiscal Policies (A. E. Holmans), Congress' Fiscal Role Is Object of Growing Concern (*Congressional Quarterly*); Alternative Budget Concepts—Some Observations of the Budget Concept (Gerhard Colm; Peter Wagner), Budgetary Reform, Notes on Principles and Strategy (F. M. Bator); Federal Expenditures: Expenditure Process (Roy E. Moor), Timing and Economic Impact of Government Spending (Murray L. Weidenbaum), "Backdoor Spending" Issue (*Congressional Quarterly Almanac*), Military Budget and Its Impact on the Economy (Charles J. Hitch), Criteria of Efficiency in Government Expenditures (Roland N. McKean), Past and Future of Public Spending (Alan T. Peacock; Jack Wiseman); Federal Taxation: Tax Problems—Income, Consumption, and Property as Bases of Taxation (Richard Goode), What's Wrong with the Federal Tax System? (Herbert Stein), Economic Effects of a Federal Value-Added Tax (Earl R. Rolph), Indirect Versus Direct Taxes—Implications for Stability Investment (Otto Eckstein), Income

Expenditure and Taxable Capacity (Nicholas Kaldor); State and Local Finance: Tax Overlapping in the United States, 1964 (Advisory Commission on Intergovernmental Relations), Centralized Versus Decentralized Finance (Harold M. Groves), Tenable Range of Functions of Local Government (George J. Stigler), Federal-State-Local Government Fiscal Relations (Joint Economic Committee). Fiscal Policy. Introduction: Fiscal Policy (Norman B. Ture) (Commission on Money and Credit), Was Fiscal Policy in the Thirties a Failure? (Alvin H. Hansen), Taxation and Economic Progress (Nicholas Kaldor); Budget Rules: In Search of a New Budget Rule and an Alternative Economic Guide for Budget Policy (Gerhard Colm; Marilyn Young), Fiscal Policy in Perspective (Economic Report of the President), Full Employment Budget Surplus (Michael Levy); Balanced Budget Theorem: Federal Expenditure and Economic Stability, The Fallacy of the Balanced Budget (Harold M. Somers), Taxes, Income Determination, and the Balanced Budget Theorem (William A. Salant), Multiplier Effect of an Increase in Government Expenditures Versus a Decrease in Tax Rates (Svend O. Hermansen); Automatic Stabilizers: Concept of Automatic Stabilizers (M. O. Clement), Federal Fiscal Policy in Postwar Recessions, Summary (Wilfred Lewis, Jr.), Personal Income Tax as an Automatic Stabilizer (E. Cary Brown); Federal Loan and Credit Programs: Study of Federal Credit Programs (House Committee on Banking and Currency), Summary of Conclusions and Recommendations (Committee on Federal Credit Programs). Debt Management. Principles: Debt Management in the United States (Warren L. Smith), Advance Refunding, Technique of Debt Management (Joseph Scherer), On Debt Policy (Henry Simons), Debt Management and Banking Reform (Milton Friedman), United States Savings Bond Program in the Postwar Period (George Hanc), Statement of the Honorable Douglas Dillon, Secretary of the Treasury, Before the House Banking and Currency Committee. Monetary Policy, Fiscal Policy, and Debt Management Policy. Interrelations: Functional Finance and the Banking System (Lawrence S. Ritter), Potential Contribution of Fiscal-Monetary Policy (Committee on Economic Development). Economic Growth. Role of Fiscal Policy: Fiscal Policy and Economic Growth (Commission on Money and Credit), Beyond Recovery to Full Employment (Alvin H. Hansen); Role of Taxes: Growth Aspects of Federal Tax Policy (Norman B. Ture), Income Tax Rates and Incentives to Work and Invest (George F. Break), Effects of Tax Policy on Private Capital Formation (Richard A. Musgrave), Effects of Depreciation Allowances for Tax Purposes (Robert Eisner), Fading Boom in Corporate Tax Depreciation (George Terborgh); Role of Expenditures: Federal Expenditure Policy for Economic Growth (Otto Eckstein), Government Expenditures and Growth (James S. Duesenberry).

844. Lichfield, Nathaniel, *Economics of Planned Development*, London, England (Estates Gazette), 1956, 460 pp.

Development and the Development Plan: Nature and General Requirements; Economics of Development: Economic Planning and Development, Consumer Demand, Selection and Securing of Sites, Capacity of the Building Industry, Investment, Financing; Financial Calculations for Development: Private Enterprise, Public, Subsidized Housing, Roads and Streets, Car Parks, Comprehensive Re-Development, Town Development, New Towns; Balance Sheet of Development and Planning: Private and Social Costs, Planning Balance Sheet; Programmes and the Development Plan: Programmes of Development, Programming and the

Development Plan; Land Values and Land Planning: Land Use and Land Values, Compensation and Betterment, Effect of a Development Plan on Land Values, Land Values and Planning for an Urban Area; Appendices; Further Reading, pp. 439-443.

845. Lindholm, Richard W. (Editor), *Property Taxation, USA*, Madison, Wis. (University of Wisconsin), 1967, 315 pp. [67-20762]

Some Fundamental Considerations: Property-Tax Development—Selected Historical Perspectives (Arthur D. Lynn, Jr.), Past and Future Growth of the Property Tax (Benjamin Bridges, Jr.), Conflict Between State Assessment Law and Local Assessment Practice (John Shannon), Henry George—Economics or Theology? (Reed R. Hansen); Business and Industry: Taxation of Agriculture (Raleigh Barlowe), Taxation of Business Personal Property (Earl E. Burkhard), Property Taxation of Intangibles (Harold M. Groves), Some Aspects of the *Ad Valorem* Taxation of Railroads (Lynn A. Stiles), Property-Tax Inducements to Attract Industry (Paul E. Alyea); Special Problems: Property-Tax-Rate Limits—View of Local Government (Irving Howards), Payments from Tax-Exempt Property (Joan E. O'Bannon), Exemption of Veterans' Homesteads (Bernard F. Sliger), Property-Tax Concessions to the Aged (Yung-Ping Chen), Broader Lessons from the History of Lake Superior Iron-Ore Taxation (Clarence W. Nelson); Conference Hour Discussions.

846. Margolis, Julius (Editor), *The Public Economy of Urban Communities*, Baltimore, Md. (Johns Hopkins), 1965, 264 pp. Paper. [65-26179]

Model of Economic and Political Decision Making (Jerome Rothenberg); Long-Run Welfare Criteria (Harvey Leibenstein); Public and Private Interaction under Reciprocal Externality (James M. Buchanan; Gordon Tullock); Voting Behavior on Municipal Public Expenditures—Study in Rationality and Self-Interest (James Q. Wilson; Edward C. Banfield); Empirical Evidence of Political Influences Upon the Expenditure Policies of Public Schools (Otto A. Davis); Majority Voting and Alternative Forms of Public Enterprise (Benjamin Ward); Urban Transportation Parables (Robert H. Strotz); Rationalizing Decisions in the Quality Management of Water Supply in Urban-Industrial Areas (Allen V. Kneese); Geographic Spillover Effects and the Allocation of Resources to Education (Burton A. Weisbrod); Spatial Externalities in Urban Public Expenditures—Case Study (Nathaniel Lichfield); Role of Cost-Benefit Analysis in the Public Sector of Metropolitan Areas (Benjamin Chinitz; Charles M. Tiebout).

847. Maxwell, James A., *Financing State and Local Governments*, Washington, D.C. (Brookings), 1965, 276 pp. Paperback. [65-26007]

Developing of the Federal System; Fiscal Performance and Capacity; Intergovernmental Transfers; State Taxes on Individual Income and Sales; Other State Taxes; Property Tax; Nonproperty Taxes and Nontax Revenue; State and Local Debt; Earmarked Revenues and Capital Budgets; Whither State and Local Finance? Bibliography, pp. 263-265.

848. Miller, Herman P., *Rich Man, Poor Man*, New York (New American Library), 1964, 255 pp. Paperback. [64-12108]

Where Do You Fit in the Income Picture?; Nearly Everyone Lies to the Census

Man; The Pie Gets Bigger, the Critics Louder; What's Happening to Our Social Revolution?; Look Around—the Poor Are Still Here; Race, Creed, and Color—the Income of Minorities; Who, Me? In the Top Income Groups?; Cash Value of Education; It's the Job that Counts; Why Don't You Work, Like Other Wives Do?; A Glance into the Crystal Ball; Appendix—Validity of Income Statistics.

849. Perloff, Harvey S.; Dodds, Vera W., *How a Region Grows—Area Development in the U.S. Economy,* New York (Committee for Economic Development), 1963, 147 pp. Paper. [63-14670]

Changing Patterns of Regional Growth—Look at the Growth Figures, 1870-1960; Factors Behind the Volume Growth of Regions; Long-Term Changes in Regional Distribution of Economic Activity; Growth of the States in Recent Years, 1939-1958; Recent Shifts in Employment Among the States, 1939-1958; Changes in: Mining Employment, Agriculture, Manufacturing Employment; Per Capita Income —Levels and Rates of Growth; Approaches to Regional Development.

850. Perloff, Harvey S.; Wingo, Lowdon, Jr. (Editors), *Issues in Urban Economics,* Baltimore, Md. (Johns Hopkins), 1968, 668 pp. Paper. [68-15454]

Introduction (Harvey S. Perloff; Lowdon Wingo, Jr.); Urban Community Within the National Economy: Internal and External Factors in the Development of Urban Economics (Wilbur R. Thompson), Appendix—Toward an Econometric Model of Urban Economic Development (John M. Mattila; Wilbur R. Thompson), Evolving System of Cities in the United States—Urbanization and Economic Development (Eric E. Lampard), Uses and Development of Regional Projections (Sidney Sonenblum), Changing Profile of Our Urban Human Resources (George J. Stolnitz), Discussion of Part I (Harold J. Barnett); Intrametropolitan Development: Evolving Form and Organization of the Metropolis (Edgar M. Hoover), Urban Residential Land and Housing Markets (Richard F. Muth), Poverty in the Cities (Oscar A. Ornati), Quantitative Models of Urban Development—Their Role in Metropolitan Policy-Making (Britton Harris), Discussion of Part II (Donald J. Bogue; Anthony Downs; Britton Harris); Urban Public Economy: Federal, State, and Local Finance in a Metropolitan Context (Dick Netzer), Supply of Urban Public Services (Werner Z. Hirsch), Demand for Urban Public Services (Julius Margolis), Discussion of Part III (Richard A. Mussgrave); Policy Issues: Public Policy for Urban America (Alan K. Campbell; Jesse Burkhead).

851. Perloff, Harvey S.; Nathan, Richard P. (Editors), *Revenue Sharing and the City,* Baltimore, Md. (Johns Hopkins), 1968, 112 pp. [68-16164]

Sympathetic Reappraisal of Revenue Sharing (Walter W. Heller); Federal Government and Federalism (Richard Ruggles); Reflections on the Case for the Heller Plan (Lyle C. Fitch); Federal Grants to Cities, Direct and Indirect (Carl S. Shoup); Comments on Block Grants to the States (Harvey E. Brazer); Rebuttal Comments (Walter W. Heller; Richard Ruggles).

852. Pickard, Jerome P., *Changing Urban Land Uses as Affected by Taxation,* Washington, D.C. (Urban Land Institute), 1962, 105 pp. Paper [62-20691]

Federal Taxation and Changing Urban Land Use; Alternative Forms of Property Taxation, Tax Abatement and Exemption; Tax Variation and Local Metropolitan Area Tax-Land Use Impacts; Tax-Land Use Complexity in Allegheny County, Pennsylvania; High Property Taxation and Varying Resources in the Boston, Massachusetts Metropolitan Area; Comparison of Tax-Land Use Factors in Four Metropolitan Areas; Property Taxation in the United States, 1957—Per Capita Levels and Regional Variation; Bibliography, pp. 73-83.

853. Reynolds, D. J., *Economics, Town Planning and Traffic*, London, England (Institute of Economic Affairs), 1966, 166 pp.

Nature of Economics and Its Application to Town Planning; Distribution of Industry, Housing and Transport—Appendix, Economics of Urban Congestion; Planning Controls, Land Values and General Survey; Traffic in Towns—The Buchanan Thesis and Its Appraisal; Traffic in Towns—Problems of Application and Alternatives—Appendix, Parking Policy; Urban Renewal; Density and Distribution of the Population; the Planning Process; Conclusions—Possible Solutions to the Urban Planning Problem; Selected Bibliography, p. 161.

854. Rostow, W. W., *The Stages of Economic Growth*, New York (Cambridge University), 1960, 179 pp. Paperback. [60-1847]

Introduction; Five Stages-of-Growth—A Summary; Preconditions for Take-Off; Take-Off; Drive to Maturity; Age of High Mass-Consumption; Russian and American Growth; Relative Stages-of-Growth and Aggression; Relative Stages of Growth and the Problem of Peace; Marxism, Communism, and the Stages-of-Growth; Appendix—Diffusion of the Private Automobile.

855. Schaeffer, Wendell G., *Modernizing Government Revenue Administration*, The Application of Technical Cooperation in Improving Revenue Administration in the Governments of Developing Countries, Prepared by Public Administration Service, Washington, D.C. (Agency for International Development), January 1961, 92 pp.

The Public and the Revenue System; National Economies and Revenue Structures; Revenue Budgeting; Major Revenue Sources; Administration of Revenue Programs; Technical Assistance in Revenue Administration; Selected Bibliography, pp. 91-92.

856. Skinner, Donald J., *Industrial Markets for Oakland County Firms*, Pontiac, Mich. (Oakland County Planning Commission), 1963, 173 pp.

"The methodology and approach used in this two-year study, examining the importance of the auto industry to other manufacturing firms, product markets, and problems of diversification for a more stable base, have valid application in other areas. Market growth projections, forces for relocation, and other factors are considered."

857. Solomon, Ezra, *The Theory of Financial Management*, New York (Columbia University), 1963, 170 pp. [63-8405]

Scope of the Finance Function; Objective of Financial Management; Concept of the Cost of Capital; Cost of External Equity Capital; Cost of Retained Earnings; Combined Cost of Debt and Equity; Cost of New Borrowing; Leverage and the

Cost of Capital; Optimal Use of Leverage; Investment Decisions; Financing Decisions; Classified Bibliography, pp. 156-160.

858.   Stone, P. A., *Housing, Town Development, Land and Costs*, London, England (Estates Gazette), 1964, 154 pp.

"This economic comparison of the cost of high-density housing in urban areas with low-density development in suburbs and new towns, supported by an abundance of charts and graphs, is set in the British Isles and costs are in English pounds, but the principles of the analysis are pertinent to the American planner."

859.   Thompson, Wilbur R., *A Preface to Urban Economics*, Baltimore, Md. (Johns Hopkins), 1968, 413 pp. Paperback. [65-19537]

Principles—Goals and Processes in the Urban Economy: Economic Growth and Development—Processes, Stages, and Determinants, Money Income and Real Income—From Labor Markets to Urban Efficiency, Income Inequality—Personal and Governmental Poverty, Patterns of Economic Instability—Preventives and Cures, Interactions Among Goals—Opportunity Cost at the Policy Level; Prescription—Problems and Policy in the Urban Economy: Urban Poverty—Employment, Employability, and Welfare, Urban Public Economy—Problems in Scale and Choice, Housing and Land Use Patterns—Renewal, Race, and Sprawl, Traffic Congestion—Price Rationing and Capital Planning, Interactions Among Problems—Problems of "Solutions."

860.   U.S. National Resources Committee, Urbanism Committee, *Our Cities: Their Role in the National Economy*, Washington, D.C. (Superintendent of Documents), 1937, 88 pp. Paper.

Foreword, Emerging Problems, Recommendations, Possible Accomplishments; Facts About Urban America; The Process of Urbanization: Underlying Forces and Emerging Trends; Problems of Urban America; Special Studies of the Urbanism Committee; Statements of General Policy and Recommendations.

861.   Vernon, Raymond, *The Changing Economic Function of the Central City*, New York (Committee For Economic Development), 1959, 81 pp. Paperback. [59-8897]

Central City Today: Place In the Economy, Costs Of Doing Business, Communication Factor, Costs of Uncertainty, "External Economies of Scale," Early Start; Central City In Transition: Movement—Population, Retail Job, Wholesale Job, Manufacturing Job, Office Job; Summary.

862.   Weston, J. Fred; Brigham, Eugene F., *Managerial Finance*, New York (Holt), Second Edition, 1966, 829 pp. [66-16957]

Financial Management and Its Environment: Finance Function, Promotion and Choice of Form of Organization, Tax Environment; Financial Analysis and Control: Financial Analysis, Financial Control, Management of Assets, Capital Budgeting; Financial Planning: Financial Forecasting, Budgeting, Profit Measurement and Planning, Capital Structure and Use of Financial Leverage, Valuation and Cost of Capital; Short-Term and Intermediate-Term Financing: Cash Cycle, Major Sources of Short-Term Financing, Forms of Secured Short-Term Financing; Long-Term Financing: Equity Funds, Preferred Stock, Long-Term Debt, Dividend

Policy and Internal Financing, Use of Rights in Financing, Warrants and Convertibles; Capital Market Institutions: Investment Banking, Financial Markets and Direct Placement, Comparison of Sources of Financing; Financial Strategies for Growth: Mergers and Holding Companies, Financial Aspects of Mergers, Failure and Financial Rehabilitation, Financial Aspects of Reorganization and Liquidation Procedures; Integrated View of Financial Management: Timing of Financial Policy, Financial Life Cycle of the Firm.

863. Zimmer, Basil G., *Rebuilding Cities: The Effects of Displacement and Relocation on Small Business*, Chicago, Ill. (Quadrangle Books), 1965, 363 pp. [64-14135]

". . . an account of the fate of displaced business firms . . . statistically documented by a survey of 350 small businesses displaced in Providence, R.I. between 1954 and 1959 by urban renewal and highway projects. The study . . . has the dual purpose of recording the effect of displacement on individual firms and on the spatial distribution of business within the area. . . . Discusses the extent of business failures, business losses to the central city, suburbanization, reaction to displacement and adjustment to new locations, and . . . concludes . . . that relocation leads to losses both to the city and to individual businessmen."

Bibliographies

864. Bain, Harry O., *California Real Estate Bookshelf*, Berkeley, Calif. (Center for Real Estate and Urban Economics, Institute of Urban and Regional Development, University of California), Revised Edition 1966, 189 pp. Paperback.

Real Estate Business: General, Selling, Advertising and Public Relations, Management, Exchanges and Taxation, Finance and Investment, Insurance; Legal Aspects; Property Valuation and Appraisal; Housing: General, for the Elderly, How and What To Buy, Housing and Race; Building; Income Properties; Urban Planning, Development, and Land Use; Land Economics; California Real Estate.

865. *C-E Newsletter*, "1967 Periodical Articles on Cost-Effectiveness," Vol. 2, No. 5, 1968, p. 6.

". . . continuing series listing some of the more significant articles on cost-effectiveness published in periodicals . . ." [Publication of the Cost-Effectiveness Section, Operations Research Society of America.]

866. Chinitz, Benjamin (Editor), *City and Suburb, The Economics of Metropolitan Growth*, Suggested Readings, pp. 179-181.

867. Henry, S. G. B., *Elementary Mathematical Economics*, Guide to Further Reading, pp. 106-108.

868. Knox, Vera H., *Public Finance—Information Sources*, Management Information Guide 3, Detroit, Mich. (Gale Research Company), 1964, 142 pp. [64-16503]

Public Finance; Public Revenues; Public Expenditures; Public Debt; Fiscal Policy;

Tax Revision; Intergovernmental Fiscal Relations; Fiscal Administration; International Public Finance; Periodicals, Services and Indexes; Appendix; Author Index; Subject Index.

869. Lichfield, Nathaniel, *Economics of Planned Development,* Further Reading, pp. 439-443.

870. Maxwell, James A., *Financing State and Local Governments,* Bibliography, pp. 263-265.

871. Pickard, Jerome P., *Changing Urban Land Uses as Affected by Taxation,* Bibliography, pp. 73-83.

872. Reiner, Thomas A., "Review Article: Economic Planning," References, pp. 118-119.

873. Rickert, John E., *Open Space Land, Planning and Taxation: A Selected Bibliography,* Washington, D.C. (Superintendent of Documents), February 1965, 58 pp. Paper.

Prepared by Urban Land Institute for the Urban Renewal Administration, U.S. Housing and Home Finance Agency.

Open-Space Land, Planning and Taxation—A Selected Bibliography; National Capital Region—Taxation, Planning, and Open Space Studies, A Review; Bibliography: Research Resources, Land Planning and Use, Regional Studies, Land Economics, Taxation, Public Finance, National Capital Region; Index of Authors.

874. Schaeffer, Wendell G., *Modernizing Government Revenue Administration,* Bibliography, pp. 91-92.

875. Solomon, Ezra, *The Theory of Financial Management,* Classified Bibliography, pp. 156-160.

General Works; Finance Function—Scope and Objectives; Cost of Capital; Investment Decisions and Capital Budgeting; Financing Decisions and Financial Structure.

Laws, Legislation, Adjudication, Regulation

Articles, Periodicals

876. American Society of Planning Officials, "The British Land Commission Bill," *Trends,* No. 1, 1967, 21 pp.

Land Planning and Development Values in Postwar Britain (Wyndham Thomas); The Taxation of Development Value in Land—The English Bill for a Land Commission (Desmond Heap); The Land Commission.

877. American Society of Planning Officials, "New Directions In Connecticut Planning Legislation," *Land-Use Controls—A Quarterly Review,* Vol. 2, No. 3, Summer 1968, pp. 21-46.

Recommendations—Local Development Powers, Community Development Program: Introduction, Program, Reasons For Development Policies, Nature of Development Policies; New Local Development Powers: Introduction, Dedication of Park and School Land and Payment in Lieu, Planned Unit Development, Conditional Rezoning, Unmapped Zones, Holding Zones, Land Bank, Oversized Improvements, Flood Plain Regulation, Amortization of Nonconformities, Official Map; Modifications of Existing Local Powers.

878. Becker, David M., "Municipal Boundaries and Zoning: Controlling Regional Land Development," *Washington University Law Quarterly*, Vol. 1966, No. 1, February 1966, pp. 1-58.

Intraterritorial Controls—Extraterritorial Considerations and the Courts; Solution—County, Metropolitan, or Regional Control?; Solution—Extraterritorial Controls?; Conclusion.

879. Bernard, Michael M., "The Comprehensive Plan Concept as a Basis for Legal Reform," *University of Detroit Journal of Urban Law*, Vol. 44, Summer 1967, pp. 611-624.

Consequences of Recognition of a Body of City Planning Law; Benefits of Codification; Planning Regulations and the Courts; Application of the "Mock Planning" Concept; Comprehensive Plan Concept; Analysis and Evaluation of Existing Plans; Functions of Reform in Planning Law; Conclusion.

880. Bernard, Michael M., "The Development of a Body of City Planning Law," *American Bar Association Journal*, Vol. 51, No. 7, July 1965, pp. 632-636.

Land Regulations; New Legal Documents; New Legal Bodies; City Planning Case Law; City Planning Legislation and Enabling Acts; Periodicals and Reporting of City Planning Law; Reference Works on City Planning Law; Law School Courses; Textbooks and Casebooks on City Planning Law; Committees of the Association Concerned with Planning Law; Conclusion.

881. Bryden, Roderick M., "Zoning: Rigid, Flexible, or Fluid?", *University of Detroit Journal of Urban Law*, Vol. 44, Winter 1966, pp. 287-326.

Purpose of Variances; Should the Variance Be Retained in the Zoning System? If So, Then in What Form?; The Area (or Bulk) Variance.

882. *Columbia Law Review*, "Comment—Zoning, Aesthetics, and the First Amendment," Vol. 64, No. 1, January 1964, pp. 81-108.

Instant Case; Zoning, Aesthetics, and the Police Power; Nonverbal Expression and the First Amendment; Accommodating Competing Constitutional Claims; Conclusion.

883. Dodds, James C.; Elenowitz, Leonard, "Legislative Review—Planning Legislation: 1966-1967," *Journal of the American Institute of Planners*, Vol. XXXIV, No. 5, September 1968, pp. 312-322.

Introduction—Legislative Trends; Federal Planning Requirements and Programs; State Planning: New Agencies, Reorganization, Legislative Reviews and Studies; Urban Development: State Departments of Urban Affairs, Metropolitan Councils of Governments, Regional and Metropolitan Planning, Membership on Regional or Metropolitan Planning Commissions; Local Planning: Zoning and Subdivision Regulation, Extraterritorial Jurisdiction, Housing and Planned Communities, State Financial Incentives, County Planning; Recreation and Open Space: Land Acquisition, Open Space Land Assessment, Financing Conservation and Recreation; Water Resource Planning; Transportation Planning; Summary of Legislative Developments.

884. Dukeminier, Jesse, Jr., "Zoning for Aesthetic Objectives: A Reappraisal," *Law and Contemporary Problems*, Vol. 20, No. 2, Spring 1955, pp. 218-237.

Existing Confusion; Clarification of Community Objectives; Implementation of Values: Control of Building Design, Billboards, Gravel Pits, and Junk Yards; Conclusion.

885. Dukeminier, Jesse, Jr., "The Zoning Board of Adjustment: A Case Study in Misrule," *Kentucky Law Journal*, Vol. 50, No. 3, Spring 1962, pp. 273-350.

Procedure Before the Lexington Board of Adjustment; Variances; Special Exceptions; Interpretation; Statistical Summary; Conclusions; The Problem Re-Examined.

886. Fortenberry, Jerry P., "Exercise of Eminent Domain by Private Bodies for Public Purposes," *University of Illinois Law Forum*, Vol. 1966, Spring 1966, pp. 131-173.

The Third Economy and Public Utilities; Condemnation Policy and Practice; Getting Into Court; Damage to the Remainder; Condemnor's Choices; Conclusion.

887. Freund, Eric C., "Past, Present, and Emergent Problems and Practices in Land Use Control," Symposium on Changing Concepts of Human Habitations, Session II, Paper 20, India, November 1965, Washington, D.C. (Building Research Institute), 1965, 19 pp.

Abstract; Introduction; Early and Medieval Land Use Control; Utopian and Model Cities; Law and Land Use Control; Physical Expression of Legal Controls: Lot Area per Room, Open Space, Bonuses for Open Space, Height and Setback Provisions; Challenge of Emergent and Future Problems.

888. Mandelker, Daniel R., "Delegation of Power and Function in Zoning Administration," *Washington University Law Quarterly*, Vol. 1963, No. 1, February 1963, pp. 60-99.

Introduction; Administrative Structure and Administrative Discretion in Zoning; The Variance Power; The Exception Power; Adequate Procedures as a Defense to Improper Delegation; Discretionary Administration; Conclusion.

889. Mandelker, Daniel R., "Inverse Condemnation: The Constitutional Limits of Public Responsibility," *Wisconsin Law Review*, Vol. 1966, No. 1, Winter 1966, pp. 3-57.

Background to Inverse Condemnation; Substantive Recovery in Inverse Condemnation; Some Solutions to the Inverse Liability Problem.

890. Mandelker, Daniel R., "Planning the Freeway: Interim Controls in Highway Programs," *Duke Law Journal,* Vol. 1964, No. 3, Summer 1964, pp. 439-476.

Setbacks Under the Police Power; Subdivision Controls; Official Map; Highway Reservation Laws Not Based on the Variance Principle; Toward a More Effective Interim Control Law; Conclusion.

891. Mandelker, Daniel R., "Some Policy Considerations in the Drafting of New Towns Legislation," *Washington University Law Quarterly,* Vol. 1965, No. 1, February 1965, pp. 71-87.

Problems of Site Selection; Planning Controls in New Town Areas; Conclusion.

892. Mandelker, Daniel R., "Standards for Municipal Incorporations at the Fringe," *Texas Law Review,* Vol. 36, No. 3, February 1958, pp. 271-298.

Political and Social Perspective on the Incorporation Problem; Standards Applicable to Metropolitan Incorporations; Conclusion.

893. Morris, Eugene J., "The Quiet Legal Revolution: Eminent Domain and Urban Redevelopment," *American Bar Association Journal,* Vol. 52, No. 4, April 1966, pp. 355-359.

". . . traces the judicial expansion of the power of eminent domain, showing that it has become the prime tool for the use of governments in bringing about urban redevelopment. The expansion of the concepts 'public use' and 'public purpose' illustrates . . . the ability of law and legal institutions to react to changing needs of society."

894. *Northwestern University Law Review,* Symposium—Apartments in Suburbia: Local Responsibility and Judicial Restraint, Vol. 59, No. 3, July-August 1964, pp. 344-432.

The Battle for Apartments in Benign Suburbia—A Case of Judicial Lethargy; Aesthetic Control of Land Use—A House Built Upon Sand?; Flexible Land Use Control—Herein of the Special Use; The Legal Significance of Cost Considerations in the Regulation of Apartments by Suburbs.

895. Reps, John W., "Requiem for Zoning," in *Planning 1964,* Chicago, Ill. (American Society of Planning Officials), November 1964, pp. 56-67.

Critical examination of zoning as an instrument of city planning.

896. *UCLA Law Review,* Land Planning and the Law: Emerging Policies and Techniques, Vol. 12, No. 3, March 1965, pp. 707-1009.

Foreword: The Coming Search for Quality (Jesse Dukeminier, Jr.); Historic Land Policies—Historic and Emergent (Robert C. Weaver); Controlling Land Values in Areas of Rapid Urban Expansion (Daniel R. Mandelker); The Single Tax and Land-Use Planning: Henry George Updated (Donald G. Hagman); Some Observations on the Role of Speculators and Speculation in Land Development (C. E. Elias, Jr.; James Gillies); Transportation Problems of the Megalopolitan (M. L. Gunzburg); Creation and Maintenance of Open Spaces in Subdivisions: Another

Approach (Richard S. Volpert); The Legal Role in Urban Development (Bernard J. Frieden); About Tomorrow's Urban America (Werner Z. Hirsch); Comments: The Use and Abuse of Contract Zoning (Bruce R. Bailey), Subdivision Regulation and the Park Problem (Harold J. Smotkin), Judicial Control Over Zoning Boards of Appeal: Suggestions for Reform (Ronald Tepper; Bruce Toor), *Coast Bank V. Minderhout*, and the Reasonable Restraint on Alienation: Creature of Commercial Ambiguity (Wilford D. Godbold, Jr.).

897. *University of Pennsylvania Law Review*, Planned Unit Development, Vol. 114, No. 1, Entire Issue, November 1965, pp. 3-170.

A Developer Looks at Planned Development (Gerald D. Lloyd); Planned Unit Development—A Challenge to Established Theory and Practice of Land Use Control (Jan Z. Krasnowiecki); Reflections on the American System of Planning Controls—Response to Professor Krasnowiecki (Daniel R. Mandelker); Village Planning in East Sussex (L. S. Jay; K. D. Fines; J. Furmidge); Planned Unit Development as Seen from City Hall (David W. Craig); The Model State Statute (Richard F. Babcock; Jan Z. Krasnowiecki; David N. McBride).

898. Williams, Norman, Jr., "Annual Judicial Review—Recent Decisions in Planning Law: 1967," *Journal of the American Institute of Planners*, Vol. XXXIV, No. 3, May 1968, pp. 180-189.

Residential Land Use Controls: Density, Building Types, "Planned Residential District," Accessory Uses, Community Facilities, Bulk Regulations, Historic Preservation, Subdivision Control; Residential v. Nonresidential Land Use; Nonresidential Controls; Other Zoning Devices: Off-Street Parking Requirements, Signs, Nonconforming Uses, Procedure; Other Types of Land Use Controls: Urban Renewal, Racial Discrimination—Housing and Schools; Notes.

Books, Reports, Pamphlets

899. American Bar Association, National Institute, *Junkyards, Geraniums and Jurisprudence: Aesthetics and the Law*, Chicago, Ill. (Section of Local Government Law), 1967, 346 pp. typewritten. Paper.

Police Power vs. Eminent Domain (Ross D. Netherton); Notes on the Lack of Aesthetic Principles As a Guide to Urban Beautification (Christopher Tunnard); Outline—Aesthetics As a Public Purpose (Anne Louise Strong); Recent Legislative Approaches and Enabling Legislation to Accomplish Aesthetic Objectives (Ruth R. Johnson); Nuisance Doctrine (James D. Billet); Law and the Environment (Russell E. Train); Preservation of Open Space of Private Arrangements (Allison Dunham); Environmental Awareness (Philip H. Lewis, Jr.); Valuation Problems Involving Aesthetic Programs (Edward B. Atherton); Brakes For the Beauty Bus (David M. Gooder); Preservation of Scenic Beauty—One Way To Get There (R. C. Leverich); Federal, State and Local Programs for Beautification (Ross D. Netherton); Problems in Condemnation of Property Rights Involving Aesthetic Controls (Joseph M. Montano); Scenic Easements—Techniques of Conveyance (Ben J. Mullen); Aesthetics and the Marketability of Title (Robert Kratovil); Human Response to the Urban Environment (Cyril Herrmann); Aesthetics in the Law

(Sidney Z. Searles); Posies, Politics and the Courts (Fred S. Farr); Planning, Zoning and Aesthetic Control (Dennis O'Harrow); Review of Current Literature, pp. 308-326 (David R. Levin); Jurisprudence of Aesthetics (Ruth R. Johnson).

900. American Society of Planning Officials, *New Directions in Connecticut Planning Legislation: A Study of Connecticut Planning, Zoning, and Related Statutes,* A Summary Report, Chicago, Ill. (1313 East 60th St., 60637), March 1967, 34 pp.

Introduction; Planning and Zoning History in Connecticut; Role of State Legislation; Recommendations: Introduction, Local Development Powers, Local Organization, Procedural Fairness and Uniformity, Regions and State; Large-Lot Zoning and Its Effect on Moderately-Priced Housing.

901. Curry, James E., *Public Regulation of the Religious Use of Land,* West Trenton, N.J. (Chandler-Davis), 1964, 429 pp. [65-2097]

General and Historical; Police Power and the Constitution; Absolutism in Church Zoning Law; Substantiality of Purpose: Balancing of Interests; Justifications; Aesthetic and General, Space, Light, Air and Ease of Access, Fiscal Considerations, Protection of Property Values, Protection of Neighbors, Traffic Control; Discrimination in General; Spot Zoning and Religious Discrimination in Reverse; Discrimination Between Religious and Public Schools; Suspicions of Prejudice; Religion as a Personal Right and as a Public Service; Role of Precedent; Presumption of Legality and Burden of Proof; Regulations and Their Interpretation; What Is a Church?; Private Opinion; Church Zoning: in State Courts, in United States Courts.

902. Dixon, Robert G., Jr., Kerstetter, John R.; Hollister, Charles A., *Adjusting Municipal Boundaries—Law and Practice,* Washington, D.C. (Department of Urban Studies, National League of Cities), 1966, 354 pp.

"A detailed summary of annexation laws and practices in the fifty states comprises the major part of this publication . . . includes [a] chapter containing 'specifications' for improving the quality of annexation legislation."

903. Haar, Charles M. (Editor), *Law and Land—Anglo-American Planning Practice,* Cambridge, Mass. (Harvard University), 1964, 290 pp. [64-11129]

Land Planning and Land Ownership: Control of the Use of Land in English Law (W. O. Hart), Property, City Planning, and Liberty (Allison Dunham); Making and Effect of the Land Plan: Development Plan and Master Plans—Comparisons (James B. Milner), English Development Plans for the Control of Land Use (Desmond Heap); Individual and Machinery of Planning: Planning Decisions and Appeals (F. H. B. Layfield), Flexibility and the Rule of Law in American Zoning Administration (Lawrence A. Sullivan), Enforcement of Planning Controls in England and Wales (J. G. Barr); Regulation and Taking of Property under Planning Laws: Regulation and Purchase—Two Governmental Ways To Attain Planned Land Use (David W. Craig), Compensation for the Compulsory Acquisition of Land in England (R. E. Megarry), Aspects of Eminent Domain Proceedings in the United States (David R. Levin); In Appraisal: Comparisons and Contrasts (Charles M. Haar).

904. Krasnowiecki, Jan; Babcock, Richard F., *Legal Aspects of Planned Unit Residential Development, with Suggested Legislation*, Technical Bulletin 52, Washington, D.C. (Urban Land Institute), 1965, 95 pp. Paper. [65-23923]

Legal Research and the Basis for Legislative Procedure: Historical, Some Fundamental Questions, Project Approval—Constitutional Limitations, Existing Statutory Limitations, Treatment of Open Space, Planning Unit—Problems of Staged Development; Model Statute and Ordinance.

905. Lawrence, Glenn, *Condemnation—Your Rights When Government Acquires Your Property*, Dobbs Ferry, N.Y. (Oceana), 1967, 123 pp. [67-14394]

Condemnation Defined; What Property May Be Condemned?; Compensable and Non-Compensable Items; How Is Property Valued?; Negotiating the Settlement; Making the Best of the Condemnation; Federal and State Constitutions and Eminent Domain.

906. Lerner, Daniel (Editor), *Evidence and Inference*, Glencoe, Ill. (Free Press), 1958, 164 pp. [59-12065]

Introduction, On Evidence and Inference (Daniel Lerner); Evidence and Inference in History (Raymond Aron); Some Aspects of Evidence and Inference in the Law (Henry M. Hart, Jr.; John T. McNaughton); Nature of Clinical Evidence (Erik H. Erikson); Evidence and Inference: in Nuclear Research (Martin Deutsch), in Social Research (Paul F. Lazarsfeld); In Search of a Poison (Jacob Fine). Guide to Further Readings, pp. 159-164.

907. Manvel, Allen D., *Local Land and Building Regulation—How Many Agencies? What Practices? How Much Personnel?* Research Report No. 6, National Commission on Urban Problems, Washington, D.C. (Superintendent of Documents), 1968, 48 pp. Paper.

Summary Highlights; Coverage of Data; Overall Findings; Cities and Towns of 5,000-Plus; Findings for the 52 Largest Cities; Data Sources and Limitations; Appendix.

908. Mayer, Martin, *The Lawyers*, New York (Dell), 1966, 575 pp. Paperback.

In General: A Number of Lawyers—The Profession from a Distance, Skills and Functions, The Law Schools—Where the Lawyers Come From, Jurisprudence—Where the Laws Come From; People: Criminal Matters—The Way It Is Now, Search for Something Better, P. I. and Other Wrongs, Idea of Justice and the Poor; Business: Pieces of Paper, Business in Washington, Small Bag of Specialties; Infrastructure: Who Has Seen the Law? Book, Binders and Bits, Business of the Courts, Personality of the Judge, Supreme Court—Concluding Unscientific Postscript.

909. Murphy, Francis C., *Regulating Flood-Plain Development*, Department of Geography Research Paper No. 56, Chicago, Ill. (University of Chicago), November 1958, 204 pp. Paperback.

Problem of Flood-Loss Reduction; Current Practices in Regulating Flood-Plain

Development; Integrated Action for Flood-Damage Reduction; Appendices; Bibliography, pp. 190-204.

910. Murphy, G. H. (Compiler), *Laws Relating To—Conservation, Planning, and Zoning, Agricultural Land Conservation, Airports, Beaches, Subdivided Lands . . .*, Sacramento, Cal. (Documents Section, State of California), 1968, 435 pp.

911. National League of Cities, *Adjusting Municipal Boundaries—Law and Practice*, Washington, D.C. (Department of Urban Studies), December 1966, 354 pp. Paper.

Purposes and Principal Methods of Annexation; Annexation Process; Effects of Annexation; Consolidations, Mergers, and Detachments; Basic Principles for a Good Annexation Law; Annexation Methods in the Fifty States.

912. Siegel, Shirley Anderson, *The Law of Open Space*, Legal Aspects of Acquiring or Otherwise Preserving Open Space in the Tri-State New York Metropolitan Region, New York (Regional Plan Association), 1960, 72 pp. Paper.

Public Use and Purpose—Important Legal Concepts; Legal Matters Respecting Land for Parks and Recreation; Legal Matters Respecting Other Essential Open Space; About Taxes; Summary and Recommendations.

913. U.S. Public Health Service, *A Compilation of Selected Air Pollution Emission Control Regulations and Ordinances,* Public Health Service Publication No. 999-AP-43, Washington, D.C. (Superintendent of Documents), Revised Edition, 1968, 146 pp. Paper. [60-60121]

Definitions Typically Included in Air Pollution Ordinances; Smoke Emissions and Equivalent Opacity Regulations; Particulate Emissions from Refuse-Burning Equipment; Particulate Emissions from Manufacturing Processes; Emissions From Asphalt Plants; Emission Control: Sulphur Compound, Organic Solvent, Hydrocarbon, Fluoride, Motor Vehicle, Odor; Zoning Ordinances.

Bibliographies

914. American Bar Association, National Institute, *Junkyards, Geraniums and Jurisprudence: Aesthetics and the Law*, Review of Current Literature (David R. Levin), pp. 308-326.

915. Jacoby, Gordon D.; Others, *The Legal Environment for Planning: Annotated Bibliography of Selected Reference Materials for Environmental Planning and Management,* Pomona, Cal. (Dr. Gerald Rigby, Department of Political Science, California State Polytechnic College), 13 May 1968, 46 pp. Paper. Mimeo.

227 references, most available only in a law library; index.

916. Murphy, Francis C., *Regulating Flood-Plain Development,* Bibliography, pp. 190-204.

Water Resources; Floods; Urban Planning; Channel Encroachment; Zoning; Flood-Plain Zoning Articles; Zoning Ordinances; Sub-Division Regulations; Building Codes; Miscellaneous Ordinances; Urban Renewal; Permanent Evacuation; Flood Insurance; Flood Risk Reports; Miscellaneous.

Politics, Public Attitudes

Articles, Periodicals

917. Babcock, Richard F.; Bosselman, Fred P., "Citizen Participation: A Suburban Suggestion for the Central City," *Law and Contemporary Problems,* Vol. 32, No. 2, Spring 1967, pp. 220-231.

Need for Decentralization; Decentralization—A Proposal; Limits of Decentralization; Advantages of Decentralization; Parallels in the Suburbs.

918. Banfield, Edward C., "The Political Implications of Metropolitan Growth," *Daedalus,* Vol. 90, No. 1, Winter 1961, pp. 61-78.

Tasks of British Local Government; Relation of Citizen to Government; Direction of Change; Contrasting American Tradition; Change in the United States; Summary and Conclusions.

919. Chinitz, Benjamin, "New York: A Metropolitan Region," *Scientific American,* Vol. 213, No. 3, September 1965, pp. 134-138, 143-146, 148; also in: Scientific American (Periodical), *Cities,* New York (Knopf), 1965, 211 pp. Paperback. [65-28177]

"Within the region of which New York is the central city are 550 separate municipal governments. How can the sometimes mutual and sometimes conflicting interests of these communities be unified?"

920. Silva, Ruth C., "Reapportionment and Redistricting," *Scientific American,* Vol. 213, No. 5, November 1965, pp. 20-27.

"The court decisions that reinforce the concept of 'one man, one vote' are now operating to change two aspects of the U.S. political structure: apportionment of elective offices and the form of election districts."

921. Thompson, Peter, "Brandywine Basin: Defeat of an Almost Perfect Plan," *Science,* Vol. 163, No. 3872, 14 March 1969, pp. 1180-1182.

"To accommodate development of the region without sacrificing scenic beauty or water quality, the Institute for Environmental Studies . . . of the University of Pennsylvania spent 3 years drawing up a plan for the river basin which is being widely acclaimed in regional planning circles . . . perhaps the lesson in human relations which came out of the confrontation between 'experts' and rural Americans is the most valuable piece of base-line data to emerge from the study."

Books, Reports, Pamphlets

922. Altshuler, Alan A., *The City Planning Process, A Political Analysis,* Ithaca, New York (Cornell University), 1965, 466 pp. [65-25498]

Introduction; Intercity Freeway; Land-Use Plan for St. Paul; Ancker Hospital Site Controversy; Plan for Central Minneapolis; Goals of Comprehensive Planning; Means of Planning, Reason and Influence in the Public Service; Political Restraints and Strategies; Opportunism vs. Professionalism in Planning; Alternate Perspectives; Bibliography, pp. 454-461.

923. Banfield, Edward C., *Big City Politics*, New York (Random House), August 1967, 149 pp. Paperback. [65-13765]

Atlanta—Strange Bedfellows; Boston—The New Hurrah; Detroit—Balancing Act; El Paso—Two Cultures; Los Angeles—Pre (Civil) War; Miami (Dade County)—Yes, But . . . ; Philadelphia—Nice While It Lasted; St. Louis—Better Than She Should Be; Seattle—Anybody in Charge.

924. Banfield, Edward C.; Wilson, James Q., *City Politics*, New York (Vintage), 1963, 362 pp. Paperback. [63-19134]

Introduction; Nature of City Politics: City As a Setting for Politics, the Political Function, Cleavages, Attachments; Structure of City Politics: City in the Federal System, Distribution of Authority Within the City, Electoral Systems, Centralization of Influence; Political Forms and Styles: Machine, Factions and Factional Alliances, Reform, Nonpartisanship, Council-Manager Form, Master Planning; Some Political Roles: City Employees, Voters, Power Structure and Civic Leadership, Businessmen in Politics, Organized Labor, Negroes, Press; Conclusion: Trend of City Politics.

925. Bullitt, Stimson, *To Be a Politician*, Garden City, N.Y. (Doubleday), 1959, 190 pp. [59-6353]

The Profession: Politics as a Calling, Motives and Incentives, Liabilities, Dilemmas; Methods: Campaign, Candidate, Some Personal Experience; Qualities: Flavor, Moderation, Honesty, Boredom, An Arsenal of Civic Assets; Leadership Among the Leisured.

926. Center for Neighborhood Renewal, *People and Neighborhood Renewal*, Chicago (Metropolitan Housing and Planning Council of Chicago), 1962, 86 pp.

"In a four-year action-research project in Chicago . . . [the Center] worked closely with over 50 neighborhood groups and a host of city-wide agencies, while also conducting a citizen education program in renewal. . . . The experience, pointed observations and recommendations in this report [are] of value in the continuing quest for achieving better rapport between citizens and government in urban renewal."

927. Coulter, Philip B. (Editor), *Politics of Metropolitan Areas: Selected Readings*, New York (Crowell), 1967, 497 pp. Paperback. [67-14291]

Urbanization and Political Conflict: On the Impact of Urbanism on Social Organization, Human Nature and the Political Order (Philip M. Hauser), New Shame of the Cities (Morton Grodzins), Community Conflict (James S. Coleman), Typology for Comparative Local Government (Oliver P. Williams); Structures of Urban Conflict Management: Changing Pattern of Urban Party Politics (Fred I. Greenstein), Mayor as Chief Executive (Duane Lockard), City Manager and His Council —Sources of Conflict (Jeptha J. Carrell), Case Study of the Legislative Process in

Municipal Government (J. Leiper Freeman), Insulation of Local Politics under the Non-Partisan Ballot (Oliver P. Williams; Charles R. Adrian), Governmental Structure and Political Environment (John H. Kessel), St. Louis Politics—Relationships Among Interests, Parties, and Governmental Structure (Robert H. Salisbury); Urban Political Behavior: Class and Party in Partisan and Non-Partisan Elections— Case of Des Moines (Robert H. Salisbury; Gordon Black); Suburbanization and Political Conflict: Suburbanization and Suburbia (Coleman Woodbury), Suburban Decision-Making (Robert C. Wood); Structures of Suburban Conflict Management: Social Structure and Political Process of Suburbia (Scott Greer), Suburban No-Party Politics (Robert C. Wood); Suburban Political Behavior: Suburbs and Shifting Party Loyalties (Fred I. Greenstein; Raymond E. Wolfinger); Metropolitanization and Political Conflict: Resistance to Unification in a Metropolitan Community (Amos H. Hawley; Basil G. Zimmer), Curse of Multiplicity (Reinhold P. Wolff), Urban and Suburban Nashville—Case Study in Metropolitanism (Daniel R. Grant), Metro and the Suburbanite (David A. Booth); Management of Metropolitan Conflict: Organization of Government in Metropolitan Areas—Theoretical Inquiry (Vincent Ostrom; Charles M. Tiebout; Robert Warren), Politics of Metropolitan Area Organization (Edward C. Banfield), Framework for Change (Winston W. Crouch; Beatrice Dinerman), Metropolitics and Professional Political Leadership (Daniel R. Grant), Governing a Metropolitan Area (John C. Bollens; Henry J. Schmandt; Others), Urban County—Study of New Approaches to Local Government in Metropolitan Areas (Mark B. Feldman; Everett L. Gassy), Metropolitan Planning Function (William C. Havard; Floyd Corty); Metropolitan Political Behavior: Urbanization, Suburbanization, and Political Change (August Campbell; Others); State Politics: Urbanization and Competitive Party Politics (Phillips Cutright), Metropolitan and Outstate Alignments in Illinois and Missouri Legislative Delegations (David R. Derge); National Politics: Urbanism and American Democracy (Francis E. Rourke), National Political Alignments and the Politics of Large Cities (Charles E. Gilbert), Equal Protection and the Urban Majority (C. Herman Pritchett).

928. Dahl, Robert A., *Modern Political Analysis*, Foundations of Modern Political Science Series, Englewood Cliffs, N.J. (Prentice-Hall), Fourth Printing, October 1964, 118 pp. Paperback. [63-11092]

Some Unavoidable Political Questions; What Is Politics?; Political Systems: Similarities, Differences; Power and Influence; Political Man; Political Conflict; Coercion, and Popular Government; Political Evaluation; To Explore Further, pp. 111-115.

929. Dahl, Robert A., *Who Governs? Democracy and Power in an American City*, New Haven, Conn. (Yale University), 1961, 355 pp. [61-16913]

". . . interprets the historical succession of occupancy of the principal governmental offices in New Haven as one leading from oligarchy to 'pluralistic democracy.' From an examination of participation in . . . political nominations, urban redevelopment, and public education, [the author] concludes that there is, except for a few high ranking officials, little overlap of active participation. . . ."

930. Danielson, Michael N. (Editor), *Metropolitan Politics—A Reader*, Boston, Mass. (Little, Brown), 1966, 400 pp. Paperback. [66-15459]

Impact of Urbanization on the Political Environment: Transformation of the Urban Community (York Willbern), In Defense of the City (Lewis Mumford), Urban Change and Public Opinion (Raymond Vernon), Racial Segregation in the Metropolis (John H. Strange); Political System of the Metropolis: Political Separation of City and Suburb—Water for Wauwatosa (David D. Gladfelter), Party Cleavages in the Metropolis (Edward C. Banfield), Machine of the Incumbents—Governance of the Central City (Scott Greer), Suburbia—Fiscal Roots of Political Fragmentation and Differentiation (Robert C. Wood), Who Makes Decisions in Metropolitan Areas? (Norton E. Long), Community Power and Metropolitan Decision-Making (Frank J. Munger); Metropolitan Problems and Role of Government: Rationale for Metropolitan Government (Luther H. Gulick), Too Many Governments (Citizens Advisory Committee, Joint Committee on Urban Development, Legislature of the State of Washington), Metropolis and Its Problems Reexamined (Roscoe C. Martin; Douglas Price), Some Flaws in the Logic of Metropolitan Reorganization (Edward C. Banfield; Morton Grodzins); Metropolitan Government—Politics of Revolution: Morality Plays of Metropolitan Reform (Scott Greer), Suburbia and the Folklore of Metropology (Charles R. Adrian), Why Mayors Oppose Metropolitan Government (Paul Ylvisaker), Campaign for Metropolitan Government in St. Louis (Henry J. Schmandt; Paul G. Steinbiker; George D. Wendel), Reflections on the Creation of Miami's Metro (Edward Sofen), Nashville's Politicians and Metro (Daniel R. Grant); Adaptive Metropolis—Politics of Accommodation: Metropolitan Authorities (Victor Jones), Special Districts in the San Francisco Bay Area (Stanley Scott; John Corzine), Differentiation and Cooperation in a Metropolitan Area (Thomas R. Dye; Charles S. Liebman; Oliver P. Williams; Harold Herman), Accommodation *Par Excellence*—Lakewood Plan (Richard M. Cion); In Defense of the Polycentric Metropolis (Vincent Ostrom; Charles M. Tiebout; Robert Warren), Metropolis Against Itself (Robert C. Wood); Metropolis and the Federal System: Halfway House of State Government (Robert C. Wood), States' Role in Urban Development (Council of State Governments), Limitations on State Action—View from Albany (Harold Herman), Federal Policy for Metropolitan Areas (Robert C. Wood), Pattern of Federal-Metropolitan Politics (Michael N. Danielson); Future Metropolis: Changing Political Pattern in the Suburbs (Clarence Dean), Washington and the Cities—Sinister Alliance (Charles H. Percy), Promise of Metropolitan Planning (Charles M. Haar and Associates), Shaping the Metropolis with Federal Money (Jerry Landauer), Natural Limits of Cooperation—Metropolitan Washington Council of Governments (Roscoe C. Martin), Toronto—Problems of Metropolitan Unity (Frank Smallwood).

931. Denver, Planning Office, *Community Directions for the City and County of Denver*, Denver, Colo. (Rm. 300, 1445 Cleveland Pl., 80202), 1964, 143 pp.

"Resumé of a cross-section of Denver citizen opinion, as expressed through key civic and community organizations, relating the city's assets to community objectives for planning."

932. East, John Porter, *Council-Manager Government—The Political Thought of Its Founder, Richard S. Childs*, Chapel Hill, N.C. (University of North Carolina), 1965, 183 pp. [65-19386]

Foundations of Childs' Philosophy; Childs and His Model of Municipal Government; Critique of the Model: Essence of the Closed System, Effect of the Closed System.

933. Finer, Herman, *Road to Reaction,* Boston (Little, Brown), 1945, 228 pp. [45-9861]

Day before Yesterday, and Tomorrow; Reactionary Manifesto; Lunacy about Planning; Rule of Law Is the Rule of Hayek; Adam Smith and Planning for Competition; "Dictatorship" Means Dictatorship; Labeled POISON; The Engineer's Dials; The Engineered Dials; Most Splendid Race; Conclusion.

934. Greer, Scott, *Metropolitics: A Study of Political Culture,* New York (Wiley), 1963, 207 pp. [63-22206]

". . . about the metropolitan reform campaigns in . . . : St. Louis, Miami, and Cleveland. . . . Documented with a wealth of carefully analyzed data leading to the somber conclusion that bringing about rational change in metropolitan government is a bleak prospect."

935. Hayek, Friedrich A., *The Road to Serfdom,* Phoenix Books, Chicago (University of Chicago), 1944, 248 pp. Paperback. [44-4381]

"Classic warning against the dangers to freedom inherent in social planning."

Abandoned Road; Great Utopia; Individualism and Collectivism; "Inevitability" of Planning; Planning and Democracy; Planning and the Rule of Law; Economic Control and Totalitarianism; Who, Whom?; Security and Freedom; Why the Worst Get on Top; The End of Truth; Socialist Roots of Naziism; Totalitarians in Our Midst; Material Conditions and Ideal Ends; Prospects of International Order; Conclusion.

936. Kaufman, Herbert, *Politics and Policies in State and Local Governments,* Foundations of Modern Political Science Series, Englewood Cliffs, N.J. (Prentice-Hall), Second Printing, 1964, 120 pp. Paper. [63-11093]

A Family of Governments; Architecture of State and Local Governments; Everyone's in Politics; Political Strategies; State and Local Government in Perspective; Bibliography, pp. 116-118.

937. Makielski, S. J., Jr., *The Politics of Zoning—The New York Experience,* New York (Columbia University), 1966, 233 pp. [65-25662]

Introduction; Zoning in New York City: Resolution of 1916, Years of Frustration—1917-1945, Rezoning New York City—1945-1960; Patterns and Analysis: Governmental Participants and the Rules of the Game, Nongovernmental Participants, Processes and Strategies; Conclusions: Politics of Zoning; Selected Bibliography, pp. 229-233.

938. Meyerson, Martin; Banfield, Edward C., *Politics, Planning, & the Public Interest,* Glencoe, Ill. (Free Press), 1955, 353 pp. [55-7335]

Background to the Case Study; Organization and Its Tasks; Politicians; Climate of Neighborhood Opinion; Development of Policy; Struggle Begins; Climax; Settlement; Politics; Planning; Public Interest.

939. Press, Charles, *Main Street Politics,* Policy Making at the Local Level,

East Lansing, Mich. (Institute for Community Development Michigan State University), 1962, 147 pp. [63-62535]

"... comprehensive survey of the literature (since 1950) ... [which] discussed four major questions: are communities governed by a conspiratorial power elite?; what is influence?; what are the roles of the low- and middle-status groups in democratic decision-making, and in particular, what is the role of labor in the community?; and, can material in the field be classified for comparative analysis?"

940. Ranney, David C., *Planning and Politics in the Metropolis,* Columbus, Ohio (Merrill), 1969, 168 pp. Paperback. [69-14520]

Planning in the Metropolis—Overview; Physical Design, Utopias, and Reform—Planners' Heritage; Local Governments and Their Planning Agencies; Intergovernmental Relations and Local Planning; Metropolitanism and the Planning Function; Conflict and the Planning Process—Politics of Planning; Determinants of Political Involvement in Planning; Some Concluding Remarks.

941. Sayre, Wallace S.; Kaufman, Herbert, *Governing New York City—Politics in the Metropolis,* New York (Norton), 1965, 777 pp. Paperback. [65-5474]

Background and Setting: Birth and Growth of the Greater City, Stakes and Prizes of the City's Politics, Contestants for the Prizes of the City's Politics; Rules of the Contest; Strategies of the Contestants—Efforts to Determine Who Gets Public Office and Employment: Structure and Operation of Nominating Machinery, Elections, Appointments and Removals; Strategies of the Contestants—Shaping Governmental Decisions: Administrators of the Line Agencies, Officials of Special Authorities; Leaders of the—Overhead Agencies, Organized City Bureaucracies, Party Leaders and Governmental Decisions, Nongovernmental Groups and Governmental Action, Courts and Politics, Officials of Other Governments in the City's Political Process, Council, Board of Estimate, Mayor; Conclusions: Risks, Rewards, and Remedies; Appendices.

942. Talbot, Allan R., *The Mayor's Game, Richard Lee of New Haven and the Politics of Change,* New York (Harper & Row), 1967, 270 pp. [67-11346]

"... intended mainly as a narrative of those events which relate to the city's effort to overcome physical blight and human poverty."

Setting; Organizing Support; Rebuilding the City; Human Resources; Reprise.

943. U.S. House of Representatives, Committee on Banking and Currency (90th Congress, 1st Session), *Basic Laws and Authorities on Housing and Urban Development,* Revised Through May 15, 1967, Washington, D.C. (Government Printing Office), 1967, 749 pp. Paper.

National Policy and Purpose; Housing; Urban Development; Related Functions and Provisions; Participation in Presidential and National Committees; Organization Acts and Orders; Appropriation Acts; General Administrative Provisions.

944. Wibdavsky, Aaron, *The Politics of the Budgetary Process,* Boston (Little, Brown), 1964, 216 pp. Paperback. [64-13972]

Budgets; Calculations: Complexity, Aids to Calculation, Fair Share and Base, the Agency—Roles and Perspectives, Deciding How Much to Ask For—to Spend, Department Versus Bureau, Bureau of the Budget—Roles and Perspectives, Deciding How Much to Recommend, Appropriations Committees—Roles and Perspectives, Deciding How Much to Give—Appropriations Committees, Summary; Strategies: Ubiquitous and Contingent Strategies, Clientele, Confidence, Congressional Committee Hearings, Results, Strategies Designed to Capitalize on the Fragmentation of Power in National Politics, Contingency and Calculation, Defending the Base—Guarding Against Cuts in the Old Programs, Expanding the Base—Inching Ahead with Existing Programs, Expanding the Base—Adding New Programs, Outcomes; Reforms: Normative Theory of Budgeting?, Politics in Budget Reform, Typical Reform, Program Budgeting Versus Traditional Budgeting, Program Budget in the Department of Defense, Efficiency, Knowledge and Reform; Appraisals: Comprehensiveness, Coordination, Neglect, Roles, PVPI Versus PVPI, Strategies, Best Case, Support, Deceit?, Merit Conclusion; Appendix; Bibliography, pp. 201-209.

945. Williams, Oliver P.; Adrian, Charles R., *Four Cities—A Study in Comparative Policy Making*, Philadelphia, Penn. (University of Pennsylvania), 1963, 334 pp. [63-7853]

Introduction: Study Approach, Four Cities; Political Process: Nonpartisanship—Recruitment of Elected Officials, Electoral Patterns, Referendums, Local Interests of Labor, Neighborhoods; Typology of Civic Policies: Types—Promoting Economic Growth, Providing Life's Amenities, Maintenance of Traditional Services, Arbitration Among Conflicting Interests; Conclusions: Community Differences and Policy Variations, Political Institutions and the Structure of Politics.

946. Wood, Robert C., *Suburbia—Its People & Their Politics*, Boston, Mass. (Houghton Mifflin), 1959, 340 pp. [58-9078]

Image of Suburbia; Roots for the Image; Rise of Metropolis; Nature of Suburbia; Politics of Suburbia; Public Programs of Suburbia; Miniature Re-Examined.

Bibliographies

947. Dahl, Robert A., *Modern Political Analysis*, To Explore Further, pp. 111-115.

948. Hawley, Willis D.; Wirt, Frederick M. (Editors), *The Search for Community Power*, Englewood Cliffs, N.J. (Prentice-Hall), 1968, 379 pp. Paperback. [68-54678]
Bibliographic Appendix, pp. 367-379.

949. Kaufman, Herbert, *Politics and Policies in State and Local Governments*, Bibliography, pp. 116-118.

950. Makielski, S. J., Jr., *The Politics of Zoning—The New York Experience*, Selected Bibliography, pp. 229-233.

951. Wibdavsky, Aaron, *The Politics of the Budgetary Process*, Bibliography, pp. 201-209.

Technology, Engineering

Articles, Periodicals

952. Brady, Frank B., "All-Weather Aircraft Landing," *Scientific American*, Vol. 210, No. 3, March 1964, pp. 25-35.

"Jet aircraft are now routinely landed with the help of instruments when visibility is only half a mile. The traditional goal of fully automatic landings under any conditionals, however, is still elusive."

953. Briggs, Asa, "Technology and Economic Development," *Scientific American*, Vol. 209, No. 3, September 1963, pp. 52-61.

". . . outlines the history of development and the division of nations into 'rich' and 'poor.'"

954. *Business Week*, "Case History of a New Product," No. 1590, 20 February 1960, pp. 105-106, 108.

"From idea to commercial product took 12 years. . . . Many of DuPont's product ideas don't make the grade, but those that do follow a well-defined logical course."

955. Davis, Louis E., "Science, Technology, and Social Problems," *California Engineer*, Vol. 44, No. 3, December 1965, pp. 26-29, 31.

Social Housekeeping; Social Studies by Aerospace Firms; Results of the Aerospace Contracts; Social Policy.

956. Etzioni, Amitai, "Agency for Technological Development for Domestic Programs," *Science*, Vol. 164, No. 3875, 4 April 1969, pp. 43-50.

"Such an agency is prerequisite for effective social 'engineering.'"
Importance of Technological Shortcuts; Technology as a Source of Shortcuts; Existing Organization of Domestic R & D; System Effects; Speed of Payoffs and Congress; Resistance to Other Agencies; Relations with Consumers of New Technologies.

957. Hoffman, George A., "The Electric Automobile," *Scientific American*, Vol. 215, No. 4, October 1966, pp. 34-40.

"Air pollution and other drawbacks inherent in the internal-combustion engine make this early kind of car seem increasingly attractive. All depends on the improvement of batteries, and here there are advances."

958. Lessing, Lawrence, "Doing Something About the Weather—In a Big Way," *Fortune*, Vol. LXXVII, No. 4, April 1968, pp. 133-137, 171-172.

Toward a Global System, Overviewer in Outer Space, Ghost Balloons in the Sky, Exploring the Earth's Heat Engine, Bomex—and Bigger, Taking in Two Billion Pieces of Data, Finding a Lever on the Weather, Darkening Sky.

959. McGlauchlin, Lawrence D., "Long-Range Technical Planning," *Harvard Business Review*, Vol. 46, No. 4, July-August 1968, pp. 54-64.

"How a large multidivision company [Honeywell Inc.] is coordinating its fundamental research effort with its engineering and marketing needs."

Enlisting Broad Support; Rating Current Research; Program Results; Conclusion.

960. MacInnis, Joseph B., "Living Under the Sea," *Scientific American*, Vol. 214, No. 3, March 1966, pp. 24-33.

"To learn more about the ocean and harvest its resources, men must be able to live and work as free divers on the continental shelf. Several research programs are currently developing this ability."

961. *Scientific American*, Technology and Economic Development, Vol. 209, No. 3, September 1963, pp. 52-265.

Technology and Economic Development (Asa Briggs); Population (Kingsley Davis); Food (Nevin S. Scrimshaw); Water (Roger Revelle); Energy (Sam H. Schurr); Minerals (Julian W. Feiss); Education for Development (Frederick Harbison); The Structure of Development (Wassily Leontief); Development of Nigeria (Wolfgang F. Stolper), of India (Pitambar Pant), of Brazil (Celso Furtado), of the U.S. South (Arthur Goldschmidt); The Planning of Development (Edward S. Mason).

962. Quarton, Gardner C., "Deliberate Efforts to Control Human Behavior and Modify Personality," *Daedalus*, Vol. 96, No. 3, Summer 1967, pp. 837-853.

Technological Possibilities; Modification of the Genetic Code; Gene Selection by Controlled Mating; Nutritional Influences; Hormones; Use of Drugs; Neurosurgical Interventions; Surgery Outside the Brain; Environmental Manipulations; Monitoring; Mixed Methods; Social Acceptance of Technological Change; Some Extreme Patterns of Social Acceptance of Behavior Manipulation.

963. Quinn, James Brian, "Technological Forecasting," *Harvard Business Review*, Vol. 45, No. 2, March-April 1967, pp. 89-106.

". . . purposes of this kind of forecasting . . . methods and approaches [which] are proving useful . . . capacities and limitations of these management techniques . . ."

Books, Reports, Pamphlets

964. DeGarmo, E. Paul, *Engineering Economy*, New York (Macmillan), Fourth Edition, 1967, 532 pp. [67-10518]

Introduction; Economic Environment; Interest and Annuity Relationships; Depreciation and Valuation; Capital Financing and Budgeting; Selection Among Designs, Materials, and Methods; Value Analysis; Solutions By Linear Programming;

Critical-Path Economy; Basic Patterns for Making Economy Studies; Basic Investment-Decision Studies; Risk, Uncertainty, and Sensitivity; Selections Between Alternatives; Fixed, Increment, and Sunk Costs; Effects of Income Taxes in Economy Studies; Replacement Studies; Minimum-Cost and Break-Even Studies; Economy Studies of Public Projects; Economy Studies in Public Utilities; Appendices; Selected References, pp. 478-479.

965. (E. I.) Du Pont de Nemours & Co., *The D of Research and Development*, Wilmington, Del., 1966, 32 pp.

Development; The Case of "Detaclad" Explosion-Bonded Metals, "Corfam" Poromeric Material, "Nomex" Nylon, Heat Exchangers of "Teflon" Resin, Chloride Pigments Process; Development Process; Research and Marketing, Interplay of Research and Marketing; People; Diversity of Disciplines in Development; Multiplication of Manpower; Broad Business Base; Management; Continuing Development.

966. Hoel, Lester A.; Perle, Eugene D.; Kansky, Karel J.; Kuehn, Alfred A.; Roszner, Ervin S.; Nesbitt, Hugh P., *Latent Demand for Urban Transportation*, Pittsburgh, Penn. (Transportation Research Institute, Carnegie-Mellon University), 1968 (?), 311 pp. Paper.

Introduction; Mobility Behavior of Special Urban Groups; Mobility Development Needs; Theory and Method of Measuring Latent Demand; Future Research Efforts; References, pp. 255-259; Appendices.

967. Hoffman, George A., *Automobiles—Today and Tomorrow*, Memorandum RM-2922-FF, Santa Monica, Cal. (RAND), November 1962, 84 pp. Paper.

Introduction; Weight Composition of Passenger Cars; Some General Characteristics of Passenger Cars; Future Possibilities in Automobile Components; Automobile of the 1970's; References, pp. 83-84.

968. Hoffman, George A., *Urban Underground Highways and Parking Facilities*, Memorandum RM-3680-RC, Santa Monica, Cal. (RAND), August 1963, 47 pp. Paper.

Introduction; Cost of Urban Surface Highways; Cost of Vehicular Tunnels and Underground Highways; Some Design Considerations of Future Urban-Highway Tunnels; A Few Extreme Examples: Los Angeles, Chicago, New York; Conclusions; Suggestions for Further Study; References, p. 47.

969. Krick, Edward V., *An Introduction to Engineering and Engineering Design*, New York (Wiley), 1965, 214 pp. [64-8767]

". . . an extensive introductory description of engineering, primarily for . . . a person considering or beginning a formal education in this field."

Introduction to Engineering; Three Basic Engineering Skills—Representation, Optimization, and Design; The Future; Suggested Readings, pp. xiii-xv.

970. Mumford, Lewis, *Technics and Civilization*, New York (Harcourt, Brace), 1934, 495 pp. [34-11680]

Cultural Preparation; Agents of Mechanization; Eotechnic Phase; Paleotechnic

Phase; Neotechnic Phase; Compensations and Reversions; Assimilation of the Machine; Orientation; Bibliography, pp. 447-474.

971. National Academy of Engineering; National Academy of Sciences, *Science, Engineering, and the City*, Publication No. 1498, Washington, D.C. (2101 Constitution Ave., 20418), 1967, 142 pp. Paperback. [67-61861]

Social Requirements for Urban Design: Introductory Remarks (Walter A. Rosenblith), Challenge to Science and Engineering (Robert C. Wood), Planning for Education (Jerrold R. Zacharias), From the Viewpoint of Business Firms (Arthur M. Weimer); Urban Transportation Problems: Introductory Remarks (Donald S. Berry), Transportation and Urban Goals (Britton Harris), Technology and Formulating Alternative Transport Systems (Sumner Myers), Evaluating Alternative Systems (Charles J. Zwick), Urban Transportation Planning Process (J. Douglas Carroll, Jr.); Urban Construction: Introductory Remarks (Ezra Ehrenkrantz), Concerns of City Governments (John G. Duba), Planning for the New City (William Finley); Urban Experimentation Program: Introductory Remarks (Robert S. Wood), Identifying Alternative Urban Environments (William L. Garrison), Urban Experimentation and Urban Sociology (Peter Orleans), The City as a System (Cyril Herrmann), Urban Research and Experiment (Richard Llewelyn-Davies).

972. Newville, Jack, *New Engineering Concepts in Community Development*, Technical Bulletin 59, Washington, D.C. (Urban Land Institute), 1967, 50 pp. [67-28256]

Examination of New Techniques and Materials: To the Engineer, Clearing and Grading, Drainage, Streets, Sewerage, Public Utilities, Epilogue; Analysis of Improved Lot Costs—Review of Past Experience: Analysis of Improved Lot Costs.

973. Nichols, Herbert L., Jr., *Moving the Earth—the Workbook of Excavation*, Greenwich, Conn. (North Castle), Second Edition, 1962, c. 1440 pp. [62-18588]

Land Clearing; Surveys and Measurements; Soil and Mud; Cellars; Ditching and Dewatering; Ponds; Landscaping and Agricultural Grading; Roads; Blasting and Tunneling; Pit Operation; Costs; Machines: Basic Information, Revolving Shovels, Conveyor Machinery, Tractors and Bulldozers, Tractor Loaders, Scrapers, Dump Trucks, Grading and Compacting Machinery, Compressors and Drills, Miscellaneous Equipment.

974. Oglesby, Clarkson H.; Hewes, Laurence I., *Highway Engineering*, New York (Wiley), Second Edition, 1963, 783 pp. [63-17482]

Introduction; Highway Systems, Types, Organizations, and Associations; Highway: Planning, Economy, Finance, Surveys, Plans, and Computations; Rights of Way; Driver, Vehicle, and Road Characteristics; Highway Design; Traffic Engineering; Highway Drainage; Roadside Development; Highway Subgrade Structure; Constructing the Roadway; Gravel and Crushed Rock Roads—Stabilized Roads; Base Courses; Macadam Surfaces, Bases; Bituminous Pavements; Portland-Cement-Concrete Pavements; Highway Maintenance.

975. Sweeney, Stephen B.; Charlesworth, James C. (Editors), *Governing*

Urban Society: New Scientific Approaches, Monograph 7, American Academy of Political and Social Science, Philadelphia, Pa. (3937 Chestnut St., 19104), May 1967, 254 pp.

Urban Goals and Urban Action: Government and the Intellectual—Necessary Alliance for Effective Action to Meet Urban Needs (Robert C. Wood), Commentaries (Robert A. Dahl; Whitney M. Young, Jr.); Relevance of Science and Technology to Urban Problems: Use of Science in Public Affairs (C. West Churchman), The City and the Computer Revolution (John G. Kemeny), Commentaries (Britton Harris, Walter Scheiber, David Kurtzman and George S. Duggar), Impacts on Urban Governmental Functions of Developments in Science and Technology (John Diebold), Commentary (Roger L. Sisson); Adapting Government for Innovative Action: Emerging Executive and Organizational Responses to Scientific and Technological Developments (Matthias E. Lukens), Roles of Public Officials and Educators in Realizing the Potentials of New Scientific Aids for Urban Society (Carl F. Stover), Commentaries (Don K. Price; Roger W. Jones; Lyle C. Fitch); Progress in Applying Science and Technology in Urban Government: Scope of Scientific Technique and Information Technology in Metropolitan Area Analysis (Herman G. Berkman), Scope of Large-Scale Computer Based Systems in Governmental Functions (Joel M. Kibbee), Scope of Management Information Systems in Governmental Administration (Edward F. R. Hearle); Developing an Urban Research Capability: Overview of Urban Observatories (Henry W. Maier), Defining and Implementing the Urban Observatories Concept (H. Ralph Taylor), Determining Priorities and Developing Basic Research on Urban Problems (Thomas J. Davy), Information Requirements for Urban Research Programs (John K. Parker).

976. U.S. Federal Power Commission, *National Power Survey,* Washington, D.C. (Superintendent of Documents), 1965, 296 pp. Paper.

Introduction—Background and Highlights of Survey; Electric Power Industry Today; Industry's Prospects for Growth; Fuels and Fuel Transport for Electric Energy; Conventional Steam-Electric Generation; Nuclear Power; Hydroelectric Power Resources; Peaking Power; Possible New Methods of Power Generation; Air and Water Pollution at Thermal-Electric Generating Plants; Transmission of Electric Power; Interconnected System Operation and Automation; General Concepts of Coordination, Load Diversity and Capacity Needs; Reduction in Reserves of Generating Capacity; Patterns of Generation and Transmission for 1980; Power Supply for Small Systems; Outlook for Cost Reductions.

Bibliographies

977. Altshuler, Alan A., *The City Planning Process, A Political Analysis,* Bibliography, pp. 454-461.

978. Caldwell, Lynton K. (Editor), *Science, Technology, and Public Policy —A Selected and Annotated Bibliography,* Bloomington, Ind. (Department of Government, Indiana University), 1968, 492 pp. Paper.

Bibliographies and Research Tools; Philosophy of Science; History of Science and

Technology; Nature and Impact of Science and Technology; Science, Politics and Government; Science, Technology, and the Law; Science, Education and the Universities; Scientific and Technical Personnel; Scientific Organizations and Institutions; Organization and Management of Research and Development; Science, the Humanities, and Religion; Science and Society.

979. DeGarmo, E. Paul, *Engineering Economy*, Selected References, pp. 478-479.

980. Mumford, Lewis, *Technics and Civilization*, Bibliography, pp. 447-474.

General Introduction; List of Books.

11. SUBSYSTEMS

Environment

Articles, Periodicals

981. Berank, Leo. L., "Noise," *Scientific American*, Vol. 215, No. 6, December 1966, pp. 66-74, 76.

"There is widespread concern that the noisy environment of modern man not only is distracting but also causes damage to the ear. What are the facts of the matter, and how can noise be brought under control?"

982. Boffey, Philip M., "Radioactive Pollution: Minnesota Finds AEC Standards Too Lax," *Science*, Vol. 163, No. 3871, 7 March 1969, pp. 1043, 1044, 1046.

History and Situation; Consultant's Recommendations; Proposals for Minnesota.

983. Boffey, Philip M., "Smog: Los Angeles Running Hard, Standing Still," *Science*, Vol. 161, No. 3845, 6 September 1968, pp. 990-992.

Regulatory Systems; Tentative New Theory of Cancer.

984. Caldwell, Lynton K., "Environment: A Focus for Public Policy?," *Public Administration Review*, Vol. XXIII, No. 1, March 1963, pp. 132-139.

Meaning of Environment; Dilemma for Decision-Makers; Emerging Environmental Concepts; A Focus for Policy?

985. Calhoun, John B., "Population Density and Social Pathology," *Scientific American*, Vol. 206, No. 2, February 1962, pp. 139-146, 148.

"When a population of laboratory rats is allowed to increase in a confined space, the rats develop acutely abnormal patterns of behavior that can even lead to the extinction of the population."

986. Cavanaugh, James H., "The Physicians Role in Areawide Planning," *Journal of the American Medical Association*, Vol. 195, No. 7, Feb. 14, 1966, pp. 153-156.

Need for [Hospital] Planning; Rising Costs; Solutions Proposed; Recent Developments in Areawide Planning; Objectives of Areawide Planning; Social Values of Planning; Conclusion.

987. Clark, John R., "Thermal Pollution and Aquatic Life," *Scientific American*, Vol. 220, No. 3, March 1969, pp. 19-27.

"Increasing use of river and lake waters for industrial cooling presents a real threat to fish and other organisms. To avoid an ecological crisis new ways must be found to get rid of waste heat."

988. Cohen, Alexander, "Location-Design Control of Transportation Noise," *Journal of the Urban Planning and Development Division*, Proceedings of the American Society of Civil Engineers, Vol. 93, No. UP4, Proc. Paper 5693, December 1967, pp. 63-86.

Introduction; Description and Evaluation of Transportation Noise Problems; Site Considerations and Noise Control; Noise Reduction by Barriers; Noise Reduction Through Modification of Transportation Facilities; Sonic Boom; Conclusions; Appendix: References, pp. 85-86.

989. *Daedalus*, America's Changing Environment, Vol. 96, No. 4, Fall 1967, pp. 1003-1223. [12-30299]

Wider Environment of Ecology and Conservation (F. Fraser Darling); Art of the Impossible (Robert W. Patterson); U.S. Resource Outlook—Quantity and Quality (Hans H. Landsberg); Some Environmental Effects of Economic Development (John V. Krutilla); American Agriculture—Its Resource Issues for the Coming Years (David Allee); Air Pollution Abatement—Economic Rationality and Reality; New Economics of Resources (Nathaniel Wollman); Aesthetic Power or the Triumph of the Sensitive Minority Over the Vulgar Mass—Political Analysis of the New Economics (Aaron Wildavsky); Experimental City (Athelstan Spilhaus); Fallacy of Single-Purpose Planning (Harold Gilliam); Architectural Schemata for Outdoor Recreation Areas of Tomorrow (Alexis Papageorgiou); Outdoor Recreation in a Hyper-Productive Society (Roger Revelle); Ecosystem Science as a Point of Synthesis (S. Dillon Ripley; Helmut K. Buechner); Reorganization of Educational Resources (George J. Maslach); Education for the Changing Field of Conservation (Robert S. Morison).

990. Drapkin, Michael K.; Ehrich, Thomas L., "The Dirty Ohio—A Major River Remains Polluted Despite Years of Effort to Clean It Up," *The Wall Street Journal*, Vol. LXXX, No. 53, Monday, 17 March 1969, pp. 1, 15.

Factories, Cities Still Pour Waste Into Water; Politics, High Costs Slow Progress, Dead Fish and Raw Sewage; Bluish-Black Goo; From Pittsburgh to Cairo; Formidable Political Obstacles; Corporate Reluctance; Invisible Poison.

991. Edwards, Clive A., "Soil Pollutants and Soil Animals," *Scientific American*, Vol. 220, No. 4, April 1969, pp. 88-92, 97-99.

"How is the ecological system of the small invertebrates that live in the soil (and play such a key role in its character) affected by the introduction of pesticides?"

992.   Goldsmith, John R.; Landaw, Stephen A., "Carbon Monoxide and Human Health," *Science,* Vol. 162, 20 December 1968, pp. 1352-1359.

Sources of Exposure; Mechanism of Carbon Monoxide Effects; Endogenous Carbon Monoxide Production—Recent Findings; Effects of Low Concentrations on the Central Nervous System; Epidemiologic Study of the Effects of Carbon Monoxide on Human Health; Methods for Studying Possible Acute Reactions; Summary; References and Notes, p. 1359.

993.   Haagen-Smit, A. J., "The Control of Air Pollution," *Scientific American,* Vol. 210, No. 1, January 1964, pp. 24-31.

"It is now clear that smog is not only annoying but also injurious to health. Los Angeles is a leading example of a city that has analyzed the sources of its smog and taken steps to bring them under control."

994.   Hardin, Garett, "The Tragedy of the Commons," *Science,* Vol. 162, No. 3859, 13 December 1968, pp. 1243-1248.

The population problem has no technical solution; it requires a fundamental extension in morality.

What Shall We Maximize?; Tragedy of Freedom In a Commons; Pollution; How To Legislate Temperance?; Freedom To Breed Is Intolerable; Conscience Is Self-Eliminating; Pathogenic Effects of Conscious; Mutual Coercion Mutually Agreed Upon; Recognition of Necessity.

995.   Keyes, Gyorgy, "Notes on Expression and Communication in the Cityscape," *Daedalus,* Vol. 90, No. 1, Winter 1961, pp. 147-165.

Juxtaposition of Symbols; Articulating the City: Process, Connections; Traffic and Symbolic Form; Light; Texture and Rhythm.

996.   Knudsen, Vern O., "Architectural Acoustics," *Scientific American,* Vol. 209, No. 5, November 1963, pp. 78-92.

"Sound is as much a part of man's man-made environment as heat or light. It can now be effectively managed, notably in rooms where music is heard, by applying the principles of acoustical physics."

997.   Knudsen, Vern O., "Community Noise," *Noise Control,* Vol. 2, No. 4, July 1956, pp. 12-13, 90.

". . . this introductory paper [is an] attempt to help delineate one or two of the fundamental issues of . . . community noise."

998.   Kryter, Karl D., "Psychological Reactions to Aircraft Noise," *Science,* Vol. 151, No. 3716, 18 March 1966, pp. 1346-1359.

"Possible methods of evaluating the acceptability of the noise from aircraft are presented."

Attributes of Sound; General Behavioral Reactions to Noise; Community Reaction to Jet Aircraft Noise; Statistical Nature of Sociological Surveys; Comparison of

Different Noise Sources; Criteria of Unacceptability; Subjective Noisiness of Sonic Booms.

999. Kryter, Karl D., "Sonic Booms from Supersonic Transport," *Science,* Vol. 163, No. 3865, 24 January 1969, pp. 359-367.

"The operation of supersonic transport is considered in the light of the effects of sonic booms on people."

Overland Supersonic Transport and Political Pressure; Conclusion; Intensity of Sonic Boom; Acceptability of Sonic Booms; Damage from Sonic Boom; Analysis of Relevant Research Studies; Behavior in Real Life and Results of Relative Judgment Tests; References and Notes, p. 367.

1000. Lowry, William P., "The Climate of Cities," *Scientific American,* Vol. 217, No. 2, August 1967, pp. 15-23.

"The variables of climate are profoundly affected by the physical characteristics and human activities of a city. Knowledge of such effects may make it possible to predict and even to control them."

1001. Lynch, Kevin, "The City as Environment," *Scientific American,* Vol. 213, No. 3, September 1965, pp. 209-214, 218-219; also in: Scientific American (Periodical), *Cities,* New York (Knopf), 1965, 211 pp. Paperback. [65-28177]

"If the world were covered with a single vast city, how would one achieve the felicitous contrasts of city and country? The metaphor dramatizes the need for making the texture of cities richer."

1002. McAllister, William, "Golden Gate Furor, Foes of More Filling in San Francisco Bay Clash with Developers," *The Wall Street Journal,* Pacific Coast Edition, Vol. LXXX, No. 73, Monday, 14 April 1969, pp. 1 & 23.

Pro-Conservation Commission Expiring; Developers Deny Peril to Beauty, Wildlife; Report to Reagan; Developer's Dream; The Other Side.

1003. Macinko, George, "Saturation: A Problem Evaded in Planning Land Use," *Science,* Vol. 149, No. 3683, 30 July 1965, pp. 516-521.

"The environmental consequences of sustained population growth have yet to be recognized by planners."

Professional Planning; Planning Process; Reconstruction of Land Planning; Other Alternatives.

1004. Van Arsdol, Maurice D., "Metropolitan Growth and Environmental Hazards: An Illustrative Case (World Population Conference)," *Ekistics,* Vol. 21, No. 122, January 1966, 4 pp.

"Once an 'optimum' ultilization of urban land has been reached, it appears that a lack of control over artificial hazards makes for differential population changes and a potential lack of use of those portions of the urban scene which are subject to such hazards."

1005. Schroeder, M. R., "Architectural Acoustics," *Science,* Vol. 151, No. 3716, 18 March 1966, pp. 1355-1359.

Architectural Acoustics in the United States and Abroad; Reverberation Time; Reverberation Theory; Sound Diffusion; Other Problems; Noise Control; Appendix.

1006. Seymour, Whitney North, "Cleaning Up Our City Air—Proposals for Combating Air Pollution Through Affirmative Government Action Programs," *Urban Affairs Quarterly,* Vol. III, No. 1, September 1967, pp. 34-45.

Power Generation; Industrial Pollution; Space Heating; Refuse Disposal; Motor Vehicles; Conclusion.

1007. Sherrill, Robert, "The Jet Noise Is Getting Awful," *The New York Times Magazine,* January 14, 1968, Section 6, pp. 24, 25, 76-81.

1008. Veneklasen, Paul S., "City Noise—Los Angeles," *Noise Control,* Vol. 2, No. 4, July 1956, pp. 14-19.

". . . these silent pages attempt to present . . . the noises of a great American city—not unusual noises . . ."

1009. Yin, Theodore P., "The Control of Vibration and Noise," *Scientific American,* Vol. 220, No. 1, January 1969, pp. 98-106.

"The addition of constrained-layer damping to the conventional techniques of isolation and absorption provides the necessary technology to control almost all excessive vibration and noise."

Books, Reports, Pamphlets

1010. Air Conservation Commission, *Air Conservation,* Publication No. 80, Washington, D.C. (American Association for the Advancement of Science), 1965, 335 pp. [65-28248]

Air Conservation and Public Policy; Summary of Facts; Background Reports: Meteorology, Pollutants and Their Effects, Metropolitan Organization for Air Conservation, Air Conservation and the Law, Air Pollution Control, Socio-Economic Factors, Air Pollution and Urban Development.

1011. American Medical Association, *Areawide Planning*—Report of the First National Conference on Areawide Health Facilities Planning, 28-29 November 1964, Chicago (535 North Dearborn St., 60610), 1965, 154 pp.

Why Areawide Planning?; Contrasts in Regional Planning: New York, California; Local Community—Center of Planning. Areawide health facilities planning viewed by medicine, hospital, government, Blue Cross, insurance, consumer, and community planning agency. View of voluntary and governmental approaches in two major areas; legislation concerned with health facilities planning. Discussion of the areawide planning process, measurement of facility need, community self-determination; examples of planning for different types of communities.

1012. American Society of Planning Officials, *The Urban Planner in Health Planning*, Public Health Service Publication No. 1888, Washington, D.C. (Superintendent of Documents), 1968, 90 pp. Paper.

Introduction; Health Care and Planning; Urban Planning and Health Planning: Present Record, Barriers To Improved Relationships, Future Possibilities; Bibliography, pp. 71-78.

1013. Appleyard, Donald; Lynch, Kevin; Myer, John R., *The View from the Road*, Cambridge, Mass. (M.I.T.), 1964, 64 pp. [63-9038]

"The conclusions of this monograph, which deals with the aesthetics of highways, are based on a series of studies of existing highways and of people's reactions to them. . . . The expressway might be one of the best means of reestablishing 'coherence and order on the metropolitan scale,' and 'is a good example of a design issue typical of a city: the problem of designing visual sequences for the observer in motion.' An interesting, beautifully illustrated book from which many design lessons may be learned."

1014. Bacon, Edmund N., *Design of Cities*, New York (Viking), 1967, 296 pp. [67-23826]

The City As an Act of Will; Awareness of Space As Experience; Nature of Design; Growth of Greek Cities; Design Order of Ancient Rome; Medieval Design; Upsurge of the Renaissance; Design Structure of Baroque Rome; Dutch Interlude; Eighteenth- and Nineteenth-Century European Design; Development of Paris; Evolution of Saint Petersburg; John Nash and London; Vitruvius Comes to the New World; Le Corbusier and the New Vision; the Great Effort—Brasilia; Color and Spatial Progression; Simultaneous Movement Systems; Putting the Ideas to Work—Philadelphia; City for Humanity—Stockholm; Looking Into the Future.

1015. Blum, Henrik and Associates, *Notes on Comprehensive Planning for Health*, Berkeley, Cal. (Comprehensive Health Planning Unit, School of Public Health, University of California), October 1968, c. 360 pp. Paper. Mimeo.

Introduction; Why We Need Comprehensive Planning for Health; Philosophy of Planning as an Instrument of Change; Meaning of Comprehensive Planning for Health; Values—Source of Goals and Standards; Political Decision-Making and Health Planning; Concepts of Community as Related to Planning; Assessment; Systems Approach to Problem Analysis; Economics and Planning; Microeconomics of Health; Planning for Individual Action and Community Implementation; Tools for Planning; Evaluation—Feedback Model; Comprehensive Health Planning Bodies: Functions and Capabilities, Representation, Relationships, Organization; Limitations and Hazards of Comprehensive Planning for Health; Planning for Health Manpower; Regional Medical Program—Comparison with Comprehensive Planning for Health Services; Hospitals—As Viewed by Health Planners; Environmental Health and Natural Resources Planning; Appendices.

1016. Carr, Donald E., *The Breath of Life*, New York (Norton), 1965, 175 pp. [64-23874]

Nature of Air; How It All Started; Disasters; Red-Faced Stiffs; Anatomy of Ruin;

Tear Gas and Spinach Killers; Villain With the Sharp Knife; Cancer; . . . and Madness?; What Is Being Done; What Could Be Done; Bibliography, pp. 159-175.

1017. Committee on Environmental Health Problems, *Environmental Health Problems*, Public Health Service Publication No. 908, Washington, D.C. (Superintendent of Documents), 1962, 288 pp. Paperback.

Conclusions and Recommendations; Review of Environmental Health Problems: Effort Needed, Resources Required, Problems and Examples; Reports of Subcommittees: Manpower Resources and Training, Applied Mathematics and Statistics, Pharmacology, Toxicology, Physiology, and Biochemistry, Analytical Methods and Instrumentation, Air Pollution, Environmental Engineering, Milk and Food, Occupational Health, Radiological Health, Water Supply and Pollution Control; Appendices.

1018. Constantine, Mildred; Jacobson, Egbert, *Sign Language, For Buildings and Landscape*, New York (Reinhold), 1961, 211 pp. [61-13197]

". . . presents in visual terms these essential elements which most planners have largely dealt with in terms of written regulations and formulas . . . importance of sign and symbol and how large a role they play in the urban scene . . . problems of communication and . . . ideas as to how good solutions can be achieved."

1019. Cullen, Gordon, *Townscape*, New York (Reinhold), 1961, 315 pp. [61-16682]

"Ingredients and principles, and possibilities of the art of townscape and their application in existing and proposed plans."

1020. Cunningham, James V., *The Resurgent Neighborhood*, Notre Dame, Ind. (Fides), 1965, 224 pp. [65-16198]

Urban Rush; What Is a Neighborhood?; Pieces of Metropolis; Hyde Park-Kenwood —Gray Area in Transformation; Racial Integration—Story of Spring Hill Gardens; Neighborhood Politics—Abner Mikva Campaign; Suburban Growth—Case of Northwood Acres; New Approaches to Creating Strong Neighborhoods; Homewood-Brushton—New Approach at Work; Note of Criticism; Emerging Allies—Church and University; Indispensable Citizen; Looking Forward.

1021. Dahir, James (Compiler), *The Neighborhood Unit Plan—Its Spread and Acceptance*, A Selected Bibliography with Interpretative Comments, New York (Russell Sage Foundation), 1947, 91 pp. Paperback. [47-12038]

Facts of City Life; Background of the Neighborhood Unit Plan; Planners Test the Plan; The Plan Makes Friends and Faces Obstacles; American Plans and Projects; Neighborhood Unit Abroad; Additional Reading References, pp. 84-91.

1022. De Wolfe, Ivor, *The Italian Townscape*, London, England (Architectural Press), 1963, 280 pp. [64-54750]

"How close can you live to your neighbor and like it? In presenting the Italian answer, De Wolfe considers townscape to mean the art of humanizing high densities after the engineers have made them hygienically possible . . . 461 photographs and drawings . . ."

1023. Dubos, René, *Man Adapting*, New Haven, Conn. (Yale University), 1965, 527 pp. [65-22317]

Man's Nature; Man in the Physical World; Man's Food; The Living World; Indigenous Microbiota; Nutrition and Infection; Evolution of Microbial Diseases; Environmental Pollution; Changing Patterns of Disease; Adaptation and Its Dangers; Population Avalanche; Hippocrates in Modern Dress; Man Meets His Environment; Eradication Versus Control; Control of Disease; Medicine Adapting; Bibliography, pp. 457-508.

1024. Dubos, René, *Man, Medicine, and Environment*, New York (Praeger), 1968, 125 pp. [68-19221]

The Living World; Man's Nature; Biomedical Philosophies; Determinants of Health and Disease; Biomedical Control of Human Life; Cultural Values of Biomedical Sciences.

1025. Eckbo, Garrett, *Urban Landscape Design*, New York (McGraw-Hill), 1964, 248 pp. [63-14578]

Discussion: Begin at the Beginning, Elements of Space Organization; Examples: Room and Patio, Building and Site, Buildings in Groups, Parks and Playgrounds, Streets and Squares, Neighborhood Community and Region; Maintenance and Design.

1026. Edelson, Edward; Warshofsky, Fred, *Poisons in the Air*, New York (Pocket Books), 1966, 160 pp. Paperback.

The Three-Thousand-Mile Sewer; A Hundred and Ten Million Americans; Breath of Death; Mephitic Fountain; Two Hundred Million Tons of Poison; Foggy, Foggy Town; The Polluters; High Cost of Breathing; Indictment; What Could Be Done.

1027. Ewald, William R., Jr. (Editor), *Environment for Man—The Next Fifty Years*, Bloomington, Ind. (Indiana University), 1967, 308 pp. Paperback. [67-14215]

Introduction (William R. Ewald, Jr.); Man Adapting— His Limitations and Potentialities (René Jules Dubos); City Planning and the Treasury of Science (John W. Dyckman); The City as a Mechanism for Sustaining Human Contact (Christopher Alexander); The Environment We See (Herbert H. Swinburne); City of Man— Social Systems Reckoning (Bertram M. Gross); Form and Structure of the Metropolitan Area (William L. C. Wheaton); City of the Mind (Stephen Carr); Planning, Politics, and Ethics (Stanley J. Hallett); Habitat '67 (Moshe Sofdie); Aerospace Systems Technology and the Creation of Environment (W. L. Rogers); Role of Large-Scale Enterprise in the Creation of Better Environment (George T. Bogard); Some Thoughts on the Future (John T. Howard).

1028. Gibberd, Frederick, *Town Design*, New York (Praeger), Fifth Edition, 1967, 372 pp. [67-13491]

Design of the Complete Town: Town and Its Raw Materials, Master Plan, Analyses; Central Areas: Town Centre, Civic Spaces, Shopping Centres, Analyses; Industry:

Light Industrial Estates, Workshop and Service Areas, Analyses; Housing: Neighbourhood, Layout with Houses, Layout with Flats, Dwellings on Steep Sites, Mixed Housing Development, Analyses.

1029. Goldman, Marshall I. (Editor), *Controlling Pollution—The Economics of a Cleaner America*, Englewood Cliffs, N.J. (Prentice-Hall), 1967, 175 pp. Paperback. [67-14840]

Pollution—The Mess Around Us (Marshall I. Goldman); The Great and Dirty Lakes (Gladwin Hill); America's Airborne Garbage (C. W. Griffin, Jr.); What Is Pollution? (Marshall I. Goldman; Robert Shoop); Role of Government in a Free Society (Milton Friedman); Social Costs of Business Enterprise (Karl William Kapp); Effluents and Affluence (Jared E. Hazelton); Economic Incentives in Air Pollution Control (Edwin S. Mills); Water Quality Management by Regional Authorities in the Ruhr Area (Allen V. Kneese); A River Dies—And Is Born Again (Fred J. Cook); Pittsburgh—How One City Did It (Ted O. Thackrey); Rebirth of a River (Lee Edson); The Vokhna River Flows (K. Iosifov); Writer's Notes—Trip to Baikal (Oleg Volkov); Let Us Protect the Water, Air, and Soil from Pollution (Yu. Danilov); Sholokov's Protest (Mikhail Sholokov); Abatement of Pollution (*Annual Report of the Council of Economic Advisers, 1966*).

1030. Hall, Edward T., *The Hidden Dimension*, Garden City, N.Y. (Doubleday), 1966, 201 pp. [66-11173]

Culture As Communication; Distance Regulation in Animals; Crowding and Social Behavior in Animals; Perception of Space—Distance Receptors, Eyes, Ears, and Nose; Perception of Space—Immediate Receptors, Skin and Muscles; Visual Space; Art As a Clue to Perception; Language of Space; Anthropology of Space—Organizing Model; Distances in Man; Proxemics in a Cross-Cultural Context—Germans, English, French, Japan, Arab World; Cities and Culture; Proxemics and the Future of Man; Bibliography, pp. 183-193.

1031. Halprin, Lawrence, *Cities*, New York (Reinhold), 1963, 224 pp. [63-19226]

"This book is about the landscape of cities . . . the open spaces, and what goes on in them."

Prologue; Urban Spaces; Gardens between Walls; Furnishing the Street; Floor of the City; Third Dimension; Water in the Square; Trees for all Seasons; View from the Roof; Choreography.

1032. Herfindahl, Orris C.; Kneese, Allen V., *Quality of the Environment: An Economic Approach to Some Problems in Using Land, Water, and Air*, Baltimore, Md. (Johns Hopkins), 1965, 96 pp. Paper. [65-24756]

Introduction—Five Environmental Problems; Water Pollution; Air Pollution; Use of Chemicals as Pesticides; Physical Environment of Urban Places; Problems Associated with the Use of Rural Areas; Research Strategy.

1033. Herman, Harold; McKay, Mary Elizabeth, *Community Health Services*, Washington, D.C. (International City Managers' Association), Second Edition, 1969, 250 pp.

Scope of Community Health Services; Growth of Community Responsibility for

Health Services; Progress toward Comprehensive Health Services; Health Status Measurement and Goal Setting; Health Department Management.

1034.   Holland, Lawrence B. (Editor), *Who Designs America?*, Garden City, N.Y. (Doubleday), 1966, 357 pp. Paperback. [66-17453]

Introduction (Lawrence B. Holland); Social Influence of Design (Susanne K. Langer); Politics of Design (John William Ward); Scale and Design in a New Environment (Boris Pushkarev); Quality in City Design (Kevin Lynch); New Forms of Community (Victor Gruen); Naked Utility and Visual Chorea (Charles Colbert); Impact of Political and Social Forces on Design in America (Edward J. Logue); Artifact As a Cultural Cipher (Richard S. Latham); Some Psychiatric Aspects of Design (Humphry Osmond); Further Reading, pp. 319-320.

1035.   Jellicoe, G. A., *Motopia—A Study in the Evolution of Urban Landscape*, New York (Praeger), 1961, 168 pp. [61-7040]

Town Geometry; Biological Landscape; Town Residence; Town Centre; Segregation of Traffic; Motopia.

1036.   Klein, Louis, *River Pollution, 3: Control*, Washington, D.C. (Butterworth), 1966, 499 pp. [66-6429]

Preface; Abbreviations; Detection and Measurement of River Pollution; Pollution of Tidal and Coastal Waters; Sewage Disposal and Purification; Disposal and Treatment of Trade Wastes; Standards for Rivers, Sewage Effluents and Trade Effluents; Present and Future Status of River Pollution. References to 1,845 written sources.

1037.   Kneese, Allen V., *Water Pollution: Economic Aspects & Research Needs*, Baltimore, Md. (Johns Hopkins), 1962, 107 pp. [62-12587]

"Viewing waste disposal as a problem in the economic theory of resource allocation, the author, using a basin-wide model, directs his attention to the orientation of the technical research effort in the framework of potential physical, economic and social 'payoff' on expanded effort."

1038.   Kostka, V. Joseph, *Neighborhood Planning*, Winnepeg, Manitoba (Community Planning, University of Manitoba), January 1957, 150 pp. Paperback.

Neighborhood Street Pattern; School and Recreation Areas; Neighborhood Shopping Center; Zoning; Density; Neighborhood Land Allocation; Mixed Developments; Layout Systems; Subdivision Design; Questions and Answers; Bibliography, p. 42.

1039.   Lewis, Howard R., *With Every Breath You Take*, The Poisons of Air Pollution, How They Are Injuring Our Health, and What We Must Do About Them, New York (Crown), 1965, 322 pp. [64-23821]

Trouble in the Air; What Price Polluted Air?; Challenges and Solutions; Plea for Tomorrow; Appendices: How They Came to Know the Air, How They Sample the Air, How to Learn More About Air Pollution.

1040.   Logan, John A.; Oppermann, Paul; Tucker, Norman E. (Editors), Proceedings of the First Conference on *Environmental Engineering &*

Metropolitan Planning, Evanston, Ill. (Northwestern University Press), 1962, 265 pp. [62-16248]

Introduction (John A. Logan); Health in Metropolitan Areas: Environmental Health Criteria for Metropolitan Areas (Robert J. Anderson), Planning Criteria for Metropolitan Areas (Dennis O'Harrow), Recreation in the Metropolitan Area (Mrs. Rollin Brown), Cost of Environmental Health (Bert W. Johnson); Environmental Problems: Surface Water Supply (Gordon E. McCallum), Ground Water Supply (William C. Ackermann), Sewerage and Drainage (A. L. Tholin), Liquid Waste Pollution (H. Loren Thompson; James R. Boydston), Solid Waste (Casimir A. Rogus), Air Pollution (John A. Maga); Metropolitan Problems: A Citizen Looks at Metropolitan Problems (Roy Sorenson), An Ex-Mayor Speaks from a New Vantage Point (Ivan A. Nestingen); Organizing the Attack: Use of Mathematics in Planning (Harold A. Thomas, Jr.), World Approach (Ernest Weissman), Metropolitan Area Approach (Paul Oppermann), Metropolitan Toronto Solution (M. V. Jones), Role for Government (Wesley E. Gilbertson), Goals and Means in Metropolitan Planning (Coleman Woodbury); Recommendations of the Conference.

1041. Lynch, Kevin, *The Image of the City*, Cambridge, Mass. (Technology Press & Harvard University Press), 1960, 194 pp. Paperback. [60-7362]

Image of the Environment: Legibility, Building the Image, Structure and Identity, Imageability; Three Cities: Boston, Jersey City, Los Angeles, Common Themes; City Image and Its Elements: Paths, Edges, Districts, Nodes, Landmarks, Element Interrelations, Shifting Image, Image Quality; City Form: Designing the Paths, Design of Other Elements, Form Qualities, Sense of the Whole, Metropolitan Form, Process of Design; A New Scale; Appendices: Some References to Orientation, Use of the Method, Two Examples of Analysis; Bibliography, pp. 182-186.

1042. McHarg, Ian, *Design With Nature*, New York (Museum of Natural History), 1969, 214 pp.

". . . how the principles of ecology can help solve the enormous environmental problems facing us today."

City and Countryside; Sea and Survival; Plight and Prospect; Step Forward; Cast and Capsule; Nature in the Metropolis; On Values; Response to Values; The World Is a Capsule; Naturalists; Processes as Values; River Basin; Metropolitan Region; Processes and Form; City, Process and Form; City, Health and Pathology.

1043. Moe, Edward O.; Ward, Delbert B.; Johnson, Morris E., *Where Not To Build—A Guide for Open Space Planning*, Bureau of Land Management, U.S. Department of the Interior, Technical Bulletin 1, Washington, D.C. (Superintendent of Documents), April 1968, 160 pp. Paper.

Theory of Open Space: Introduction, Case for Open Space, Concepts of Open Space, Classification of Open Spaces, Guides and Standards for Open Space Planning, Application of Open Space Planning; Washington County Case Study: Verification, Application of the Open Space Concepts to Washington County, Utah, Data Collection, Usefulness of Open Space Planning to Field Personnel—An Evaluation; Appendices.

1044. Northeastern Illinois Planning Commission, *Managing the Air Resource in Northeastern Illinois,* Technical Report No. 6, Chicago (400 West Madison St., 60606), 1967, 111 pp.

Summary of Findings and Recommendations; Air Pollution Sources and Volumes; Effects of Air Pollution; Transport of Pollution; Air Quality; Public Concern; Search for Solutions; Air Quality and Comprehensive Planning; Recommended Air Resource Management Program.

1045. Peterson, Arnold P. G.; Gross, Ervin E., Jr., *Handbook of Noise Measurement,* West Concord, Mass. (General Radio Company), Fifth Edition, 1963, 250 pp. Paperback.

Introduction; The Decibel—What Is It?; Man as a Noise-Measuring Instrument; Description of a Sound-Measuring System; Applications for a Sound-Measuring System; Measurement of Sound Level and Sound-Pressure Level; Noise Source Characteristics; Loudness, Speech Interference, Hearing Damage, and Neighborhood Reaction to Noise; Noise Control; Introduction to Vibration Measurements; Description of Vibration-Measuring Instruments; Applications for Vibration-Measuring Equipment; Vibration Measurement Techniques; Effects of Mechanical Vibration; Appendices.

1046. President's Science Advisory Committee, Environmental Pollution Panel, *Restoring the Quality of Our Environment,* Washington, D.C. (Superintendent of Documents), November 1965, 317 pp. Paper.

Introduction; Effects of Pollution: Health, on Other Living Organisms, Impairment of Water and Soil Resources, Polluting Effects of Detergents, Deterioration of Materials and Urban Environments, Climatic Effects of Pollution; Sources of Pollution: Municipal and Industrial Sewage, Wastes—Animal, Urban Solid, Mining, Consumer Goods, Unintentional Releases; In Which Directions Should We Go?; Recommendations: Principles, Actions, Coordination and Systems Studies, Baseline Measurement Programs, Development and Demonstration, Research, Manpower, Incompleteness of Recommendations; Appendices.

1047. Ridker, Ronald G., *Economic Costs of Air Pollution—Studies in Measurement,* New York (Praeger), 1967, 215 pp. [66-26571]

Introduction—Standard-Setting As a Frame of Reference; Strategies for Measuring the Cost of Air Pollution; Economic Costs of Diseases Associated With Air Pollution; Soiling and Materials-Damage Studies; Case Study of a Pollution Episode; Property Values and Air Pollution—Cross Section Study of One City, Time-Series Study; Some Lessons for Future Research; Appendices.

1048. Simonds, John O., *Landscape Architecture,* The Shaping of Man's Natural Environment, New York (Dodge), 1961, 244 pp. [60-53454]

"Outlines the landscape planning process from selection of the site to the completed project. A beautifully illustrated book and an articulate plea for intelligent site planning."

1049. Smithsonian Institution, *The Fitness of Man's Environment,* Publication No. 4728, Washington, D.C. (1000 Jefferson Dr., S.W., 20560), 1968, 250 pp. [68-20988]

Two Points of Philosophy and an Example (Paul Goodman); Natural History of Urbanism (Robert McC. Adams); Institutions and Their Corresponding Ideals (Wolfgang Braunfels); Sense of Place (Asa Briggs); Stewardship of the Earth (Bertrand de Jouvenel); Pastoral Ideals and City Troubles (Leo Marx); Why We Want Our Cities Ugly (Philip Johnson); Human Needs and Inhuman Cities (Edward T. Hall); Science and the City (Robert C. Wood); Conservation of Cultural Property (Hiroshi Daifuku); Values, Process and Form (Ian L. McHarg); Man and His Environment (René Dubos).

1050. Solow, Anatole; Copperman, Ann, *Planning the Neighborhood,* Chicago, Ill. (Public Administration Service), 1948, 90 pp. Paper.

Basic Requirements for Site Selection; Development of Land, Utilities and Services; Planning for Residential Facilities; Provision of Neighborhood Community Facilities; Layout For Vehicular and Pedestrian Circulation; Neighborhood Density—Coordination of Housing Elements; Appendices, Bibliography, pp. 80-86.

1051. Souder, James J.; Clark, Welden E.; Elkind, Jerome I.; Brown, Madison B., *Planning For Hospitals—A Systems Approach Using Computer-Aided Techniques,* Chicago, Ill. (American Hospital Association), 1964, 167 pp. [64-56469]

Planning Process; Conceptual Framework; Commerce Subsystem; Computer-Aided Planning.

1052. Spreiregen, Paul D., *Urban Design: The Architecture of Towns and Cities,* New York (McGraw-Hill), 1965, 256 pp. [65-25520]

". . . based on a series of 12 articles that appeared in the Journal of the American Institute of Architects from 1962 to 1964. . . . By far the best and most complete discussion of the principles, and practice of urban design yet to appear. While the classical examples of design in European cities and elsewhere are not neglected, the emphasis is on American cities. . . ."

Heritage of Urban Design; Roots of Our Modern Concepts; Making a Visual Survey; Some Basic Principles and Techniques; Examples and Scope; Urban Esthetics; Designing the Parts of the City; Residential Areas; Circulation; Regulation and Control; Government Programs; Comprehensive Role for Urban Design; Bibliography, pp. 231-238.

1053. Stewart, George, *Not So Rich As You Think,* Boston, Mass. (Houghton, Mifflin), 1967, 248 pp. [67-25450]

". . . about raw sewage and factory effluents flowing into lakes, rivers, and tidal waters; about mountains of garbage, junk, and litter that rise fastest where affluence is greatest; about agricultural and mineral refuse that is offensive to the eye and often prejudicial to health; about irritating smoke and killing smog; about incessant noise and 'the ultimates: CO_2 and atomic wastes.' "

1054. Stipe, Robert E. (Editor), *Perception and Environment: Foundations of Urban Design,* Chapel Hill, N.C. (Institute of Government, University of North Carolina), 1966, 111 pp.

". . . papers presented at [a seminar of professional people whose opinions about the design of cities and their impact on man as an animal are rarely sought] and a tightly-edited version of the dialogue that ensued among the various participants: a psychologist, a psychiatrist, an art historian, a painter, a musician, a cultural anthropologist, an architect, an urban designer, and a landscape architect . . . makes quite clear that responsibility for the creation of an amenable and more human urban environment lies not only upon the shoulders of those professionals traditionally associated with city design, but with many other disciplines as well."

1055. Tunnard, Christopher; Pushkarev, Boris, *Man-Made America: Chaos or Control?*, New Haven, Conn. (Yale University), 1963, 479 pp. [62-16243]

". . . discusses visual principles and aesthetic values of low-density residential areas, freeway design, industrial plants and commercial facilities, open space and recreational area design, and preservation of historic sites and areas."

1056. University of Georgia, Landscape Architecture Department, with Northeast Georgia Area Planning and Development Commission, *Madison: A Visual Survey and Civic Design Study*, Athens, Ga. (30601), 1964, 70 pp.

"This attempt to reconcile visually and functionally a historic remnant of the ante-bellum South with the mid-twentieth century involved a careful block-by-block survey, with consideration of individual houses, structures, and landmarks. . . . The study group drew up a detailed set of recommendations—graphic, technical and administrative. . . . A good example of the product of a group possessed with a 'feel' for its task."

1057. U.S. Department of Commerce, Panel on Electrically Powered Vehicles, *The Automobile & Air Pollution—A Program for Progress, Part II*, Washington, D.C. (Superintendent of Documents), December 1967, 160 pp. Paper.

Air Pollution; Current Automotive Systems; Energy Storage and Conversion Systems; The Automobile and the Economy; Automotive Energy Sources; Transportation System Requirements; Bibliography, pp. 151-160.

1058. U.S. Department of Health, Education, and Welfare, Task Force on Environmental Health and Related Problems, *A Strategy for a Livable Environment*, Washington, D.C. (Government Printing Office, 20402), June 1967, 90 pp.

Preface; Summary; Recommendations; Environment, Goals, Strategy; Environmental Protection Act; Appendices.

1059. U.S. Department of the Interior, Bureau of Mines, *Automobile Disposal: A National Problem*, Washington, D.C. (Superintendent of Documehts), 1967, 569 pp.

". . . information relating to the automobile wrecking industry, scrap processing and metal recovery, classes and grades of scrap, automobile sales and scrap markets, scrap iron and steel industry, independent collectors, prices, laws and regulations."

1060. U.S. Office of Science and Technology, Executive Office of the President, *Alleviation of Jet Aircraft Noise Near Airports,* Report of the Jet Aircraft Noise Panel, Washington, D.C. (Superintendent of Documents), March 1966, 167 pp.

Report of the Panel; Principal—Conclusions, Recommendations; Appendices: Physical and Psycho-Acoustic Measurements, Federally Sponsored Research, Developments in Engines, Airframes and Aircraft Utilization, Operational Procedures, General Economic Considerations, Land Utilization and Legal Problems in Noise Abatement.

1061. U.S. Public Health Service, *Environmental Health Planning Guide,* Public Health Service Publication No. 823, Washington, D.C. (Superintendent of Documents), September 1967, 99 pp. Paper.

Preparing for the Study; Collecting and Evaluating the Data; Using the Data; Implementation; Appendix.

1062. U.S. Public Health Service, *Planning for Health: As They See It— the Role of Health and Welfare Councils in Comprehensive Community Health Planning,* Publication No. 1488, Washington, D.C. (Division of Community Health Services, U.S. Department of Health, Education, and Welfare), June 1966, 16 pp.

Comprehensive Community Health Planning; Community Health and Welfare Council; Comprehensive Community Health Planning: Essentials, Organizational Framework, Geographic Area To Be Served, Authority, Role of Citizens, Manpower Training and Staff Experience, Techniques and Methods, Inherent Problems; Guidelines for Action.

1063. U.S. Public Health Service, *Procedures for Areawide Health Facility Planning—A Guide for Planning Agencies,* Public Health Service Publication No. 930-B-3, Washington, D.C. (Superintendent of Documents), September 1963, 118 pp. Paper.

Organization for Planning; Beginning To Function; Data Collection; Estimating Needs; Later Planning Activities; Appendices; References, pp. 115-116; Additional Bibliography, p. 117.

1064. Weddle, A. E. (Editor), *Techniques of Landscape Architecture,* New York (American Elsevier), 1967, 226 pp. [67-25058]

Site Planning (Sylvia Crowe); Practice of Landscape Architecture (Geoffrey Collens); Site Survey and Appreciation (Norman Clarke); Earthworks and Ground Modelling (Brian Hackett); Hard Surfaces (Timothy Cochrane); Enclosure (Frederick Gibbard); Outdoor Fittings and Furniture (Ian Purdy); Water (G. A. Jellicoe); General Planting (Patricia Booth); Grass (Ian Greenfield); Tree Planting (Brenda Colvin); Administration and Maintenance (J. T. Connell); Appendices.

1065. White House Conference on Natural Beauty, *Beauty for America,*

Washington, D.C. (Government Printing Office), 1965, 782 pp. Paperback. [65-65700]

Conference Call, The President's Message; General Session; Recreation Advisory Council, Open Meeting; Conference Panels: Federal-State-Local Partnership, Townscape, Parks and Open Spaces, Water and Waterfronts, Design of the Highway, Scenic Roads and Parkways, Roadside Control, Farm Landscape, Reclamation of the Landscape, Underground Installation of Utilities, Automobile Junkyards, New Suburbia, Landscape Action Program, Education, Citizen Action; Further Statements for the Record; Reports of Panel Chairmen; Response of the President.

1066. Whyte, William H., *The Last Landscape,* New York (Doubleday), 1968, 376 pp. [67-15350]

Introduction; Politics of Open Space; Devices: Police Power, Fee Simple, Easements, Tax Approach, Defending Open Space; Plans: Year 2000 Plans, Green Belts, Linkage, Design of Nature; Development: Cluster Development, New Towns, Project Look, Play Areas and Small Spaces; Landscape Action: Plan of the Landscape, Scenic Roads, Roadsides, Townscape; Design and Density: Case for Crowding, Last Landscape; Bibliography, pp. 355-363.

1067. Wolozin, Harold (Editor), *The Economics of Air Pollution,* New York (Norton), 1966, 318 pp. Paperback. [66-15319]

Preface (Gardner Ackley); Foreword (V. G. Mackenzie); Introduction (Harold Wolozin); Air Pollution—General Background and Some Economic Aspects (Allen V. Kneese); Economic Incentives in Air-Pollution Control (Edwin S. Mills); Risks Versus Costs in Environmental Health (Leslie A. Chambers); Structuring of Atmospheric Pollution Control Systems (Thomas D. Crocker); Strategies for Measuring the Cost of Air Pollution (Ronald G. Ridker); Use of Government Statistics in Air-Pollution Control (Edward T. Crowder); Use of Consumer-Expenditure Data in Air-Pollution Control (Helen H. Lamale); Air-Pollution Control in the Metropolitan Boston Area—Case Study in Public Policy Formation (Lester Goldner); Setting Criteria for Public Expenditures on Air-Pollution Abatement—Theoretical Foundations and Limitations (Harold Wolozin); Study of Pollution—Air, Staff Report to the Committee on Public Works, United States Senate, September, 1963; Bibliography, pp. 303-310.

1068. Wood, Samuel E.; Lembke, Daryl, *The Federal Threats to the California Landscape,* Sacramento, Cal. (California Tomorrow), 1967, 66 pp. Paper.

Federal Cities; Federal Landlords; Federally Subsidized Automobile; Grab for Water; Spread of Poison Over the Land; Water Pollution, Smog, Pesticides; Conclusions.

1069. Wood, Samuel E.; Heller, Alfred E., *The Phantom Cities of California,* Sacramento, Cal. (California Tomorrow), 1963, 66 pp. Paper.

Local Battle for Beauty—Small Efforts and Large Failures: Highway, Subdivision, Hill and Plain, Main Street and Elm Street; Local Battle for Beauty (Continued)— Pollution: Outdoor Advertising, Buildings and Freeways, Sewage, Smog and Insecticides; Phantom Cities of California: Unincorporated, Special-Interest, Contract, Seasonal, Legitimate, Regional; Home Rule or Home Ruin?

Bibliographies

1070. American Society of Planning Officials, *The Urban Planner in Health Planning*, Bibliography, pp. 71-78.

Health Care System: General Background and Major Policy Issues, Medical Manpower, Health Agencies—Public and Private, Hospitals and Health Facilities, Economics of Health, Health Legislation; Health Planning: Background, Theory and Methods, Mechanisms; Urban Planning and Health: Relationships, Selected Health Reports Prepared By Urban Planning Agencies; Annotated Bibliographies; Journals and Periodicals.

1071. Anderson, Stanford, *Planning for Diversity and Choice*, Bibliography: Environment, pp. 332-333.

1072. Carr, Donald E., *The Breath of Life*, Bibliography, pp. 159-175.

1073. Cohen, Alexander, "Location-Design Control of Transportation Noise," References, pp. 85-86.

1074. Dahir, James (Compiler), *The Neighborhood Unit Plan—Its Spread and Acceptance*, Additional Reading References, pp. 84-91.

Social Factor in City Planning; Changing City Patterns; Action For Cities; Further British and Other References.

1075. Dubos, René, *Man Adapting*, Bibliography, pp. 457-508.

1076. Goldsmith, John R.; Landow, Stephen A., "Carbon Monoxide and Human Health," References and Notes, p. 1359.

1077. Hall, Edward T., *The Hidden Dimension*, Bibliography, pp. 183-193.

Distance, space, perception, proximity.

1078. Holland, Lawrence B. (Editor), *Who Designs America?*, Further Reading, pp. 319-320.

1079. Klein, Louis, *River Pollution, 3: Control*, References to 1,845 written sources.

1080. Kryter, Karl D., "Sonic Booms from Supersonic Transport," References and Notes, p. 367.

1081. Loewenstein, Louis K., *The Location of Residences and Work Places in Urban Areas*, Bibliography, pp. 312-324.

1082. Lynch, Kevin, *The Image of the City*, Bibliography, pp. 182-186.

1083. Morgan, Candace D., "Aircraft and Industrial Noise, A Selected Bibliography," Chicago, Ill. (Municipal Reference Library), October 1968, 9 pp. Mimeo.

Noise Problem: General, Industrial and Transportation; Biological, Psychological and Structural Effects of Noise: General, Aircraft and Sonic Boom, Industrial; Noise Measurement: General, Aircraft and Sonic Boom, Industrial and Transportation;

Noise Control: General, Aircraft and Sonic Boom—Technical Aspects, Legal Aspects, Industrial and Transportation; Bibliographies.

1084. Spreiregan, Paul D., *Urban Design: The Architecture of Towns and Cities*, Bibliography, pp. 231-238.

1085. U.S. Department of Commerce, Panel on Electrically Powered Vehicles, *The Automobile & Air Pollution—A Program for Progress, Part II*, Bibliography, pp. 151-160.

1086. Wolozin, Harold (Editor), *The Economics of Air Pollution*, Bibliography, pp. 303-310.

Housing

Articles, Periodicals

1087. *American Builder* Special Report, "America's New Dream House Is Illegal," Vol. 88, September 1966, pp. 59-66.

Where the law allows, a great deal of useful outdoor living space can be developed on even a modest lot.

1088. Buel, Ronald A., "Creating a Slum—Banks, City, Landlords Contribute to Decay in Ghetto in Oakland," *The Wall Street Journal*, Pacific Coast Edition, Vol. LXXX, No. 21, Thursday, 30 January 1969, pp. 1 & 13.

Curbless Streets, Rundown Housing Deter Lenders; Absentee Owners Assailed; Tenants Draw Blame, Too; Worst Houses; No Loan Money; Streets Stay Dirty; Why No Riots, but . . . . ; Skepticism Remains; Ministers As Landlords; Southern Pacific's Role; Ned Reed's Tactics; Eviction Procedure; Lawsuit Is Filed.

1089. Carlson, David B., "Randhurst Center: Big Pinwheel on the Prairie," *Architectural Forum*, Vol. 117, No. 5, November 1962, pp. 107-110.

"Gracious shopping, with visual excitement, is Randhurst's lure for Chicago's fastest growing suburban areas."

1090. Everett, Robinson O.; Johnston, John D., Jr. (Special Editors), Housing, Part I: Perspectives and Problems, *Law and Contemporary Problems*, Vol. 32, No. 2, Spring 1967, pp. 187-370.

Population Pressure, Housing, and Habitat (Joseph J. Spengler); Assessment of National Housing Needs (Nathaniel S. Keith); Citizen Participation—Suburban Suggestion for the Central City (Richard F. Babcock; Fred P. Bosselman); Local Public Policy and the Residential Development Process (Edward J. Kaiser; Shirley F. Weiss); Public/Private Approaches to Urban Mortgage and Housing Problems (Saul B. Klaman); Influence of Mortgage Lenders on Building Design (John B. Halper); Housing, Computer, and Architectural Process (Arthur R. Cogswell); Mobile Homes—New Challenge (Frederick H. Bair, Jr.); Perfecting the Condominium as a Housing Tool—Innovations in Tort Liability and Insurance (Patrick J.

Rohan); Unions, Housing Costs, and National Labor Policy (Sylvester Petro); Slum Housing—Functional Analysis (George Sternlieb); Government and Slum Housing —Some General Considerations (Lawrence M. Friedman).

1091. Hartman, Chester, "The Housing of Relocated Families," *Journal of the American Institute of Planners,* Vol. XXX, No. 4, November 1964, pp. 266-286.

Where Did They Go?; What Kind of Housing Did They Move Into?: Type, Tenure; How Much Space Did They Secure?: Indoor Space, Outdoor Space; What Was the Quality of the Housing They Moved Into?; What Rents Are They Paying?; Racial Factors; Relocation Aid; Conclusion.

1092. Hartman, Chester, "The Limitations of Public Housing—Relocation Choices in a Working-Class Community," *Journal of the American Institute of Planners,* Vol. XXIX, No. 4, November 1963, pp. 283-296.

"Public housing is regarded as a major resource for rehousing families displaced by the urban renewal and highway programs. Yet a study of some 500 families relocated from Boston's West End reveals that the overwhelming majority refused to consider the possibility of living in a housing project. . . ."

1093. McQuade, Walter, "An Assembly-Line Answer to the Housing Crisis," *Fortune,* Vol. LXXIX, No. 5, 1 May 1969, pp. 98-103, 136, 138, 140.

"The cities are desperate for a way to get more housing built. Advocates of 'systems building' say they have one."

Like Stitching a Tent in a Windstorm; Less Need For Skill; Plans Across the Sea; Like Gigantic Shoe Boxes; Ahead of the Europeans Someday; An Enormous Housing Machine; To Overcome Fragmentation; Under Siege by Salesmen.

1094. Rainwater, Lee, "Fear and the House-As-Haven In the Lower Class," *Journal of the American Institute of Planners,* Vol. XXXII, No. 1, January 1966, pp. 23-31.

"The focus of interests and goals in relation to housing vary by social status even within the large group of families called working class. The most disadvantaged groups are concerned with shelter per se; traditional working class with opportunities to elaborate their dwellings in personally expressive ways; and the more prosperous working class with buying the 'standard all-American package.' The lower class seeks shelter from a wide variety of human and nonhuman threats, from which they fear consequences that combine elements of physical threat, disruption of familial and other interpersonal relations, and threats of moral damage to the self. This bears implications for the design of public housing."

1095. Ruth, Herman D.; Walker, Peter, "The Case for Planned Unit Development," *House & Home,* Vol. 22, No. 3, September 1962, pp. 123-129.

PUD Offers Developers and Builders New Design Opportunities; Everyone Gains from Planned Unit Development; PUD Experience in Two Cities in California Reveals Several Problems; A Case Study in Planned Unit Development: Westborough Homes.

1096. Schmitt, Robert C., "Implications of Density in Hong Kong," *Journal of the American Institute of Planners*, Vol. XXIX, No. 3, August 1963, pp. 210-217.

". . . Over-all density is 13 persons per gross acre, and individual neighborhoods exceed 2,800. Residential floor space averages 155 square feet per household or 32 square feet per occupant. Unlike congested areas in the United States, however, Hong Kong has relatively low death, disease, and social disorganization rates . . . density standards recommended by American planners may be unrealistic as requirements for public health and social welfare."

1097. Teicholz, Eric D., "Architecture and the Computer," *Architectural Forum*, Vol. 129, No. 2, September 1968, pp. 58-59.

". . . why the profession is not utilizing computer technology; and a description of one new program in the advancing technology."

1098. Turner, John C., "Barriers and Channels for Housing Development in Modernizing Countries," *Journal of the American Institute of Planners*, Vol. XXXIII, No. 3, May 1967, pp. 167-181.

"Many of the squatter communities of Latin America offer uniquely satisfactory opportunities for low income settlers. They are characterized by 'progressive development' by which families build their housing and their community in stages as their resources permit. . . . Official housing policies and projects . . . attempt to telescope the development process. . . . Such 'instant development' procedures aggravate the housing problem. . . ."

1099. Turner, John C., "Housing Priorities, Settlement Patterns, and Urban Development in Modernizing Countries," *Journal of the American Institute of Planners*, Vol. XXXIV, No. 6, November 1968, pp. 354-363.

"Attempts to plan the development of cities in modernizing countries and to improve housing standards are often confounded by the autonomous action of low-income squatters and clandestine developers. . . . A preliminary model of the settlement process in transitional countries is presented to illustrate the dynamics of this problem."

1100. U.S. Housing Administration, "Site Planning," Chap. III. in: _____, *Minimum Property Standards for Multifamily Housing*, Washington, D.C. (Superintendent of Documents), November 1963, pp. 35-65.

Land-Use Intensity; Floor Area Ratio; Open Space Ratio; Livability (Non-Vehicular) Space Ratio; Car Ratios; Land-Use Pattern; Yards; Courts; Grading Design; Driveways; Parking Areas and Courts; Walks; Steps and Stepped Ramps; Recreation Areas; Shopping Facilities; Other Outdoor Facilities; Outdoor Lighting; Planting.

1101. Vigier, François C., "An Experimental Approach to Urban Design," *Journal of the American Institute of Planners*, Vol. XXXI, No. 1, February 1965, pp. 21-31.

". . . attempts to clarify the role of urban design . . . through an examination of the applicability of standard cognitive testing techniques. . . . Drawing from the theory of perception as well as from a recent experiment, the author demonstrates what appears to be a close correlation between the spatial qualities of 'streets' and 'squares' and the response they elicit."

Perceptual Research and Theory; Experiment with Photographs of City Scenes; Observed Results; Implications for Urban Design.

Books, Reports, Pamphlets

1102. Abrams, Charles, *Man's Struggle for Shelter in an Urbanizing World,* Cambridge, Mass. (M.I.T.), 1964, 352 pp. [64-16506]

Population Inflation and Urban Invasion; Squatting and Squatters; The Urban Land Problem; Growth of Government Power and Policy; Obstacles to Progress in Housing; Problems of Administration and Personnel; Aid-Experts and "Inperts"; Economics and Housing; Background of Public Policies; Proposals for Solving the Problem; Problem of Finance; Self-Help, Core Housing, and Installment Construction; Roof-Loan Scheme; Education and Research: A University is Born in the Middle East; Development Planning and Housing; Role of Aid Programs; Land Idealogies and Land Policies; Communism, *Lebensraum,* Housing, and Cities.

1103. American Institute of Architects; North Georgia Chapter, American Society of Landscape Architects; Georgia Section, American Institute of Planners, *Improving the Mess We Live In,* Atlanta, Ga. (230 Spring St., N.W., 30303), 1965, 125 pp. Paper.

Utility Poles and Wires; Signs and Billboards; Service Stations; Commercial Areas; Landscaping; Art; Implementation.

1104. Aregger, Hans; Glaus, Otto, *Highrise Building and Urban Design,* New York (Praeger), 1967, 200 pp. [67-18826]

Highrise Building—Symbol of Our Time (Hans Aregger); Documentation—29 Examples; Analysis—22 Residential Highrise Blocks; Thoughts on Future Development (Otto Glaus).

1105. Bartley, Ernest R.; Bair, Frederick H., Jr., *Mobile Home Parks and Comprehensive Community Planning,* Gainesville, Fla. (University of Florida), 1960, 147 pp. Paperback. [60-62513]

Introduction; Mobile Home As a Living Unit; Mobile Home Park in the Comprehensive Plan; Regulatory Problem—Generally Considered; Zoning and the Mobile Home; Taxation and the Mobile Home; Summary and Conclusions.

1106. Bastlund, Knud, *José Luis Sert—Architecture, City Planning, Urban Design,* New York (Praeger), 1967, 244 pp. [66-21768]

1927-1939; 1939-1953; 1953-1965.

1107. Beazley, Elizabeth, *Design and Detail of the Space Between Buildings,* London, England (Architectural Press), 1960, 230 pp. [BNB 61-678]

"This very practical and useful manual presents and evaluates detailed information and design principles for paving materials, trim of paved surfaces, surface drainage, walls, fences, gates, parking, steps and ramps."

1108.   Beyer, Glenn H., *Housing and Society*, New York (Macmillan), 1965, 595 pp. [65-12480]

Background and Introduction: History of American Housing, American Family—Yesterday and Today, American Cities—Perspective of Their Growth; Housing Market and Production: Market, Residential Financing, Production; Acquisition and Consumption of Housing: Ownership, Design, Neighborhood; Housing Problems and Progress: Central Cities, Suburbia, Rural, Aged, History of Government's Role in Housing, Future Need and Housing Research; International Housing: Western Europe and United Kingdom, Developing Countries.

1109.   Chermayeff, Serge; Alexander, Christopher, *Community and Privacy —Toward a New Architecture of Humanism*, Garden City, N.Y. (Doubleday), 1963, 255 pp. Paperback. [63-10704]

Mass Culture: Background, Erosion of the Human Habitat, Vanishing Nature, Dissolving City, Suburban Flop, In Search of the Small, Enemy Number One—Car, Enemy Number Two—Noise, Faith and Reason; Urban Dwelling: Anatomy of Urbanism, Which Hierarchies, Problem Defined, Critical Appraisal, Anatomy of Privacy, New Planning Blocks.

1110.   Conrads, Ulrich; Sperlich, Hans G., *The Architecture of Fantasy: Utopian Building and Planning in Modern Times*, New York (Praeger), 1962, 187 pp. [62-8740]

"A study of the visionary in twentieth-century architecture that includes extensive analysis, documentation of critical appraisals, detailed descriptions, biographic data and bibliographic references, as well as copious illustrations."

1111.   Crosby, Theo, *Architecture: City Sense*, New York (Reinhold), 1965, 96 pp. Paperback. [65-14036]

Visual Order and Disorder; Cities; Order and Responsibility; More People with More Money; Traffic; Towards a New Urban Form; Value; Action Planning; Action Building; Towards a New Environment; Selected Bibliography, pp. 95-96.

1112.   Doxiadis, Constantinos A., *Architecture in Transition*, New York (Oxford), 1963, 199 pp. [63-6915]

Architectural Confusion; Epoch of Transition; Causes of the Crisis; Architects and Architecture; New Solutions for New Problems; Return to Universal Architecture; Laying the Foundations; Out of the Darkness.

1113.   Friedman, Lawrence M., *Government and Slum Housing—A Century of Frustration*, Chicago, Ill. (Rand McNally), 1968, 206 pp. [67-21414]

Some General Concepts; Housing Reform—Negative Style; Replacing the Slums—Alternatives; Urban Redevelopment and Renewal—Ultimate Weapon? Current Scene and Future Prospects; Some Thoughts on Welfare Legislation.

1114.   Gillies, James, *An Appraisal of Mobile Home Living—The Parks and the Residents*, Los Angeles, Calif. (Trailer Coach Association), undated, 15 pp. Paper.

Factors Influencing Social Patterns in Mobile Home Parks, Cross Section of Mobile Home Residents, Basic Characteristics of Mobile Home Residents; Mobile Home Parks and Municipal Costs and Revenues, Revenues and Costs—Mobile Home Parks as Compared to Single Family Residential Subdivisions; Typical Examples of Mobile Home Park Design; Facts and Figures.

1115. Grier, George and Eunice, *Equality and Beyond—Housing Segregation and the Goals of the Great Society*, Chicago (Quadrangle), 1966, 115 pp. Paperback. [65-27930]

Task at Hand; Spread of Segregation; Cost of Segregation; Upsurge of Civic Concern—Governmental Policy, Private Citizen Action; Beyond Non-Discrimination.

1116. Grigsby, William G., *Housing Markets and Public Policy*, Philadelphia, Penn. (University of Pennsylvania), 1963, 346 pp. [63-15010]

". . . examines the relationship between the private housing market and public renewal programs. . . . 'No one really knows whether a given [renewal] project will improve the total housing situation or merely create a better environment at one point while shifting and perhaps compounding the difficulty elsewhere.' . . . Public renewal efforts have focused on arresting and reversing the flight to the suburbs, rather than . . . the goal of the Federal Housing Act to provide a decent home for every American family. . . . Emphasizes the important role of the private housing market, and outlines public policies . . . essential if public action is to sustain and enlarge this market so that it meets a broader spectrum of American housing needs."

1117. Hoffmann, Hubert, *Row Houses and Cluster Houses: An International Survey*, New York (Praeger), 1967, 176 pp. [68-18903]

Introduction; Examples: Terrace Houses, Patio Houses, Terraced Houses; Key to Plans; Index to Architects. (English and German)

1118. Hoskin, Frank P., *The Language of Cities*, New York (Macmillan), 1968, 126 pp. [68-23635]

"The purpose of this book is to supply some of the tools for seeing the urban world in which most of us live today and to show . . . why it is important that we understand the urban environment which we see."

Introduction; How We See; What We See; Conclusion; Bibliography, pp. 125-126.

1119. Jensen, Rolf, *High Density Living*, London, England (Whitefriars Press), 1966, 245 pp. [66-12527]

Preface; General Statement and Town-Planning Considerations; Density; Economic Factors; Social Factors; Grouping, Layout and Detail Planning; Structure; Services; Tenancy and Ownership; Conclusions; Illustrated Schemes and Notes: Australia, Austria, Belgium, Canada, Denmark, Finland, France, Germany, Holland, Italy, Norway, South Africa, South America, Sweden, Switzerland, United Kingdom, United States; Tables and Diagrams.

1120. Katz, Robert D., *Design of the Housing Site—A Critique of American Practice*, Urbana, Ill. (Small Homes Council—Building Research Council, University of Illinois), 1966, 223 pp. Paper.

Nature of the Study; Housing Types and Residential Density; The Site; The Plan: Private and Communal Aspects, Functional Aspects; Innovations; Site Planning Regulations and Design Incentives; Conclusions.

1121. Katz, Robert D., *Intensity of Development and Livability of Multi-Family Housing Projects: Design Qualities of European and American Housing Projects*, Washington, D.C. (Government Printing Office), 1963, 115 pp.

". . . the study has as its purposes: to identify aspects of site design quality, and to determine the influence of housing intensity on these specific aspects of design and on project livability in general."

1122. Maisel, Sherman J.; Foley, Donald L. and Others, *California Housing Studies: Land Costs for Single-Family Housing, Housing Trends and Related Problems*, Berkeley, Cal. (Center for Planning and Development Research, University of California), 1963, 230 pp. Paper.

Background Information on Costs of Land for Single-Family Housing (Sherman J. Maisel): Changes in the Costs of Developed Lots, Relationships Between Lot Sizes (Density) and Costs, Outline Theory of the Factors Influencing the Prices of Raw Lands, Possible Reasons for Government Intervention in the Land Development Process, Demand for Houses and Lots, Future Land Requirements and Supply, Two Case Studies of Suburban Areas in the Process of Development; Housing Trends and Related Problems: Housing Supply—Progress and Limitations (Catherine Bauer Wurster), Shifting Metropolitan Spatial Patterns (Donald L. Foley), Housing of Special Groups (Wallace F. Smith), Future Trends and Housing Needs.

1123. Meyerson, Martin; Terrett, Barbara; Wheaton, William L. C., *Housing, People, and Cities*, New York (McGraw-Hill), 1962, 386 pp. [61-16532]

Setting: Housing—and Community Development, the National Economy, the Market; Consumer: Economic Behavior of Housing Consumers, The Disfranchised Consumer, What the Consumer Wants; The Producer: Housing Production—Actors and Actions, Labor's Role in Home Building; Industrializing the Housing Industry—Design, Materials, and Methods; The Investor: Residential Finance, Rehabilitating the Housing Stock, Potential Contribution of Rental Housing; The Federal Government: Federal Credit, Future of Federal Policy, Politics and Pressure Groups; The Community: Municipal Programs for Urban Development, Metropolitan Organization and Residential Expansion, Opportunities.

1124. Museum of Modern Art, *The New City: Architecture and Urban Renewal*, An Exhibition, Greenwich, Conn. (New York Graphic Society), 1967, 48 pp. [67-28526]

Perspective on Planning (Sidney J. Frigand); New Towns, New Cities (Elizabeth Kassler); Architecture and Urban Renewal (Arthur Drexler); Commissioned Project Proposals for Upper Manhattan; Cornell University, Columbia University, Princeton University, Massachusetts Institute of Technology.

1125. Musson, Noverre; Heusinkveld, Helen, *Building for the Elderly*, New York (Reinhold), 1963, 216 pp. [62-19489]

"In an authoritative survey of the housing requirements for the aged, using photographic examples from noteworthy projects throughout the U.S., the authors analyze the problems of both builders and occupants. They show how factors that are essentially architectural grow out of financial, sociological and philosophic aspects of the needs of this 10 per cent of the population. A section by Swedish architect Bo Boustedt suggests ways of de-institutionalizing homes for the aged. The book concludes with a check list of more than 275 items for consideration when planning this type of structure."

1126. Nulsen, Robert H., *All About Parks for Mobile Homes and Trailers,* Beverly Hills, Cal. (Trail-R-Club of America), 1968, 199 pp. Paperback.

Industry Facts; Classification of Mobile Home Parks; How To Select a Mobile Home Park; Parks That Sell Lots; Parks For the Trailer Traveler; How To Live in a Mobile Home Park.

1127. Paul, Samuel, *Apartments—Their Design and Development,* New York (Reinhold), 1967, 308 pp. [67-27430]

Introduction—Need and Purpose; Background; Building the Team; Search for Land; Types of Apartments; Financing (Abraham D. Leavitt); Program of Requirements; Design Process; Controls and Limitations; Site Planning; Designing the Buildings; Apartment Dwellings in Western Europe; Structural Design (Nicholas Farkas; Julian Karp); Mechanical and Electrical Engineering (Charles J. Wurmfeld); Sound Control (David J. Paul); Future Trends; Bibliography, pp. 297-298.

1128. President's Committee on Urban Housing, *A Decent Home,* Washington, D.C. (Superintendent of Documents), 1968, 252 pp. Paperback.

Introduction and Summary; Committee Report—Shape of the Nation's Housing Problems; Staff Studies: Shape of the Problem, Federal Housing Programs, Making Better Use of the Housing We Have, Building Houses; Summary of Committee and Staff Recommendations; Appendices.

1129. Rasmussen, Steen E., *Experiencing Architecture,* Cambridge, Mass. (M.I.T.), Second Edition, 1962, 245 pp. [62-21637]

"Profusely illustrated tour of some less familiar triumphs in architecture manages to convey much of the intellectual excitement of supurb design."

Basic Observations; Solids and Cavities in Architecture; Contrasting Effects of Solids and Cavities; Architecture Experienced as Color Planes; Scale and Proportion; Rhythm in Architecture; Textural Effects; Daylight in Architecture; Color in Architecture; Hearing Architecture.

1130. Schoor, Alvin L., *Slums and Social Insecurity,* Research Report No. 1, Division of Research and Statistics, Social Security Administration, U.S. Department of Health, Education, and Welfare, Washington, D.C. (Superintendent of Documents), 1963, 168 pp. Paper. [65-9373]

"Examines the effects of housing and blighted neighborhoods on the people who live in them, candidly discusses research findings on the social and psychological effects of inadequate housing, and comes to disturbing conclusions."

1131. Slitor, Richard E., *The Federal Income Tax in Relation to Housing,* Research Report No. 5, Washington, D.C. (National Commission on Urban Problems), 1968, 162 pp. Paper.

Introduction; Major Income Tax Mechanisms Affecting Housing Investment and Urban Development; Effect of Major Tax Mechanisms on Housing Investment and Urban Development; Major Proposals for Change in Federal Tax Provisions Affecting Urban Housing; Tax Incentives Versus Direct Approaches to Achieving Urban Housing Objectives; Conclusions and Recommendations; Appendix.

1132. Smith, Wallace F., *Housing for the Elderly in California,* Berkeley, Cal. (Real Estate Research Program, Institute of Business and Economic Research, University of California), 1961, 38 pp. Paper.

Housing Arrangements—Present and Future; Housing Preferences of the Elderly; Evaluation of Existing and Proposed Programs; Types of Housing Subsidies; Appendices: Determining the Number of Elderly California Households in 1950, Estimating 1950 Income Distributions, Methods of Projection, California Health Survey 1956, Relocation of Elderly People in an Urban Redevelopment Area.

1133. Smith, Wallace F., *The Low-Rise Speculative Apartment,* Research Report No. 25, Berkeley, Calif. (Center for Real Estate and Urban Economics, Institute of Urban and Regional Development, University of California), 1964, 136 pp. Paper.

"This study of recent apartment construction in a part of the San Francisco-Oakland metropolitan area was designed to clarify the interaction of three factors which the author believes 'explain' the . . . jump in apartment building: (1) a demographic shift, producing a larger demand for rental units; (2) conditions in the mortgage market, including a surfeit of funds available for loans coupled with a competitive struggle for high-yield loans; and (3) a complex of income tax provisions favorable to investment in multi-family dwellings. . . . A generalized case study of equity investment in multi-family housing."

1134. Sternlieb, George, *The Tenement Landlord,* New Brunswick, N.J. (Urban Studies Center, Rutgers University), 1966, 269 pp. Paper.

Introduction—Nature of the Problem; The City and Its Analysis; Sampling Methodology; Tenantry; Slum Operations and Profitability; Tenement Trading; Who Owns the Slums? A Profile: Single Parcel Owner, Attitudes of Owners, "Bad" and "Good Owners"; Financing Rehabilitation; Taxes and Slums; Government Policies for Action; Appendices: Area Tract Racial Changes 1940-50-60, Deteriorating or Dilapidated Dwelling Units 1960, Census Terminology and Sources, Parcel Choice Methodology, Interview Questionnaire, Interviewer's Manual.

1135. Wendt, Paul F., *Housing Policy: The Search for Solutions,* A Comparison of the United Kingdom, Sweden, West Germany, and the United States since World War II, Berkeley, Cal. (University of California), 1963, 283 pp. [62-11497]

"In addition to describing programs and policies . . . evaluates and compares each country's attempts to satisfy the needs and wants of its people, changes in housing standards, and the effects of housing policies on residential construction industries and postwar economies. . . . Government housing policies have focused on easing consumer cost, rather than reducing the cost of construction; policies are governed as much by politics as by economic factors. . . ."

1136. Wheaton, William L. C.; Milgram, Grace; Meyerson, Margy E. (Editors), *Urban Housing*, First Volume in Reader Series on Urban Planning and Development, New York (Free Press), 1966, 523 pp. [66-12082]

"The series . . . is primarily designed for students of city and regional planning. . . . The separate volumes should also serve as introductory volumes for specialists. . . ." Background; Housing: in the Neighborhood, Market; Requirements of Special Groups; Housing: Industry, Finance, Standards and Controls; Residential Renewal; Bibliography, pp. 487-500.

1137. Whiffen, Marcus (Editor), *The Architect and the City*, Cambridge, Mass. (M.I.T.), 1966, 173 pp. [66-23801]

Architect and the City (G. Holmes Perkins); Purpose of the City (John B. Jackson); Technology and Urban Form (Aaron Fleisher); Ecology of the City (Ian McHarg); Environmental Architecture (Victor Gruen); Urban Renewal Architect—Dr. Purist and Mr. Compromise (James J. Hurley); Architecture in Change (Romaldo Giurgola); Education for Urban Design (Jaqueline Tyrwhitt); Design from Knowledge, Not Belief (Barclay Jones); Education for Designing (Jesse Reichek); Summation for the Seminar (Thomas W. Mackesey).

1138. Wilner, Daniel M.; Walkley, Rosabell Price; Pinkerton, Thomas C.; Tayback, Matthew, *The Housing Environment and Family Life*, A Longitudinal Study of the Effects of Housing on Morbidity and Mental Health, Baltimore (Johns Hopkins), 1962, 338 pp. [62-10310]

". . . clouds some old theories about the strong relationship between housing and physical-social pathology. Six years and $500,000 were devoted to a meticulous analysis of what a better housing environment means in terms of physical and mental health, attitudes and morale, and performance of children at school."

1139. Woods, Shadrach, *Candilis • Josic • Woods—Building for People*, New York (Praeger), 1968, 226 pp. [67-29465]

Articulation of Function: Moroccan Housing, European Housing in North Africa, Moslem Housing, Low-Cost Housing in France, Various Buildings; Articulation of the Limits of Space: Elements of Buildings—Moslem Dwellings, European Housing in Low-Cost Apartments, Climatic and Technological Influence, Articulation of Specific Functions in Buildings, Limits of Buildings in Specific Situations, Limits of Space Formed by Buildings; Articulation of Volumes and Spaces: Number and Scale, Geometric Systems and Structures, External Expression of Cellular Units —Housing, External Expressions of Units in Various Buildings, Expression of Unity; Articulation of Public and Private Domains: within Buildings, in Spaces Formed by Buildings, in Large Scale Development, of the Public Domain, 'Organic' Systems and Structures; Biobliography.

Bibliographies

1140. Lynch, Kevin, *Site Planning*, References, pp. 241-243.

1141. Paul, Samuel, *Apartments—Their Design and Development*, Bibliography, pp. 297-298.

1142. Robinson, Ira M., *A Bibliography on Housing, Renewal and Community Development (Including New Communities)*, Los Angeles, Cal. (Graduate Program in Urban and Regional Planning, University of Southern California), Fall 1968, 18 pp. Mimeo.

General: Bibliography, Periodicals, Overviews, Some "Classics" in the Field; Setting: Housing in the National Economy, Housing Markets—Mechanism, Needs, Demand, Supply; Consumer: Behavior and Housing Preferences, Disfranchised—Minorities and Low-Income; Housing for the Elderly; Producer: General, Building, Labor, Technology, Mobile Homes; Investor: General, California; Government: History, Legislation and Proposed Legislation, Descriptions of and Comments on Specific Federal Programs, Guide to Other Federal Programs . . . , State and Regional Housing Programs, Local Housing Programs . . . ; Community: Public Housing and Relocation Housing, Urban Renewal—Purposes, Programs and Problems, Housing, Renewal and Social Planning, Housing Market and Needs Analysis—Survey Techniques, Housing Design and Functions, Planning and Programming, New Cities; The Future.

1143. United Nations, *Cumulative List of United Nations Documents and Publications Related to the Field of Housing, Building and Planning*, New York (Research, Training and Information Section, Centre for Housing, Building and Planning), November 1968, 49 pp.

General Assembly; Economic and Social Council; Social Development Commission; Committee on Housing, Building and Planning; Regional Economic Commissions; Specialized Agencies; Conferences, Seminars, Expert Groups' Meetings; Technical Assistance Experts' Reports.

1144. United Nations, Department of Economic and Social Affairs, *Housing, Building, Planning—An International Film Catalogue*, Publication Sales No.: 1956.IV.8, New York (Room 1059, 10017), 17 May 1956, 246 pp. Paper.

Analytical Subject Index of Films; Geographical Index of Films by Major World Regions and Countries; Alphabetical Catalogue of Films; Alphabetical Indexes: Film Producers and Their Addresses, Film Distributors and Their Addresses, Sources of Information, Film Titles.

1145. U.S. Department of Housing and Urban Development, Federal Housing Administration, *List of Technical Studies and Experimental Housing Projects*, FT/TS-1, Washington, D.C. 20410, 53 pp. Paper.

Technical Studies; Experimental Housing.

1146. Vance, Mary, *Housing for the Elderly*, Bibliography No. 27, Council of Planning Librarians, Monticello, Ill. (Exchange Bibliographies, P.O. Box 229, 61856), November 1963, 153 pp.

Bibliographies; Indexes; Lists, Reviews and Abstracts; Directories; Background on Aging and the Aged; Housing for the Elderly—General References; Community Facilities and Services; Economic Aspects; Design Requirements and Standards; Housing Types; Geographic Index; Author Index.

1147. Wheaton, William L. C.; Milgram, Grace; Meyerson, Margy E. (Editors), *Urban Housing*, Bibliography, pp. 487-500.

1148. Wilcoxen, Ralph, "A Short Bibliography On Megastructures," Exchange Bibliography No. 66, Council of Planning Librarians, Monticello, Ill. (P.O. Box 229, 61856), 1969, 18 pp. Mimeo.

Introduction; Bibliography: Beginnings, Items—American, Archigram, British, Canadian, French, German, Indian, Japanese, Russian, Popular Science; Science Fiction; Yona Friedman; Miscellany.

Land Use

Articles, Periodicals

1149. Abrams, Charles, "The Uses of Land in Cities," *Scientific American,* Vol. 213, No. 3, September 1965, pp. 150-156, 158, 160; also in: Scientific American (Periodical), *Cities,* New York (Knopf), 1965, 211 pp. Paperback. [65-28177]

"In cities all over the world land is used for specialized purposes such as housing and industry. One of the main problems of any city is how to control these uses to enable the city to function and evolve."

1150. Blumenfeld, Hans, "Are Land Use Patterns Predictable?", *Journal of the American Institute of Planners,* Vol. XXV, No. 2, May 1959, pp. 61-66.

". . . summarizes some of the studies that have identified key variables affecting urban growth patterns. The most consistent factor relating to land use distribution is sheer distance, and especially distance from the [Central Business District]. In Philadelphia and Toronto urban density patterns have followed almost identical patterns . . . giving support to the 'crest of the wave' theory of urban growth. . . ."

1151. *House and Home,* "Better Land Use: This Hillside Project Makes Another Strong Case for Cluster Planning," Vol. 29, No. 4, April 1966, pp. 94-99.

"Had it been developed like most other tracts in the same area, [this] Fullerton, Calif. project . . . would have been built on tiers of bare hillside pads . . . the project would have been economically unfeasible."

1152. Krasnowiecki, Jan; Strong, Ann Louise, "Compensable Regulations for Open Space—A Means of Controlling Urban Growth," *Journal of the American Institute of Planners*, Vol. XXIX, No. 2, May 1963, pp. 87-97.

". . . a novel technique for controlling the location and timing of urban development through the preservation of temporary and permanent open space."

1153. Levine, Lawrence, "Land Conservation in Metropolitan Areas," *Journal of the American Institute of Planners*, Vol. XXX, No. 3, August 1964, pp. 204-216.

Metropolitan Development Requires Open Spaces; Open Space Objectives; Conservation As a Planning Priority; Wetlands—An Important Conservation Requirement; Legislative Support for Conservation; Implementing a Conservation-Based Policy; Deficiencies of Local Governmental Processes; Fragmented Government Structure As an Obstacle; Shifting the Responsibility for Regulating Land Use; State Land Use Control; County Applications of the Hawaiian Approach; Districts or Authorities for Preserving Open Space; Case for State Regulatory Powers; Conclusion.

1154. Sawits, Murray, "Model for Branch Store Planning," *Harvard Business Review*, Vol. 45, No. 4, July-August 1967, pp. 140-143.

". . . determination of how much of a new suburban branch store's volume will represent new business."
Search for Solution; Prediction Model; Concluding Note.

1155. Wingo, Lowden, Jr., "The Use of Urban Land: Past, Present, and Future," in: *Land Use Policy and Problems in the United States*, Proceedings of the Homestead Centennial Symposium, Lincoln, Nebr. (University of Nebraska), 1963, pp. 231-254; also: Reprint Series, No. 39, Resources for the Future, Washington, D.C.

Urbanization—Pattern and Process; Urban Structure as a Variable; Changing Parameters; Some Reflections on Policy.

Books, Reports, Pamphlets

1156. Barton-Aschman Associates, *Highway and Land-Use Relationships in Interchange Areas*, Springfield, Ill. (Illinois Division of Highways, 62706), January 1968, 61 pp. Paper.

Introduction; Problems and Opportunities in Interchange Area Development; Approach to Interchange Area Planning and Development; Recommended Action Program; Appendix; Bibliography, pp. 56-61.

1157. Bauer, Anthony M., *Simultaneous Excavation and Rehabilitation of Sand and Gravel Sites*, Silver Spring, Md. (National Sand and Gravel Association), 1965, 60 pp.

Introduction; Survey: Deposit, Site Location, Operational Characteristics, Current Rehabilitation Practices; Planning Procedures: Analysis, Procedures and Recommendations, Case Studies, Conclusions; Bibliography, pp. 58-59.

1158. Berry, Brian J. L., *Commercial Structure and Commercial Blight*, Chicago, Ill. (Department of Geography, University of Chicago), 1963, 235 pp. Paper. [63-21862]

Overview; Structure; Models; Change; Blight; Planning; Bibliography, p. 221.

1159. Boley, Robert E., *Industrial Districts, Principles in Practice*, Technical Bulletin 44, Washington, D.C. (Urban Land Institute), December 1962, 197 pp. Paper.

Development of Industrial Districts: Role of the Developer, Other Factors, Industrial District Layout Considerations, Controls Imposed in Planned Industrial Districts, Summary—Development Considerations; Case Studies of Representative Industrial Districts in the U.S. and Canada: 1900-1929—Pioneer Districts, 1930-1949 —The Concept Grows, 1950-1960—Contemporary Developments: Industrial Estates, Districts, and "Parks," Balanced Community Developments, Research-Oriented "Parks," Industrial Urban Renewal Projects; Selected Bibliography, p. 197.

1160. Brush, John E.; Gauthier, Howard L., Jr., *Service Centers and Consumer Trips*, Studies on the Philadelphia Metropolitan Fringe, Department of Geography, University of Chicago, Chicago (Research Papers, 1101 E. 58th St., 60637), 1968, 182 pp. [67-25274]

Introduction; Origin and Evolution of Centers; Analysis of Service Centers; Analysis of Consumer Trips; Conclusions.

1161. Bucklin, Louis P., *Shopping Patterns in an Urban Area*, Research Program in Marketing, Berkeley, Cal. (Institute of Business and Economic Research, University of California), 1967, 154 pp. Paperback. [67-64131]

Urban Trading Area Problem; Market Shares for the Urban Retail Facilities; Demographic Influences; Some Aspects of Motivation—Products and Stores; Shopping Plan; Multivariate Analysis of Shopping Behavior; Trade Area Determination; A Theory of Urban Shopping; Appendices; References, pp. 152-154.

1162. Butler, George D., *Standards for Municipal Recreational Areas*, New York (National Recreation Association), Revised 1962, 38 pp.

"This study, based on an examination of more than 60 planning reports issued by national, state and local planning and recreation agencies in the past decade, presents a handy and concise summary of standards for recreation areas and facilities."

1163. Clawson, Marion, *Land for Americans*, Trends, Prospects, and Problems, Chicago (Rand McNally), 1963, 141 pp. Paperback. [63-17447]

Land: and You, to Live On, for Recreation; Cropland; Forests; On the Range; Odds and Ends; So What?

1164. Clawson, Marion; Held, R. Burnell; Stoddard, Charles H., *Land for the Future*, Publication of Resources for the Future, Inc., Baltimore (Johns Hopkins), 1960, 570 pp. [60-9917]

Introduction; Land in Time and Space; Urban Uses of Land; Land for Recreation;

Agricultural Land Use; Forestry as a Land Use; Land for Grazing; Miscellaneous Uses of Land; Future Land Use in the United States; Appendices: Review of Population Projection, Some General Urban Relationships, Classification of Urban-Like Areas, Statistics.

1165. Clawson, Marion; Stewart, Charles L., *Land Use Information: A Critical Survey of the U.S. Statistics Including Possibilities for Greater Uniformity*, Baltimore (Johns Hopkins), 1966, 402 pp. [66-14380]

". . . provides a history of land-use information gathering in the United States, describes the weaknesses in our present land-use data gathering and distribution system, and outlines the essential features of an improved system. . . . . suggests that progress . . . will require a permanent national organization to work on the problem."

1166. Deem, Warren H.; Reed, John S., *Airport Land Needs*, Cambridge, Mass. (Arthur D. Little, Inc.), 1966, 85 pp. Paper.

Summary; Consequences of Suburbanization for Airport Development; Financing Airport Development and Land Acquisition 1950-65; Airport Land Needs and Costs in the Next Decade; Acquiring Land to Meet Airport Needs During the Next Decade; Appendices: Cost Estimates of Land Required for Airports, NAP Cost Estimates of Land Required for Airport Development, Method for Calculating Ten-Year Cost Advantage of Advance Acquisition, Funds Obligated Under the Federal Grant in Aid Airport Program.

1167. Gruen, Victor; Smith, Larry, *Shopping Towns USA, The Planning of Shopping Centers*, New York (Reinhold), 1960, 288 pp. [60-8527]

Introduction; Prologue; Prerequisites: Developer, Location, Site, Zoning, Tenants, Financing; Planning: Team, Schedule, Site, Surrounding Areas, Growth, Traffic, Merchandising; Designing, Engineering, Leasing, Budgeting the Shopping Center; Case Studies; Completed Center: Opening and Promotion, Merchants Associations, Use of Public Areas; Future of Shopping Centers; Glossary; Bibliography, pp. 279-281.

1168. Haar, Charles M., *Land-Use Planning: A Casebook on the Use, Misuse, and Re-Use of Urban Land*, Boston, Mass. (Little, Brown), 1959, 790 pp. [58-13261]

"Examines the assumptions, doctrines, and implications of city planning law." The Land Around Us: Environment of City Planning; "Sic Utere Tuo . . .": Reconciliation by the Judiciary of Discordant Land Uses; Legislative Districting of Permissible Land Uses: "Euclidean" Zoning and Beyond; Regulating the Tempo and Sequence of Growth by Subdivision and Street Controls: The Rural-Urban Fringe; Government as Landowner and Redistributor: Eminent Domain, Planning, and the Central City; Public Role of Private Land Arrangements: Covenants, Easements, Conditions, and Other Restrictions; Planning, Housing and Real Estate Finance: Money Franchise and the Shift to Federal Dominance; Planning For What, How, and By Whom? The Sources of Decision-Making For the Urban Environment.

1169. International Council of Shopping Centers, Inc., *Enclosed Mall Shopping Centers*, New York (445 Park Ave., 10022), 1965, 60 pp.

"... in the form of questions and answers given at a conference session in 1964 and the answers are quite specific. The discussions cover conversion, construction, and operation."

1170. Jacksonville-Duval Area Planning Board, *The Land and Its Uses,* An Analysis of the Use of Land in Jacksonville and Duval County, Florida, Jacksonville, Fla. (830 American Heritage Bldg., 32202), April 1968, 57 pp. Paper.

Historical Perspective; Geography; Transportation; Economy; Duval County; Housing Conditions; Land Use; Land Use Methodology; Ring Sector Data; Statistical Tables.

1171. Jensen, David R., *Selecting Land Use for Sand and Gravel Sites,* Silver Spring, Md. (National Sand & Gravel Association), 1967, 65 pp. Paper.

Introduction; Land Use Selection and Site Development; Land Use Potentials; Summary; Appendix; References, pp. 64-65.

1172. Kane, Bernard J., Jr., *A Systematic Guide to Supermarket Location Analysis,* New York (Fairchild), 1966, 166 pp. [65-27050]

Urban Area Analysis: Urban Area, Urban Area Economy, Supermarket Trading Areas, Population, Income Levels and Food Store Expenditures, Street and Highway System, City Planning and Supermarket Development; Supermarket Competition, Location Types, and Site Characteristics: Competition Inventory and Analysis, Supermarket Saturation, Supermarket Location Types, Site Characteristics; Sales Volume Projection: Trading Area Data: Analogies and Volume Projections, Sales Volume Projections in Small Cities (Population under 25,000), ... in Large Cities (... over 25,000), Store Size and Parking Space Requirements; Broader Uses of Location Analysis Techniques: Location Analysis for Non-Food Retailers; Appendices.

1173. Kinnard, William N., *Industrial Real Estate,* Washington, D.C. (Society of Industrial Realtors, National Association of Real Estate Boards), 1967, 615 pp. [67-29938]

Economics of Industrial Real Estate: Field of Industrial Real Estate, Industrial Growth and the Demand for Industrial Space, Site Selection and Space Requirements of Industry, Meeting the Demand for Industrial Space; Industrial Real Estate as an Investment; Major Functions of the Industrial Real Estate Broker: Marketing Industrial Space, Industrial Real Estate Office Operation and Management; Supplementary Industrial Real Estate Activities: Industrial Real Estate Credit, Non-Credit Financing of Industrial Real Estate, Industrial Real Estate Leases, Rehabilitation and Conversion of Industrial Real Estate, Industrial Real Estate Valuation—General Principles, Industrial Real Estate Appraisal—Applications; Industrial Real Estate Development: Industrial Property Development and Pioneering, Development of Planned Industrial Districts and Parks, Marketing and Operating Planned Industrial Districts, Aids to Industrial Development; Selected References, pp. 600-601.

1174. Lederman, Alfred; Trachsel, Alfred, *Creative Playgrounds and Recreation Centers,* New York (Praeger), 1959, 176 pp. [59-7455]

"Numerous examples of how twelve countries have tackled the problem of building imaginative play centers. Deals with all aspects of creating modern recreation centers."

1175. Levin, Melvin R.; Grossman, David A., *Industrial Land Needs Through 1980*, Land Use Report 2, Boston, Mass. (Greater Boston Economic Study Committee), May 1962, 87 pp. Paper.

Patterns and Trends of Industrial Activity; Industrial Land Use and Zoning in Greater Boston; Industrial Employment in Greater Boston, 1970 and 1980, and Factors Influencing Its Distribution; Industrial Land Needs in Greater Boston Through 1980; Appendices.

1176. Loewenstein, Louis K., *The Location of Residences and Work Places in Urban Areas*, New York (Scarecrow), 1965, 331 pp. [65-13561]

Introduction; Location of Work Places in Urban Areas; Location of Residences in Urban Areas; Spatial Relationship Between Place of Work and Place of Residence in Urban Areas; Interzonal Trip Distribution Approach to a Study of Current Journey-to-Work Patterns; Summary and Conclusions; Bibliography, pp. 312-324.

1177. Los Angeles, Department of City Planning, *Analysis of the Land Use Characteristics*, Los Angeles, Cal. (City Hall, 90012), March 1968, 53 pp. Paper.

General Land Use Characteristics: Proportions, Densities, Absorption Coefficients, Conclusion, List of Cities Used in Analysis; Analysis of Land Characteristics in the City of Los Angeles: Population, Land Use—Residential, Commercial, Industrial, Agricultural, Recreation, Street Acreage, Other Developed Land Uses, Vacant or Undeveloped Land, Conclusion; Appendices.

1178. Nassau-Suffolk Regional Planning Board, *Existing Land Use*, Hauppauge, L.I., N.Y. (Veterans Memorial Highway, 11787), February 1968, 33 pp. Paper.

Introduction; Physical Characteristics; Analysis by Municipality; Bi-County Land Use Profile; Analysis of Major Land Uses: Residential, Commercial, Industrial Recreational.

1179. National Association of Home Builders; Urban Land Institute, *Innovations vs. Traditions in Community Development*, Technical Bulletin No. 47, Washington, D.C. (1200 18th St., N.W., 20036), 1963, 111 pp.

". . . a comparative analysis of innovative and conventional forms of residential building types and land development patterns. It cites specific case histories, analyzes alternative designs, and demonstrates the merits of cluster grouping, differing lot sizes and shapes, and density control zoning coupled with performance standards."

1180. Nelson, Richard L., *The Selection of Retail Locations*, New York (Dodge), 1958, 422 pp. [58-10539]

Influence of Location on Retailing; Selection of a Location; Technique of Estimating

Business Volume; What About Shopping Centers?; New Trends in the Economics of Location.

1181. Neutze, Max, *The Suburban Apartment Boom—Case Study of a Land Use Problem*, Baltimore, Md. (Johns Hopkins), 1968, 170 pp. Paperback. [69-10889]

Introduction; Some Reasons for the Apartment Boom; Apartment in the Suburbs; Apartments on the Suburban Fringe; Comparisons between Metropolitan Areas; Land Use Decisions and Imperfections in the Land Market; Optimal Role for Government in the Urban Land Market; Effects of Non-Optimal Public Policy; Summary.

1182. Niedercorn, John H.; Hearle, Edward F. R., *Recent Land-Use Trends in Forty-Eight Large American Cities*, Memorandum RM-3664-FF, Santa Monica, Cal. (RAND), June 1963, 37 pp. Paper.

Introduction; Land-Use Proportions; Land-Use Densities; Land-Absorption Coefficients; Conclusions; Appendices: Land-Use Data, Population and Employment Estimates.

1183. Norcross, Carl; Goodkin, Sanford, *Open Space Communities in the Market Place—A Survey of Public Acceptance*, Technical Bulletin 57, Washington, D.C. (Urban Land Institute), 1966, 97 pp. Paper. [66-30604]

Introduction; Overview of the Study; Who Buys or Rents in Open Space Communities; Why People Buy in Open Space Communities; Street Patterns and Land Planning; Open Space; Recreation; Golf; Other Community Features; Lakes and Water-Front Communities; Townhouses; Home Owners' Associations; Marketing Lessons; Brief Descriptions of the Communities Studied.

1184. Real Estate Research Corporation, *A Study of Airspace Utilization*, U.S. Department of Transportation, Federal Highway Administration, Bureau of Public Roads, Washington, D.C. (Superintendent of Documents), June 1968, 91 pp. Paper.

Assignment and Summary; Bases of Deduction; General Discussion; Economic Considerations; Major Issues; Policy and Procedure; Engineering Concepts; Legal Aspects; Conclusion; Architecture and Planning; Bibliography on Use of Freeway Airspace, pp. 69-71.

1185. San Mateo County, Regional Planning Committee, *Parks and Open Space*, A Program for San Mateo County, Redwood City (County Government Center), June 1968, 102 pp. Paper.

What We Need . . . and Why; Program; How To Make It Work; Maps; Appendices: RPC Parks and Open Space Program, Summaries of Subcommittee Reports, Elements of the Parks and Open Space Program, Inventory of Existing Recreational Facilities, Present and Future Recreation Needs, Sources of Funds, Program Cost Estimates.

1186. Traffic Research Corporation, *Review of Existing Land Use Forecast-*

ing Techniques, New York (441 Lexington Ave., 10017), 29 July 1963, 107 pp. Paper.

Summary and Introduction; Glossary of Technical Terms; Review of Land Use Forecasting Techniques: Operational or Quasi-Operational, Research-Oriented, Conceptual; Comments and Conclusions; References Pertaining to Existing Land Use Forecasting Techniques, pp. 83-93.

1187. TRW Systems Group, *California Regional Land Use Information System,* Redondo Beach, Cal. (One Space Park), 1968, 21 pp. Paper.

Objectives and Approach; Land Data Environment; Statewide Land Data Exchange; Development Program; Costs and Benefits; Look Into the Future; Principal Conclusions.

1188. Urban Land Institute, *The Dollars and Cents of Shopping Centers— Part Two: Tenant Characteristics,* Washington, D.C. (1200 18th St., N.W., 20036), 1962, 63 pp.

". . . an accurate study of the significance of the various kinds of tenants in shopping centers . . . based on a survey of 119 shopping centers of all sizes and included nearly 3,700 tenants units."

1189. Urban Land Institute, *New Approaches to Residential Land Development—A Study of Concepts and Innovations,* Technical Bulletin No. 40, Washington, D.C. (1200 18th St., N.W., 20036), January 1961, 151 pp. Paper.

Fields of Major Significance—Planned Unit Developments; Cluster Method of Residential Development; Town, Group, or Patio House; Flexible Zoning Controls; Improved Subdivision Regulations; Miscellaneous Innovations—Plans and Design; Miscellaneous Innovations.

1190. U.S. Federal Aviation Agency, *Planning the Airport Industrial Park,* AC NO:AC 150/5070-3, Washington, D.C. (Distribution Section, HQ-438, 20553), 1965, 57 pp. Paper.

Airport's Attraction for Industry; Historical Development of Airport Industrial Parks; Organization for Development; Physical Planning; Land Use Controls; Management and Operations; Federal and State Assistance; Bibliography of Reference Material; Appendix—Air Industrial Park Survey.

1191. U.S. National Aeronautics and Space Administration, *Study of the Optimum Use of Land Exposed to Aircraft Landing and Takeoff Noise,* Springfield, Va. (Clearinghouse for Federal Scientific and Technical Information, 22151), 1966, 140 pp.

". . . standards for regulating the aircraft noise problem—from building design criteria to noise level measurement . . . discusses in some detail the legal and administrative mechanisms which can be employed by a community to assure a harmonious relationship between the airport and surrounding uses. . . ."

1192. Whyte, William H., *Cluster Development,* New York (American Conservation Association), 1964, 130 pp. Paper. [64-18592]

Introduction; Economics of Cluster; Community's Attitude; Market Place; Legal Base; Homeowners Associations; Town House Developments; Linkage; Handling of Space; Appendix; A. Tabulation of Developments, B. New York State Cluster Enabling Act, C. Local Ordinances, D. Homeowners Association Forms.

1193. Williams, Edward A. (Director), *Open Space: The Choices Before California, The Urban Metropolitan Open Space Study*, San Francisco, Cal. (Diablo), 1969, 187 pp. Paperback.

Urban-Metropolitan Open-Space Study, 1965: Objectives and Scope of Study, Nature of the Open Space Problem in California, Study Areas, Methods of Preserving Open Space, Conduct of the Study; Appendices; Bibliography, pp. 179-187.

1194. Wingo, Lowden, Jr. (Editor), *Cities and Space—The Future Use of Urban Land*, Baltimore, Md. (Johns Hopkins), 1963, 261 pp. Paperback. [63-18694]

Urban Space in a Policy Perspective (Lowden Wingo, Jr.); Order in Diversity: Community Without Propinquity (Melvin M. Webber); The Importance of Open Space in the Urban Pattern (Stanley B. Tankel); The Form and Structure of the Future Urban Complex (Catherine Bauer Wurster); Urban Space and Urban Design (Frederick Gutheim); The Human Measure: Man and Family in Megalopolis (Leonard J. Duhl); Public Policy and the Space Economy of the City (Roland Artle); The Social Control of Urban Space (Charles M. Haar); Social Foresight and the Use of Urban Space (Henry Fagin).

Bibliographies

1195. Berry, Brian J. L., *Commercial Structure and Commercial Blight*, Bibliography, p. 221.

1196. Karl, Kenyon F., "Industrial Parks and Districts: An Annotated Bibliography," Exchange Bibliography No. 63, Council of Planning Librarians, Monticello, Ill. (P.O. Box 229, 61856), 1968, 14 pp. Mimeo.

General Discussion; Bibliography; Directories; Examples; Particular Aspects; Planning and Development; Planning and Development—International; Special Types of Industrial Districts; Foreign Trade Zones; Office Parks; Research Parks; Appendix.

1197. Real Estate Research Corporation, *A Study of Airspace Utilization*, Bibliography on Use of Freeway Airspace, pp. 69-71.

1198. Stevens, Benjamin H.; Brackett, Carolyn A., *Industrial Location—A Review and Annotated Bibliography of Theoretical, Empirical and Case Studies*, Bibliography Series No. 3, Philadelphia, Penn. (Regional Science Research Institute), 1967, 199 pp. Paperback.

User's Guide; Review of Recent Literature; Bibliography.

1199. Traffic Research Corporation, *Review of Existing Land Use Forecasting Techniques*, References Pertaining to Existing Land Use Forecasting Techniques, pp. 83-93.

Penn-Jersey Transportation Study, Philadelphia, Penn.; Rand Corporation, Santa Monica, Cal.; Chicago Area Transportation Study, Chicago, Ill.; Alan M. Voorhees, Charles F. Barnes, Jr., and Walter G. Hansen, Washington, D.C. and Hartford, Conn.; F. Stuart Chapin, Jr. and Shirley F. Weiss, University of North Carolina, Chapel Hill, N.C.; William L. Garrison, Northwestern University, Evanston, Ill.; Ira S. Lowry, Rand Corporation, Santa Monica, Cal.; Lowden Wingo, Jr., Resources for the Future, Inc., Washington, D.C.; William Alonso, Harvard University, Cambridge, Mass.; William B. Hansen, Boston Regional Planning Project, Boston, Mass.; Donald J. Bogue, University of Chicago, Chicago, Ill.; John S. de Cani, University of Pennsylvania, Philadelphia, Penn.; Edgar M. Horwood, University of Washington, Seattle, Wash.; William C. Pendleton, University of Pittsburgh, Pittsburgh, Penn.; Additional References.

1200. University of California at Los Angeles, Real Estate Research Program, *Industrial Location Bibliography*, Los Angeles, Cal. (Division of Research, Graduate School of Business Administration), July 1959, 82 pp.

Area Planning and Industrial Location; Bibliographies; Industrial Location: Factors, Financial Aspects, Foreign; General; Geography, Economic and Other; Industrial: Decentralization and Dispersion, Growth and Expansion; Industrial Location: Community and Community Promotion, Resources, Surveys, U.S., Regional, State, and City; Location Theory; National Security and Defense; Planned Industrial Communities and Districts; Relocation of Industry; Site Selection and Development; Small Business; Small Communities, Suburbanization and Non-Urban Areas; Specific Industries and Industrial Location; Taxation and Subsidization of Industry; Transportation and Industrial Location; Urban Redevelopment and Industrial Location; Zoning; Miscellaneous.

1201. Williams, Edward A. (Director), *Open Space: The Choices Before California, The Metropolitan Open Space Study*, Bibliography, pp. 179-187.

Transportation

Articles, Periodicals

1202. *Architectural Forum*, "Urbane Freeway," Vol. 129, No. 2, September 1968, pp. 68-73.

"For the first time in the history of the interstate program, the Crosstown [Chicago] has been conceived not just as a transportation artery, but as a tool for the immediate enhancement of the neighborhoods through which it will pass and as a framework for the rational development of an entire urban corridor in the future."

1203. Beller, William S., "Megalopolis Transportation: Attacking the Systems Problem," *Space/Aeronautics*, Vol. 48, No. 4, September 1967, pp. 72-84.

Needed: Quality, Not Quantity, Solutions; How Relevant Is Aero Space Experience?; Perspective of the Transit Problem; . . . and More Perspective; Systems Analysis Applied to Transportation; Too Long to Walk, Too Short to Drive; Urbmobile at Home on Rails or Asphalt; From Downtown to Airport in a "People Pod"; ACV Economics Call for Bigger Testbeds; Finally, a Concern for the Traffic Blight; Not Much Help from Detroit; Don't Just Build Roads, Improve 'Em; Electronics for Highway "Navigation."

1204. Bendtsen, P. H. "Traffic Generation," *Socio-Economic Planning Sciences*, An International Journal, Vol. 1, September 1967, pp. 33-42.

". . . the number of cars per 1,000 inhabitants in a new suburb depends on density and the income level . . . number of trips per day varies in different cities . . . number of trips attracted by 100 m^2 of shops and offices seems to be more constant . . . traffic to the central areas at least in American cities counted in persons has been decreasing since the car density reached 300 cars per 1,000 inhabitants. . . ."

1205. Chapman, John L., "How to Plan An Auto Accident," *The New York Times Magazine*, January 21, 1968, Section 6, pp. 26-27, 72, 74, 76.

". . . accident-research program being carried out . . . at the Institute of Transportation and Traffic Engineering, University of California at Los Angeles . . . combined medical-engineering study of actual accident cases . . . conducted . . . at the U.C.L.A. School of Medicine."

1206. Dyckman, John W., "Transportation in Cities," *Scientific American*, Vol. 213, No. 3, September 1965, pp. 162-173, 174; also in: Scientific American (Periodical), *Cities*, New York (Knopf), 1965, 211 pp. Paperback. [65-28177]

"Urban transportation has to do not only with moving people and goods into, out of and through the city but also with the spatial organization of all human activities within it."

1207. Edwards, L. K., "High-Speed Tube Transportation," *Scientific American*, Vol. 213, No. 2, August 1965, pp. 30-40.

"In which a system is proposed that would carry passengers from Boston to Washington in 90 minutes. Its vehicles would travel at speeds up to 300 miles an hour through dual evacuated tubes."

1208. Foster, Joseph A., "Impact of the Jumbo Jet on Community Planning," in: American Society of Planning Officials, *Planning 1967*, Selected Papers from the ASPO National Planning Conference, Houston, Texas, April 1-6, 1967, Chicago, Ill. (1313 East 60th St., 60637), pp. 246-252. Paperback. [39-3313]

". . . background on the jumbo jet aircraft and its position in future airline service plans . . . total planning of the airport and its community . . . establishment of land-use environments that are inherent to the airport's role as a center of community development . . ."

1209. Gessow, Alfred, "The Changing Helicopter," *Scientific American*, Vol. 216, No. 4, April 1967, pp. 38-46.

"The unique capabilities of this improbable flying machine make it adaptable to many purposes. Because of recent improvements future models will have more speed, maneuverability and carrying capacity."

1210. Gibbons, John W. (Editor), Highway Safety and Traffic Control, *The Annals*, Vol. 320, Philadelphia, Penn. (American Academy of Political and Social Science), 1958, pp. 1-141.

Automotive Transport in the United States (Wilfred Owen); Dimensions of the Traffic Safety Problem (David M. Baldwin); Action Program for Highway Safety (Norman Damon); Inventory and Appraisal (George C. Stewart); The Law Must Hurry to Catch Up (Louis R. Morony); Police, Prosecutors, and Judges (Franklin M. Kreml); Driver Licensing (James Stannard Baker); Education and the Motor Age (William C. Carr); Automotive Design Contribution to Highway Safety (Charles A. Chayne); Increased Safety—By Design (Edward H. Holmes); Human Behavior—Factor X (James L. Malfetti); Traffic and Safety in Tomorrow's Urban Areas (Sergei N. Grimm); The Military Looks at Traffic Safety (Stephen S. Jackson); Organizing for Safety (J. W. Bethea); Traffic Accident Trends in Europe and the British Isles (Denis O'Neil).

1211. Herman, Robert; Gardels, Keith, "Vehicular Traffic Flow," *Scientific American*, Vol. 209, No. 6, December 1963, pp. 35-43.

"The interaction of driver-car units on roads is now being studied by physical and mathematical methods. The results have already been used in the solution of traffic problems."

1212. Howard, Thomas E., "Rapid Excavation," *Scientific American*, Vol. 217, No. 5, November 1967, pp. 74-76, 81-85.

"Excavating from the surface is normally cheaper than tunneling underground, but its undesirable side effects and the evolution of excavating machines make tunneling increasingly attractive."

1213. Jensen, E. J.; Ellis, H. S., "Pipelines," *Scientific American*, Vol. 216, No. 1, January 1967, pp. 62-70, 72.

"A vast and rapidly growing network transports oil and gas in many parts of the world. A recent trend is the use of pipelines to move solids both in slurries and in capsules containing solid products."

1214. Leary, Frank, "Megalopolis Transportation: High-Speed Ground Transportation," *Space/Aeronautics*, Vol. 48, No. 4, September 1967, pp. 85-101.

Symptoms of Circulatory Disease; For HUD and DOT, a Critical Interface; Strip Corridor, Dumbbell, Polygon, and Hub; Technical Solutions: Only the beginning; the Steel-Wheel Railcar Will Stay; Grab a Minibus or a Maxitaxi; Vying for the 200-600-mi Haul; Is 200 mph on the Ground Too Fast?; Automatic Train Control;

One-Second Headway at 120 mph; Promise of the Linear Motor; Power Transmission Across the Air Gap; Travel in a Pneumatic Tube; Possibly a Psychological Block; Rejuvenation in Tunnel Technology; GM Still Backs its Hovair; Guideway Critical for ACVs; Transcontinental Tubeflight; Nontechnical Quicksands.

1215. Levin, Stuart M.; London, Michael P., "Megalopolis Transportation: Aircraft for the Short Haul," *Space/Aeronautics*, Vol. 48, No. 1, September 1967, pp. 102-115.

They Want To Fly; How Many V/STOLS?; Rough Test in the California Corridor; Best Mix: V/STOLS plus CTOLS; Big Planes Cheaper, but Fewer; From City Center to City Center, Fast; 'Copter Maintenance Can Be Curbed; Vulnerability During Hover; Smartest Money May Be on the Lift Engine; Less Noisy, but Closer to the Ear; Fans Moving from Wings to Fuselage; Help for the Pilot During Transition; Boosting the Lift; Today's ILS Can't Handle V/STOL Approach; No Sure Enroute Navigation, Either; By V/STOL to a Jetport—in the Ocean.

1216. *Life*, "Crisis of the Cluttered Air, Planes and Passengers Hit a Peak— and a Breaking Point," Vol. 65, No. 6, 9 August 1968, pp. 38-47.

1217. McCormick, Donald W., Jr., "A Lift in Life," *Steelways*, Vol. 24, No. 1, January-February 1968, pp. 18-21.

"Carrying 32 billion passengers a year, the elevator is one form of transportation that is truly mass and rapid in performance."

1218. Martin, James E., "San Diego Freeway," *California Highways and Public Works*, Vol. 42, No's. 1, 2, January-February 1962, pp. 45-59.

1965 Completion; Right-of-Way Cost; Construction Cost; Completed Projects; Oil Well Abandonment; Long Beach Oil Field; First Contract; Junk Removed; Projects Under Construction . . . ; Operation Sky-Hook; Fill Stockpiling; Material Sources Listed; Freeway Agreements; Additional Moneys Voted; Freeway Financed.

1219. Ministry of Transport, "Traffic in Towns," *Ekistics*, Vol. 18, No. 105, August 1964, pp. 49-64.

Abstract from the report of the Working Group (Colin Buchanan) to the Ministry of Transport, Great Britain.

Introductory: Slovenly Disregard; Theoretical Basis: Road Patterns, Traffic Architecture; Practical Studies: Newbury, Leeds Central Area, Norwich, Accessibility Limited, Tottenham Court Road (London) Redevelopment—Total, Partial, Minimum; Conclusions: Question of Investment, Static or Expanding Towns, Parking Policy, Peak Periods, Scale of Primary Roads, Basis for an Integrated Policy, Comprehensive Redevelopment, Professional Collaboration, Further Research Required.

1220. Morris, Robert L.; Zisman, S. B., "The Pedestrian, Downtown, and the Planner," *Journal of the American Institute of Planners*, Vol. XXVII, No. 3, August 1962, pp. 152-158.

Pedestrian Downtown: Why Do People Walk?; Pedestrian Patterns; Planning for the Pedestrian: Routes and Breaks, Boulevard and Vista, Pedestrian and Vehicle, Sidewalk, Street; Challenge to the Planner of Downtown.

1221. *Operations Research,* Special Transportation Science Issue, Vol. 12, No. 6, November-December 1964, pp. 807-1076.

Twelve articles. The largest number deal with delays and congestion characteristics of four-arm intersections (Kleinecke; Gazis), merging intersections (Evans-Herman- Weiss; Jewell), vehicular-controlled intersections (Dunne-Potts; Darroch-Newell-Morris), and the control of a sequence of synchronized signals (Morgan-Little). Remaining papers include results of bus-following experiments (Rothery-Silver-Herman-Torner), the flow reduction aspects of removing arcs in a transportation network (Wollmer), economic justification for establishing runway service priorities in an air traffic control system (Pestalozzi), mathematical characteristics of two traffic systems whose service times increase with their occupancy (Helly), and the distribution of certain measures of travel in a circular city (Haight). Bibliography (Haight), pp. 976-1039.

1222. Perraton, Jean K., "Planning for the Cyclist in Urban Areas," *The Town Planning Review,* Vol. 39, No. 2, July 1968, pp. 149-162.

Cycling and Cycleways; Cycleway Planning; Conclusion.

1223. Quarmby, D. A., "Relating Public Transport to Urban Planning," *Journal of the Institute of Transport,* Vol. 30, No. 12, 1964, pp. 435-440.

"The role of public transport in relieving traffic congestion is examined. . . . A Study of congestion behavior shows that the acceptable level of congestion (to the commuter) depends to some extent on the relative attractiveness and convenience of other forms of transport. Based on this hypothesis, the derivation of a model of congestion behavior relative to public transport system is explained."

1224. Rice, Richard A., "Fast New Trains Could Unclog Megalopolis," *University,* A Princeton Quarterly, No. 30, Fall 1966, pp. 19-22.

". . . true high-speed rail service will only become feasible when the roadway and the trains become technologically integrated and when schedules and costs make it convenient and economical."

1225. Schwartz, Robert L., "The Case for Fast Drivers," *Harpers,* Vol. 227, No. 1360, September 1963, pp. 65-70.

"Contrary to common belief, they are not the worst menace on the road . . . low speed limits don't really reduce accidents . . . and 'safe-driving' campaigns may actually make our highways more dangerous."

1226. Turpin, Robert D., "Evaluation of Photogrammetry and Photographic Interpretation for Use in Transportation Planning," *Photogrammetric Engineering,* Vol. 30, No. 1, January 1964, pp. 124-130.

"The complexity of planning emphasizes the strong need for a method of evaluation . . . comprehensive rather than merely a means of producing masses of uncorrelated, individual studies, each based on a different set of conditions or even on different premises. . . . Ground-inventory methods of data collection are hampered by the problem of correlation and the impossibility of re-creation of conditions in existence at the time the data were collected. . . ."

1227. Warburton, Ralph (Guest Editor), New Concepts in Urban Transportation Systems, *Journal of The Franklin Institute*, Special Issue, Vol. 286, No. 5, November 1968, pp. 377-552.

Preface (Athelstan Spilhaus); Federal Role (Ralph Warburton); System Design: Impact of Rapid Transit on Atlanta (Leon S. Eplan), Transit Architecture—Seattle and New York (Rai Okamoto), Design Procedures in the Bay Area (Tallie B. Maule, John E. Burchard), Market-East Transportation Mall (R. Damon Childs), Central City Aerial Distribution—Pittsburgh (Walter A. Netsch, Jr.); Component Design: Orientation in the Transit Environment (Peter Chermayeff), Hub of the New Philadelphia (Vincent G. Kling), Los Angeles La Brea Station (P. J. Iovin), Moving Masses (Carl W. Sundberg), Transit Expressway (George Prytula), Puget Sound Ferries (A. F. Eikum), Urban Transporters as Human Environments (Robert Gutman); Conclusion: Urban Systems Design (Ralph Warburton).

Books, Reports, Pamphlets

1228. Ashton, Winifred D., *The Theory of Road Traffic Flow*, New York (Wiley), 1966, 178 pp. [66-72680]

"A general mathematical background including some knowledge of statistical distributions and queueing theory is desirable. . . ."

Fundamental Diagram of Traffic I; Dynamical Theories—Follow-the-Leader Models of Traffic Flow; Kinematic Theories and Fluid Analogies; Fundamental Diagram of Traffic II; From Traffic Signals to Traffic Cybernetics; Statistical Distributions and Queueing Theory; Stochastic Approach to Static Delay Problems; Simulation of Traffic Problems; Accident Statistics.

1229. Automotive Safety Foundation, *Urban Transit Development—In Twenty Major Cities*, Washington, D.C. (200 Ring Building, 20036), March 1968, 73 pp. Paper.

Introduction; Changes Affecting Transit; Transit Vehicle Development; Future Plans for Transit; New York; Los Angeles; Chicago; Philadelphia; Detroit; San Francisco; Boston; Montreal; Washington, D.C.; Pittsburgh; Cleveland; Toronto; St. Louis; Baltimore; Houston; Buffalo; Dallas; Seattle; New Orleans; Atlanta; Summary Table.

1230. Baerwald, John E. (Editor), *Traffic Engineering Handbook*, Washington, D.C. (Institute of Traffic Engineers), 1965, 770 pp. [65-17560]

Vehicle, Highway and Travel Facts; Vehicle Operating Characteristics; Driver; Pedestrian; Traffic: Characteristics, Accidents, Studies; Highway Capacity and Levels of Service; Traffic: Markings and Markers, Signing, Signalization; Parking; Loading and Unloading; Speed Regulation; Other Operational Controls; Roadway Lighting; Geometric Design of Roadways; Long Range Urban Traffic Planning; Traffic Engineering Administration.

1231. Barry, Walter A. (Committee Chairman), *Traffic Planning and Other Considerations for Pedestrian Malls*, Washington, D.C. (Institute of Traffic Engineers), October 1966, 66 pp. Paper.

Traffic Improvements for Pedestrians; Planning a Successful Mall; Traffic Circulation; Parking; Transit; Trucks and Freight Distribution; Pedestrians and Pedestrian Features; Police and Fire Protection and Utilities; Additional Guidelines; Implementation; Selected References, pp. 65-66.

1232. Berry, Donald S.; Blomme, George W.; Shuldiner, Paul W.; Jones, John Hugh, *The Technology of Urban Transportation*, Evanston, Ill. (The Transportation Center, Northwestern University), 1963, 145 pp. [63-11943]

Automotive Transportation; Transit Operations; Transit Equipment; Rapid Transit Innovations; Other Possible Innovations; Central Area Transportation Facilities; Comparisons of Systems; Conclusions.

1233. Bethlehem Steel Corporation, *Fabrication and Erection of Steel Elevated Structures for the Westinghouse Transit Expressway*, South Park Demonstration Project, Allegheny County, Pennsylvania, Bethlehem, Penn., June 1966, 12 pp. Paper.

Typical Structural Drawings; Superstructure Shop Fabrication; Erection of Steelwork; Plan of the Experimental Loop; Transit Expressway in Operation.

1234. Branch, Melville C., *Transportation Developments, Cities and Planning*, Chicago (American Society of Planning Officials), 1965, 29 pp.

Changing Context of Urban Transportation; Transportation Developments: Land—Railroads, Railroad-Automobile Combination, Automotive, Air—Superspeed Aircraft and Missiles, Supersize Aircraft, Vertical-Lift Aircraft, Winged Ground-Effect Machines, Individual Flight, Water—Hydrofoil Vessels, Air-Cushion Craft, Subsurface Vessels, Underground—Pipelines, Tunnels; Other Transportation Developments; Comprehensive Urban Analysis.

1235. Buchanan, Colin, *Traffic in Towns*, The Specially Shortened Edition of the Buchanan Report, S-228, Middlesex, England (Penguin), 1964, 263 pp.

Working Context: Frustration in the Use of Vehicles, Accidents, Deterioration of Environment, Future of the Motor Vehicle, Future Growth of Traffic, Form of Urban Areas; Theoretical Basis: Nature of Urban Traffic, Essence of the Problem, Design, Conclusion; Practical Studies: Small Town—Newbury, Large Town—Leeds, Historic Town—Norwich, Central London Block; Some Lessons from Current Practice: Britain, Europe, United States; General Conclusions.

1236. California Division of Highways, *Los Angeles Regional Transportation Study, Volume 1, Base Year Report 1960*, Los Angeles, Cal. (Advance Planning Department), December 1963, 61 pp.

Introduction and Summary; Organization and Financing; Present and Historical Background of the LARTS Study Area: Topography, Climate, Transportation, People and Economy of the LARTS Area; Methods and Results: Study Methods, Study Results.

1237. Carroll, J. Douglas, Jr., *Chicago Area Transportation Study*, Final

Report, Vol. 1, *Survey Findings,* Chicago (City of Chicago), December 1959, 126 pp.

Introduction; Design of the Study; Land Use; Amounts and Characteristics of Travel; Trip Generation; Supply and Use of Transportation Facilities; Summary and Conclusion; Appendices.

1238.   Ciampi, Mario J. & Associates; Warnecke, John Carl & Associates, *Market Street Design Plan,* Summary Report—November 6, 1967, San Francisco, Cal. (City and County of San Francisco), 1968, 27 pp.

Introduction; Design Framework: Historical, Regional, and Urban Setting; Design Plan: Design Principles, Plan, Pedestrian Environment, Powell Station Plaza, Civic Center Station Plaza, Street Furniture and Equipment; Action Plan; Phasing Actions, Financing, Costs.

1239.   Citizens Advisory Council on Public Transportation, *Improving Public Transportation in Los Angeles,* Los Angeles, Calif. (1226 South Flower St., 90015), 1967, 153 pp. Paper.

Introduction; Summary and Conclusions; Is There a Need for Improved Public Transportation in Los Angeles?: Economic Growth Trends, Motor Vehicle Trends, County Freeway Program, Problems with the Motor Vehicle System, Conclusions; How Should Public Transportation Be Improved?: Alternative No. 1—Wait for a Technological "Breakthrough," Alternative No. 2—Proceed with "Known" Solution, Conclusions; Who Should Pay for Improved Public Transportation?: Measuring the Benefits of Improved Public Transportation, Allocating the Costs of the Proposed 64-Mile System, Conclusions; What Steps Are Required to Implement an Improvement Program?; Appendices.

1240.   Clare, Kenneth G., *Southern California Regional Airport Study,* South Pasadena, Cal. (Stanford Research Institute), 1964, 274 pp.

Introduction; Summary and Conclusions; Determinants of Future Airport Development Requirements; Projection of Future Air Carrier Activity; Projection of Future General Aviation Activity; Assessment of the Adequacy of Existing Air Carrier Airports and Specification of Future Development Requirements; Assessment of the Adequacy of Existing General Aviation Airports and Identification of Future Development Requirements; Capital Needs and Possible Sources of Funds; Organization and Administration of Airports.

1241.   Cohen, Lawrence B., *Work Staggering for Traffic Relief—An Analysis of Manhattan's Central Business District,* New York (Praeger), 1968, 646 pp. [67-25240]

Introduction to Work Staggering; How Much Congestion? How Much Relief?; CBD Industry and Work Schedules; Modifiability of Work Schedules—Functional Capability; Acceptability of Schedule Change; Forecasting Cordon Counts; Feasible Work Schedules; Feasibility of Work Staggering; Appendices.

1242.   Cole, Leon Monroe (Editor); Merritt, Harold W. (Technical Editor), *Tomorrow's Transportation, New Systems for the Urban Future,* Washington, D.C. (Office of Metropolitan Development, Urban

Transportation Administration, U.S. Department of Housing and Urban Development), 1968, 100 pp. Paperback. [68-61300]

Summary; Urbanization and Urban Transportation; Federal Role and Responsibility; What Should Be Done: Findings and Conclusions, Strategy for Action, Immediate Improvements for Present Urban Transportation Systems, New Systems for the Future, Summary—Urban Problems and Program Recommendations; Recommended Research and Development Program; Bibliography of New Systems Study Project Reports, pp. 92-95.

1243. Cook, Walter L., *Bike Trails and Facilities: A Guide to Their Design, Construction, and Operation*, Wheeling, W.Va. (American Institute of Park Executives), May 1965, 52 pp.

Introduction; Bicycle Facilities; Where We Stand; Getting Started; Bikeways; Development Ideas; Construction; Rental Service; Programs and Activities.

1244. Detroit Metropolitan Area Regional Planning Commission, *Environs Study and Plan, Detroit Metropolitan Wayne County Airport*, Detroit, Mich. (800 Cadillac Square Bldg., 48226), 66 pp. Paper.

Introduction; Existing Condition and Growth Potentials; Aircraft Noise Study and the Determination of Affected Areas in the Environs of Detroit Metropolitan Airport; Development of the Land Use Plan; Summary and Conclusions; Appendices; Tables; Maps; Figures.

1245. District of Columbia, Government; Washington Metropolitan Area Transit Commission, *The Minibus in Washington, D.C.*, Final Report on a Mass Transportation Demonstration Project DC-MTD-2, Washington, D.C. (1634 Eye St., N.W., 20006), May 1965, 71 pp. Paper.

Summary; Background; Selection of a Vehicle; Minibus Operations; Impact on Traffic; Impact on Behavior and Attitudes; Exhibits; Tables.

1246. Fitch, Lyle C., *Urban Transportation and Public Policy*, San Francisco, Calif. (Chandler), 1964, 288 pp. [64-15743]

Summary and Recommendations; Urban-Transportation Problem; Urban Travel and Mass Transportation; Urban-Transportation Planning, Policy-Making, and Implementation; Economic Considerations in Urban-Transportation Planning; Technological Potentialities for Improving Urban Transportation; Forms of Financial Assistance for Urban-Transportation Development; Development of Federal Policy on Urban Transportation.

1247. Goldstein, Sidney; Thiel, Floyd, *Highways and Economic and Social Changes*, Washington, D.C. (Economics and Requirements Division, Office of Research and Development, Bureau of Public Roads, U.S. Department of Commerce), November 1964, 221 pp. Paper.

Introduction; Highway Influence on Urban Land Development and Land Values; Highway Effects on Industry, Commerce, and Services; Rural Effects of Highways; Highways and Recreation; Additional Groups Benefiting from Highway Improvements.

1248. Haar, Charles M. (Director), *Tomorrow's Transportation, New Systems for the Urban Future*, Washington, D.C. (U.S. Department of Housing and Urban Development), 1968, 100 pp. [68-61300]

Summary; Urbanization and Urban Transportation: Urban Framework, Public Transportation Today, Trends in Automobile Use, Urban Traveler, Time for Action; Federal Role and Responsibility; What Should Be Done: Findings and Conclusions, Strategy for Action, Immediate Improvements for Present Urban Transportation Systems, New Systems for the Future, Summary—Urban Problems and Program, Recommendations; Recommended Research and Development Program.

1249. Haight, Frank A., *Mathematical Theories of Traffic Flow*, New York (Academic Press), 1963, 242 pp. [63-15033]

Probability and Statistics; Theory of Queues; Fundamental Characteristics of Road Traffic; Arrangement of Cars on a Road; Simple Delay Problem; Miscellaneous Problems; Two Lane Road.

1250. Halprin, Lawrence, *Freeways*, New York (Reinhold), 1966, 160 pp. [66-22687]

My Kingdom for a Horse; Highwayman in the Country; Freeway in the New Town; Confrontation; Functions of Urban Freeways; Evaluation of Freeway Types; Change and Movement in the City; Transportation and Urban Design.

1251. Highway Research Board, Traffic Origin-and-Destination Studies— Appraisal of Methods, *Highway Research Board Bulletin 253*, Washington, D.C. (2101 Constitution Ave.), 1960, pp. 1-188.

Estimating Efficient Spacing for Arterials and Expressways (Roger L. Creighton; Irving Hoch; Morton Schneider; Hyman Joseph); Generation of Person Trips by Areas Within the Central Business District (B. C. S. Harper; H. M. Edwards); Tests of Interactance Formulas Derived from O-D Data (F. Houston Wynn; C. Eric Linder); Interpretation of Desire Line Charts Made on a Cartographatron (J. Douglas Carroll, Jr.; Garred P. Jones); Continuing Traffic Study: Methods of Keeping O-D Data Up-to-Date (Albert J. Mayer; Robert B. Smock); Appraisal of Sample Size Based on Phoenix O-D Survey Data (Arthur B. Sosslau; Glenn E. Brokke); Panel Discussion on Inter-Area Travel Formulas (John T. Lynch; Glenn E. Brokke; Alan M. Voorhees; Morton Schneider); Multiple Screenline Study to Determine Statewide Traffic Patterns (Mark K. Green); Land Use Forecasting for Transportation Planning (Walter G. Hansen); New York Port Authority's 1958 O-D Survey Using Continuous Sampling (Warren B. Lovejoy); Theoretical Prediction of Work-Trip Patterns (Robert T. Howe); Critique of Home-Interview Type O-D Surveys in Urban Areas (Nathan Cherniack).

1252. Hoel, Lester A.; Lepper, Robert L.; Anderson, Robert B.; Thiers, Gerald R.; DiCesare, Frank; Strauss, Jon; Parkinson, Tom E., *Urban Rapid Transit Concepts and Evaluation*, Pittsburgh, Penn. (Transportation Research Institute, Carnegie-Mellon University), 1968, 241 pp. Paper.

Systems Research: Vehicle Design, Guideway Superstructure and Substructure, Interface Simulation; Systems Review: Vehicle Technology—Guidance, Adhesion,

Performance, Automatic Train Control, Noise, Conclusions, Transit Systems: Metro, Transit Expressway, Monorail, Limited Tramway, Small Vehicle Systems, Future Technology.

1253.   Horton, Thomas R. (Editor), *Traffic Control Theory and Instrumentation*, New York (Plenum), 1965, 218 pp. [65-26915]

Traffic Control Instrumentation, Theory and Practice: Instrumentation for the Traffic Engineer (John L. Barker), Traffic Control, Time-Space Diagrams, and Networks (Denos C. Gazis), Computer Programming for Traffic Problems and Flow Characteristics (Matthew J. Huber), Simulation as a Tool in Traffic Control System Evaluation (D. L. Gerlough), Practicality in Traffic Control (L. J. Pignataro); Case Studies of Large-Scale Integrated Traffic Control Systems: Some Theoretical Considerations of Peak-Hour Control for Arterial Street Systems (Donald R. Draw; Charles Pinnell), Installation of a Tunnel Traffic Surveillance and Control System (Robert S. Foote), Use of Vehicle Presence Detectors in Metropolitan Traffic-Control Systems (J. H. Auer, Jr.), Experimentation with Manual and Automatic Ramp Control (Adolf D. May, Jr.), Toronto Computer-Controlled Traffic Signal System (Neal A. Irwin).

1254.   Kennedy, Norman; Kell, James H.; Homburger, Wolfgang S., *Fundamentals of Traffic Engineering*, Syllabus, Berkeley, Cal. (Institute of Transportation and Traffic Engineering, University of California), Sixth Edition, 1966, 228 pp.

"The purpose of these notes is to provide an outline for University courses in the fundamentals of traffic engineering."

1255.   Klose, Dietrich, *Metropolitan Parking Structures*, A Survey of Architectural Problems and Solutions, New York (Praeger), 1965, 247 pp. [65-19576]

Introduction; Growth of Motorization; Revolutionary Changes of the Motor Age; Space Requirements for Standing Cars; Reorganizing the Structure of Our Cities; Venues for Motorists and Pedestrians; High-Speed Traffic and the Redevelopment of City Centres; Central Area Parking Requirements, American and European Projects; Concept of a Comprehensive Parking Plan; Parking Authorities and Parking Programmes; Mechanical Parking Installations; Ramp Garages; Parking and Checking Procedure in Ramp Garages; Straight Ramps; Helical Ramps and Ramped Floors; Capacity of Parking Structures; Pedestrians and Parking Garages; Development and Design; Underground Garages; Department Stores and Shopping Centres; Office and Bank Buildings; Hotels; Housing Estates; Examples: Mechanical Parking Facilities, Ramp Garages, Underground Garages below Public Squares, Parking Garages Associated with Department Stores, Administrative Buildings, Banks, Hotels, Housing Estates. Text in English and German.

1256.   Kuhn, Tillo E., *Public Enterprise Economics and Transport Problems*, Berkeley, Calif. (University of California), 1962, 243 pp. [62-15956]

". . . discusses such topics as the cost-benefit analyses that make a project appear self-liquidating, and the 'backwards' reasoning used to decide on the necessity of a project already begun. . . . Calls for rejection of public competition (state-supported freeways v. local public rapid transit systems, for example) and the substitution of cooperative planning to arrive at solutions more in the total public interest."

1257. Lambert, John L., *Air Cushion Vehicle—Mass Transportation Demonstration Project*, CAL-MTD-3, Oakland, Cal. (Port of Oakland), April 1967, 72 pp. Paper.

Background; Governmental Jurisdiction and Regulation of ACV's; Preparation; Data Collection and Evaluation; Operational Analysis; Economic Analysis; Passenger Acceptance Analysis; Special Operational Features; Conclusions.

1258. Lang, A. Scheffer; Soberman, Richard M., *Urban Rail Transit: Its Economics and Technology*, Publication of the Joint Center for Urban Studies, Cambridge, Mass. (M.I.T.), 1963, 139 pp. [63-23379]

Introduction; The Supporting Way; Stations; Rail Transit Vehicles; Capacity; Rail Transit Costs; Rail Transit and the Demand for Urban Transportation; Future of Rail Transit; Some Considerations of Minimum Headway; Determination of Transit Costs.

1259. Lapin, Howard S., *Structuring the Journey to Work*, Philadelphia, Penn. (University of Pennsylvania), 1963, 227 pp. [63-15012]

Previous Studies; Character of Work Travel; Work-Trip Origins—Persons and Places; Confluence at Workplace Destinations; Factors of Change in Patterns of Work Travel; Suggestions and Conclusions; Appendices; Selected Bibliography, pp. 216-221.

1260. Lewis, David (Editor), *The Pedestrian in the City*, Princeton, N.J. (Van Nostrand), 1965, 300 pp. [66-3836]

Introduction; Mies' Urban Spaces; New Urban Design Concepts, Greenways and Movement Structures, The Philadelphia Plan; Constructed Art; Planning in Liverpool Today; Sheffield; Projective Relief; Humanistic City of Camillo Sitte; Curvilinear Space Form; Two Projects: Fort Worth, Texas, 1956; East Island, New York, 1961; Do Not Segregate Pedestrians and Automobiles; Pittsburgh Northside; London, The Living City; History of Nothing; Matmata; The Dogon and the Tellem; Two Sculptures; Four Reliefs; Tepotzlan: Native Genius in Town Planning; Squatters in Peru; Markets in Morocco; Cities: Stasis or Process; Collage; Recent Thoughts in Town Planning and Urban Design; Erith; Sculpture; The Linear City; Clyde City; Urban Renewal and Central Area Redevelopment in Great Britain—Participation of Private Enterprise; Planning and Designing the Central Areas of Cumbernauld New Town; Pedestrian's Experience of the Landscape of Cumbernauld; Joy Box; Chandigarh: Punjab Scene, Realized, Housing; Relevance of Greek Planning Today; Doxiadis' Contribution to the Pedestrian View of the City; Fascia Signs in Development Schemes in London and Basle; U.S. Gas Station.

1261. Little, Arthur D., Inc., *Cost-Effectiveness in Traffic Safety*, New York (Praeger), 1968, 180 pp. [68-18920]

Structure and Elements of the Process: Introduction, Analytical Approach, Traffic Safety Data, Traffic Safety Measurement; Methods and Applications: Optimization Techniques, Pilot Study, Accident Costs, Data Evaluation, Measurement Examples.

1262. Mayo, Robert S.; Others, *Tunneling, The State of the Art*, A Review and Evaluation of Current Tunneling Techniques and Costs, with Emphasis on Their Application to Urban Rapid-Transit Systems in the U.S.A., PB 178036, Springfield, Va. (Clearinghouse for Federal Scientific & Technical Information), January 1968, 263 pp. Paper.

Prepared for U.S. Department of Housing and Urban Development.
Introduction; State of the Art; Rock Tunneling; Soft-Ground Tunneling, Secondary Linings of Concrete; Shafts and Hoisting; Control of Soft Ground; Alternatives to Tunneling; Safety in Tunnels; Recommendations and Conclusions; Appendices.

1263. Meyer, J. R.; Kain, J. F.; Wohl, M., *The Urban Transportation Problem*, Cambridge, Mass. (Harvard University), 1966, 427 pp. [65-13848]

Introduction—Problem in Search of a Solution; Context of Urban Change: Economic Change and the City—Qualitative Evaluation and Some Hypotheses, Recent Trends in Urban Location, Recent Trends in the Supply of Urban Transport and Highway Financing, Trip Pattern and Demand, Interrelationship of Housing and Urban Transportation, Race and the Urban Transportation Problem; Comparative Costs: Costing Procedures and Assumptions, Line-Haul Systems, Residential Collection and Distribution, Downtown Distribution; Solutions and Public Policy: Role of Technology, Pricing, Subsidies, Market Structure, and Regulatory Institutions, Urban Transportation in Summary and Perspective; Appendices: A. Summary Description of Rail Line-Haul Systems, B. Bus Line-Haul Systems, C. Evaluation of Residential Service Costs Under Different Hypothesized Conditions.

1264. Ministry of Transport, Scottish Development Department, *Urban Traffic Engineering Techniques*, New York (British Information Service), 1965, 92 pp.

". . . deals with the principles of applying traffic engineering techniques to British urban road traffic problems. It stresses the need for making preliminary studies before any traffic or parking survey is planned, explains how many types of surveys can be undertaken and their results presented, and then shows how these results can be used for the formulation of proposals."

1265. Mitchell, Robert B.; Rapkin, Chester, *Urban Traffic—A Function of Land Use*, New York (Columbia University), 1954, 226 pp. [54-6483]

Orientation—Land Use and Traffic Problems in City Planning; Relating Land Use to Traffic; Structure of Movement—Spatial and Temporal Organization; Components of Movement Structures—Individual Movements of People; Systems of Movements of Persons; Systems of Movement of Goods and Materials; Influence of Movement on Land Use Patterns; Relating Movement of Persons and Goods to Land Use— Selected Explorations; Toward Improved Methods of Traffic Analysis; Appendices: Methods and Procedures of the Field Investigation, Classification of Items Pertinent to the Study of Movement, Channels of Marketing.

1266. Munby, Denys (Editor), *Transport*, Selected Readings, Baltimore, Md. (Penguin), 1968, 334 pp. Paperback.

Consumers' Surplus and Pricing: Public Works and the Consumer (J. Dupuit,

1844); Costs and Prices: Fixed Costs (W. A. Lewis, 1949), Marginal Cost Pricing (W. Vickrey, 1955), Costs and Rail Charges (W. J. Baumol, 1962), Railway Track Costs (S. Joy, 1964); Traffic and Cities: Traffic in Towns (C. Buchanan; Others, 1963), Road Pricing (A. A. Walters, 1961), Urban Transport Costs (J. R. Meyer; J. F. Kain; M. Wohl, 1965); Project Appraisal: Victoria Line (C. D. Foster; M. E. Beasley, 1963), Channel Tunnel (Ministry of Transport, 1963), Port Investment (R. O. Gross, 1967); Further Reading, pp. 313-318.

1267. National Academy of Sciences, Developing Transportation Plans—
        6 Reports, *Highway Research Record,* No. 240, Washington, D.C.
        (Highway Research Board), 1968, 99 pp. Paper.

Factors Influencing Traffic Generation at Rural Highway Service Areas (Herman A. J. Kuhn); Trip Generation in the Transportation Planning Process (Christopher R. Fleet; Harold D. Deutschman); Journey to Work—Singular Basis for Travel Pattern Surveys (G. A. Shunk; W. L. Grecco; V. L. Anderson); Income and Related Transportation and Land-Use Planning Implications (Harold D. Deutschman; Nathan L. Jaschik); Simplified Techniques for Developing Transportation Plans— Trip Generation in Small Urban Areas (Wilbur R. Jefferies; Everett C. Carter); Rand Classification—Procedure for Determining Future Trip Ends (John R. Walker).

1268. National Academy of Sciences, Origin and Destination—Advances in
        Transportation Planning—8 Reports, *Highway Research Record,* No.
        165, Washington, D.C. (Highway Research Board), 1967, 128 pp.
        Paper.

Improvements in Understanding, Calibrating, and Applying the Opportunity Model (Earl R. Ruiter); Effect of Zone Size on Zonal Interchange Calculations Based on the Opportunity Model in a Homogeneous Region (W. Stearns Caswell); Model Split Model in the Penn-Jersey Transportation Study Area (Anthony R. Tomazinis); Effect of Trip Direction on Interzonal Trip Volumes—Test of a Basic Assumption of Trip Distribution Models (David E. Boyce); Tri-State Transportation Commission's Freight Study Program (Robert T. Wood); Analysis of Land-Use Linkages (Frank E. Horton; Paul W. Shuldiner); Direct Estimation of Traffic Volume at a Point (Morton Schneider); Covariance Analysis of Manufacturing Trip Generation (Michael Kolifrath; Paul Shuldiner).

1269. National Academy of Sciences, Transportation System Planning and
        Current Census Techniques for Planning, *Highway Research Record,*
        No. 229, Washington, D.C. (Highway Research Board), 1968, 98 pp.
        Paper.

Measurements and the Regional Planning Process (Roger L. Creighton); Canberra —Toward a Scheme for Continuous Growth (Ian W. Morison; Walter G. Hansen); Emerging Patterns of Urban Growth and Travel (Alexander Ganz); Toward Measurement of the Community Consequences of Urban Freeways (Raymond H. Ellis); Measurement of Community Values—Spokane Experiment (Edward L. Falk); Survey of Citizens' Opinions of the Effectiveness, Needs, and Techniques of Urban Transportation Planning (Martin Wachs); Implications of the New Haven Census Use Test for Transportation and Land-Use Planning (George P. Leyland).

1270. National Academy of Sciences–National Research Council, *The Art and Science of Roadside Development–A Summary of Current Knowledge*, Special Report 88, NAS-NRC Publication 1329, Washington, D.C. (Division of Engineering and Industrial Research, Highway Research Board), 1966, 81 pp. Paper.

Historical Background; Aesthetics and Roadside Development in Highway Location and Design; Right-of-Way, Scenic Areas and Adjacent Land Use; Conservation of Natural Resources in Highway Design and Construction; Erosion Control; Landscape Plantings; Roadside Rest Areas; Scenic Turnouts and Lookouts; Safety in Roadside Development; Maintenance; Future of Roadside Development; Legal Authority and Techniques for Roadside Development.

1271. Nelson, Robert S.; Johnson, Edward M. (Editors), *Technological Change and the Future of the Railways*, Evanston, Ill. (Transportation Center, Northwestern University), 1961, 239 pp. [61-17929]

Comparison of the Advantages and Disadvantages of the Various Modes of Transport (John R. Meyer); Engineering Characteristics and National Policy (W. W. Hay); Is Nationalization of Common Carriers Inevitable? (E. G. Plowman); Importance of Technological Change to the Railway Industry (Alfred E. Perlman); Technological Change and the Railway's Need for Capital (James C. Nelson); Sociological Barriers to Technological Change (W. F. Cottrell); Technological Innovation and Labor in the Railway Industry (William Haber); Attitude of Labor Toward Technological Change in the Railway Industry (Guy L. Brown); Management and Technological Change (Kent T. Healy); Possibilities of Automation in the Railway Industry (Peter B. Wilson); Technological Changes in Transportation and Public Policy (Anthony F. Arpaia); Technological Change and the Future of Passenger Traffic (W. W. Patchell); New Equipment Technology (John D. Loftis); Railroads, Laws, and the New Congress (George P. Baker); New Motive Power Technology (Ray McBrian); Developments in Terminal Operation Control Technology (J. W. Burrows); Concluding Remarks (Norton E. Long).

1272. Netherton, Ross D., *Control of Highway Access*, Madison, Wis. (University of Wisconsin), 1963, 518 pp. [63-13744]

"Public Highway" as a Legal Concept; Servitude of the Roadside to the Roadway; Doctrine of Abutters' Rights; Twentieth Century Evolution of Access Control; Access-Control Legislation in the United States; Control of Access: Nature of the Public's Powers; Control of Access Through the Law of Nuisance, Through Regulation of Highway and Land Uses; Acquisition of Access Rights Through Eminent Domain, By Negotiated Purchase; Access Control in Land-Use Planning, The Interchange Area; Process of Evaluation; Land Use, Access, and Value; A Final Appraisal; Appendices: Model Controlled-Access Highway Act, Glossary, Diagrams of Leading Cases on Control of Highway Access.

1273. Oi, Walter I.; Shuldiner, Paul W., *An Analysis of Urban Travel Demands*, Evanston, Ill. (Northwestern University), 1962, 281 pp. [62-14643]

"The authors of this study of urban residential trip generation question the statistical validity of various aspects of existing origin-destination data collecting and traffic forecasting techniques. They argue that funds now devoted to individual city surveys would be better spent on research using accumulated O-D data to develop a reliable urban travel model that could be employed on a general basis throughout the country."

1274. Owen, Wilfred, *Cities in the Motor Age*, New York (Viking), 1959, 176 pp. [59-5645]

"An expert on traffic and transportation realistically appraises the dangers posed by automobile traffic in urban centers. A highly readable book."

1275. Pell, Senator Claiborne, *Megalopolis Unbound—The Supercity and the Transportation of Tomorrow*, Washington, D.C. (Praeger), 1966, 233 pp. [66-18917]

Fogbound; Megalopolis—Model of America To Come; Mobility—Basic Human Freedom; Automobile—Personal Mobility; Preserving Highway Mobility; Mobility Aloft; Railroads—Mobility Unused; Pell Plan—Review and Prospect; Translating Words Into Action; Supermobility—Challenge for the Future, pp. 222-226.

1276. Penn Jersey Transportation Study, *Volume 1, PJ Reports, The State of the Region*, Philadelphia, Pa. (51st St. and Parkside Ave., 19131), April 1964, 157 pp.

History and Background; Study Design; Introduction to Region; Distribution of Activities in the Region; Travel Demand in the Region; Transportation Facilities in the Region; Appendix.

1277. Port of New York Authority, Comprehensive Planning Office, *Metropolitan Transportation—1980*, A Framework for the Long-Range Planning of Transportation Facilities to Serve the New York-New Jersey Metropolitan Region, New York (111 8th Ave., 10011), 1963, 380 pp. [63-16002]

Metropolitan Background; Transporting the Goods of Commerce and Industry; Transporting People.

1278. Ritter, Paul, *Planning for Man and Motor*, New York (Macmillan), 1964, 384 pp. [63-22671]

". . . detailed analysis of the nature and needs of man and vehicle documents the need for traffic segregation, provides the principles necessary to carry out a separation program, and examines examples of traffic separation in 17 new towns, 33 urban renewal areas, and 46 residential areas. . . . Approaches the subject from an interdisciplinary point of view."

Context and Introduction; Man-Vehicle Relationship; Needs of Man as an Organism; Needs and Nature of Vehicles; New Towns with Traffic Segregation; Urban Renewal; Traffic Segregation in Residential Areas; Residential Renewal; History of Traffic Segregation; Bibliography; Appendices.

1279. Roberts, Paul O.; Suhrbier, John H., *Highway Location Analysis: An Example Problem*, M.I.T. Report No. 5, Cambridge, Mass. (M.I.T.), 1966, 94 pp. Paper. [66-29139]

Introduction; Prelocation Planning; Terrain Data Procurement; DTM Location System; DTM Design System; Volumetric Quantities; Structures; Pavement; Drainage; Right-of-Way Acquisition; Highway Maintenance Cost; Vehicle Simulation and Operating Cost System; Economic Analysis; Sensitivity Analysis; Presentation of Results; Conclusion.

1280.   Robinson, Carlton C.; Koltnow, Peter G., *Progress in Traffic Management in Los Angeles,* Los Angeles, Cal. (Automotive Safety Foundation; Automobile Club of Southern California), 1966, 49 pp. Paper.

About This Report; Transportation Planning; Program Development; Traffic Surveillance; Street Use; Installation and Maintenance; Design; Management; Research; Traffic Commission; In Conclusion.

1281.   Schneider, Lewis M., *Marketing Urban Mass Transit,* Boston, Mass. (Harvard Business School), 1965, 217 pp. [65-13254]

"A series of case studies is used to demonstrate that transit companies can increase patronage if they will devote more time and effort to marketing their product. Suggested marketing techniques for attracting new riders include: seats-for-all policy, reduced fares for senior citizens, higher fares for express service, air conditioning, heavy promotional campaigns, and distribution of timetables."

1282.   Schriever, Bernard A.; Seifert, William W. (Cochairmen), *Air Transportation 1975 and Beyond—A Systems Approach,* Report on the Transportation Workshop 1967, Cambridge, Mass. (M.I.T.), 1968, 516 pp. [68-20050]

Overview; Socioeconomic Trends and Their Potential Impact on Air Transportation; Air Vehicle; Air Traffic Control; Airports and Terminals; Collection and Distribution of Passengers and Air Freight; Government Policies and Trends; Bibliography; Glossary.

1283.   Sharp, Clifford, *Problems of Urban Passenger Transport,* With Special Reference to Leicester, Leicester, England (Leicester University), 1967, 118 pp. Paperback. [SBN 7185 1074 7]

General Survey of the Problems of Urban Passenger Transport; Passenger Transport in Leicester; Costs and Prices; Boundary System; Criteria for Seeking a Transfer of Passengers from Car to Bus; Methods of Bringing About a Car-to-Bus Passenger Transfer; Railways; Conclusions.

1284.   Smerk, George M. (Editor), *Readings in Urban Transportation,* Bloomington, Ind. (Indiana University), 1968, 336 pp. Paperback. [68-14613]

Decline of Public Transportation; Automobile Usage and Its Cost; Problem of Peak-Hour Demand; Pricing in Urban Transportation; Mass Transport or Private Transport?; Federal Aid for Urban Highways—Commitment and Challenge; Mass Transportation.

1285.   Smith, Wilbur and Associates, *Baltimore Metropolitan Area Transportation Study,* Vol. 1, *Findings and Recommendations,* Baltimore, Md. (Maryland State Roads Commission), 1964, 219 pp.

Summary; Introduction; Existing Traffic Conditions; Origin-Destination Surveys; Development Characteristics of the Baltimore Region; Travel Characteristics; Trip Distribution and Projection; Future Growth in the Baltimore Region; Future Traffic Facility Needs; Evaluation of Alternate Highway Plans; Recommended Highway Plan; Plan Implementation; Continuing the Transportation Planning Process; Appendices.

1286. Smith, Wilbur and Associates, *Parking in the City Center,* New Haven, Conn. (495 Orange St., 06511), May 1965, 143 pp. Paperback.

Evaluating Downtown Parking Needs: Supply and Characteristics, Demands and Space Needs in Typical Urban Areas, Generalized Calibration of Downtown Parking Demands; Downtown Parking Economics: Benefits to CBD Land, Meeting Downtown Needs, Cost and Revenues of Downtown Parking Developments; Recent Trends in Off-Street Parking: Typical Development Programs, Design Concepts, Garage Types, Design Standards; Parking Policy and Downtown Transportation Planning: Encouraging Short-Term Parking, Coordinating Parking with Transport Terminals, Zoning for Parking, Cooperative Parking Developments, Enhancing Parking Facility Design; Case Studies in Downtown Parking: Los Angeles, Regional Setting, Economic and Historical Development Influences, Off-Street Parking Trends and Characteristics, Parking As An Economic Use of Downtown Land, Parking and the Future of Downtown, Hartford, Regional Setting, Off-Street Parking Trends and Developments, Parking and the Future of Downtown, Summary of Similarities and Differences, Generalization and Extension; Appendices; Bibliography, pp. 139-141.

1287. Smith, Wilbur and Associates, *Transportation and Parking for Tomorrow's Cities,* New Haven, Conn. (495 Orange St., 06511), 1966, 393 pp.

Urban Transportation: Problems and Perspectives; Transportation in the Changing City; Dimensioning Downtown's Transportation Requirements; Characteristics of Urban Transport Systems; Choice of Urban Travel Mode; Strategy for System Planning; Planning Freeways and Rapid Transit; Parking in the City Center; Metropolitan Transportation Terminals; Economic and Policy Considerations; Appendices: A Macro-Approach to Calibrating the Urban Transport System, Tabulations, Selected References, pp. 383-393.

1288. Southern California Rapid Transit District, *Final Report* to the People of Los Angeles Metropolitan Area Regarding a First-Stage System of Rapid Transit, Los Angeles, Calif. (1060 South Broadway, 90015), May 1968, 111 pp.

Preface; Statement of Policy; Summary and Findings; Public Transportation History of Los Angeles Urban Area; SCRTD Planning Process; Introduction to Recommended System; Planning and Preliminary Engineering; Estimates of Traffic, Revenues and Expenses; Financing Rapid Transit; Benefit-Cost Analysis.

1289. Strakosch, George R., *Vertical Transportation, Elevators, and Escalators,* New York (Wiley), 1967, 365 pp. [67-26637]

Essentials of Elevating; Basis of Elevatoring a Building; Passenger Service Requirements; Incoming Traffic; Two Way and Interfloor Traffic; Outgoing Traffic,

Elevation Operation and Control; Space and Physical Requirements; Escalators and Moving Ramps; Elevatoring Commercial Buildings; Material Handling and Service Elevators; Multipurpose Buildings and Special Elevator Applications; Elevator Specifying and Contracting; Elevator Modernization; Economics of Elevatoring.

1290. Telford, Edward T. (Study Director), *Los Angeles Regional Transportation Study, Volume 1, Base Year Report 1960*, Los Angeles, Cal. (LARTS, P.O. Box 2304, Terminal Annex, 90054), December 1963, 61 pp.

Introduction and Summary; Organization and Financing; Present and Historical Background of the LARTS Study Area: Topography, Climate, Transportation, People and Economy; Methods and Results.

1291. Tri-State Transportation Committee, Connecticut-New Jersey-New York, *General Aviation—Airports for the Future*, New York (100 Church St., 10007), March 1965, 62 pp. Paper.

Background; Recommendations; Present Conditions; Future Conditions; Appendix.

1292. Urban Advisors to the Federal Highway Administrator, *The Freeway in the City—Principles of Planning and Design*, Washington, D.C. (Superintendent of Documents), 1968, 141 pp. [68-60072]

Comprehensive Planning and Community Values; View from the Freeway; Location of the Freeway; Roadway; Highway Structures; Multiple Use of the Corridor; Systems Approach; Bibliography, pp. 130-134.

1293. U.S. Bureau of Public Roads, *Traffic Surveillance, Simulation & Control*, Proceedings of Highway Conference on the Future of Research and Development, Washington, D.C. (Office of Research and Development), 1964, 391 pp. Paper.

Progress in Traffic: Data Instrumentation, Simulation, Surveillance, Control.

1294. Wingo, Lowden, Jr., *Transportation and Urban Land*, Baltimore, Md. (Johns Hopkins), 1961, 144 pp. [61-13662]

Introduction; Characterizing Technology in Urban Transportation Systems; Economics of the Journey-to-Work; Economic Dimensions of Urban Space; Demand for Space and the Basic Model; Applications of the Model; Summary and Index of Important Expressions and Equations; Appendices.

1295. Wohl, Martin; Martin, Brian V., *Traffic System Analysis for Engineers and Planners*, New York (McGraw-Hill), 1967, 558 pp. [67-26355]

Role of Traffic Engineering and Planning—Some Pertinent Aspects; Toward a Formalized Basis for Traffic Engineering Design and Planning; Introduction to Probability and Statistics for Traffic Engineering Design; Introductory Sampling Theory and Linear Regression Analysis; General Travel Forecasting Principles and Techniques; Predicting Travel Flow and Determining Design Hourly Volume for Varying Demand Conditions; Framework for Evaluating Alternative Traffic Engineering Designs; Evaluation of Alternative Highway and Traffic Engineering

Designs—Methods of Analysis; Application of Economic Analysis Methods—
Practical Example; Introduction to Highway and Congestion Cost and Pricing
Principles; Theoretical Techniques for Describing Traffic Flow; Highway Capacity
and Performance Characteristics; Capacity-Performance Relationships for Non-
signalized Intersections; Capacity-Performance Relationships for Time-Sharing or
Signalized Intersections; Use of Simulation in Traffic Engineering Design.

1296. Zettel, Richard M.; Carll, Richard R., *Summary Review of Major
Metropolitan Area Transportation Studies in the United States*, Ber-
keley, Calif. (Institute of Transportation and Traffic Engineering,
University of California), 1962, 66 pp. Paper.

". . . brings together for the first time information about the basic characteristics of
the dozen or so largest transportation studies. . . . Discusses the transportation
problem and study approach; the design, purposes and study methods; and the
organization, financing, and staffing of the study team.

Bibliographies

1297. Ad Hoc Committee on Librarians, *Sources of Information in Transportation*, Evanston, Ill. (Northwestern University), 1964, 262 pp.
Paper, typescript.

General Sources (Ruth F. Blaisdell); Highways (Beverly Hickok); Motor Carriers
(Marianne Yates); Metropolitan Transportation (Brigitte L. Kenney); Railroads
(Helene Dechief); Pipelines (Mary L. Roy); Merchant Marine (Dorothy V.
Ramm); Inland Waterways (Margaret L. Sullivan); Air Transportation (Marion R.
Herzog); Missiles and Rockets (Ronald J. Booser).

1298. Barry, Walter A. (Committee Chairman), *Traffic Planning and Other
Considerations for Pedestrian Malls*, Selected References, pp. 65-66.

1299. Cole, Leon Monroe (Editor); Merritt, Harold W. (Technical Editor),
Tomorrow's Transportation, New Systems for the Urban Future,
Bibliography of New Systems Study Project Reports, pp. 92-95.

1300. Columbian Research Corporation, "Computation Of Highway Eco-
nomic Impacts," Based on a Bibliography by Alan M. Voorhees and
Associates, Inc., Exchange Bibliography No. 67, Council of Planning
Librarians, Monticello, Ill. (P.O. Box 229, 61856), 1969, 27 pp. Mimeo.

1301. Edin, Nancy J., *Residential Location and Mode of Transportation to
Work, A Model of Choice*, Publication 311,012—VI, Chicago, Ill.
(Chicago Area Transportation Study), October 1966, 64 pp. Paper.

Bibliography, pp. 59-64.

1302. Grauman, Jacob; Bethel, Walter; Carr, Carol; Doran, Mary; Manch,
Steven, *A Selected and Annotated Survey of the Literature on Trans-
portation— Status, Structure, Characteristics, Problems, and Proposed
Solutions*, Research Laboratories, Science Information Services, Bib-

liographic Series No. 1, Philadelphia, Penn. (Franklin Institute), April 1968, 202 pp. Paper, typescript.

Preface; Subject Index; Abstracts.

1303. Haight, Frank A., "Annotated Bibliography of Scientific Research in Road Traffic and Safety," *Operations Research,* Vol. 12, No. 6, November-December 1964, pp. 976-1039.

Entries are listed alphabetically by author and then by date of publication. Titles published in English, French, or German are given in the original language; others in English translation. Annotations are in three columns at the right of the entries: an arabic number indicates the approximate mathematical texture of the work, a roman numeral and letter show the author's method, a capital letter shows the general subject area of the publication.

1304. Lapin, Howard S., *Structuring the Journey to Work,* Bibliography, pp. 216-221.

1305. Metcalf, Kenneth N., *Transportation—Information Sources,* Management Information Guide 8, Detroit, Mich. (Gale Research Company), 1965, 307 pp. [65-24657]

Library Facilities; Sources: Government, Statistical; Fact Source Books; Periodicals, Indexes, Abstracts, and Services; Research and University Programs, Literature and Bibliography; Professional and Trade Organizations and Associations; Transportation Industry Literature; Transportation Library Collections; Organization Sourcebooks; University Transportation Activities; Author-Title-Source Index; Subject Index.

1306. Munby, Denys (Editor), *Transport,* Further Reading, pp. 313-318.

Books and Articles: General, Costs and Prices, Investment Appraisal, Railways, Roads, Road Transport, Urban Transport, Ports and Shipping, Air Transport; Journals.

1307. Northwestern University, Transportation Center Library, *Current Literature in Traffic and Transportation,* Evanston, Ill. (1810 Hinman Ave., 60204), Regularly Issued. Paper, typescript.

1308. Northwestern University, Transportation Center Library, *The Journey to Work, Selected References 1960-67,* Bibliography No. 40, Council of Planning Librarians, Monticello, Ill. (Exchange Bibliographies, P.O. Box 229, 61856), 1968, 8 pp.

1309. Pollack, Leslie S., *Driver Distraction As Related to Physical Development Abutting Urban Streets: An Empirical Inquiry into the Design of the Motorist's Visual Environment,* Bibliography No. 8, Council of Planning Librarians, Monticello, Ill. (Exchange Bibliographies, P.O. Box 229, 61856), 1968, 4 pp.

Thesis abstract and bibliography.

1310. Robbins, Jane B., "Access to Airports: Selected References," Exchange

Bibliography No. 61, Council of Planning Librarians, Monticello, Ill. (P.O. Box 229, 61856), 1968, 21 pp. Mimeo.

General; Ground Transportation; Access by Specific Mode: Ground Effect Machines, Helicopters, Local Transit, Monorail, Motor Bus, Railroads, Vertical Take Off Aircraft; Bibliography: Geographic Index, Key to Periodical Abbreviations.

1311. Smith, Wilbur and Associates, *Parking in the City Center,* Bibliography, pp. 139-141.

1312. Smith, Wilbur and Associates, *Transportation and Parking for Tomorrow's Cities,* Selected References, pp. 383-393.

1313. Urban Advisors to the Federal Highway Administrator, *The Freeway in the City—Principles of Planning and Design,* Bibliography, pp. 130-134.

1314. Wheeler, James O., "Research On the Journey to Work: Introduction and Bibliography," Exchange Bibliography No. 65, Council of Planning Librarians, Monticello, Ill. (P.O. Box 229, 61856), 1969, 21 pp. Mimeo.

Introduction; Articles; Books; Reports; Theses and Dissertations.

Utilities, Services

Articles, Periodicals

1315. Alexander, Tom, "Where Will We Put All That Garbage?," *Fortune,* October 1967, pp. 149-151, 189-190, 192, 194.

1316. Chitwood, Jack, "Library and Community: Community Analysis, Population Characteristics, Community Growth, Governmental Relationships, Library Site Criteria," in: Shaw, Robert J., *Libraries— Building for the Future,* Chicago, Ill. (American Library Association), 1967, pp. 23-28. [67-23001]

1317. Faraci, Piero, "Planning the Public Library," ASPO Planning Advisory Service, Report No. 241, Chicago, Ill. (American Society of Planning Officials), December 1968, 20 pp.

Improving Library Service; Understanding Library Needs; Library Location; Planning Recommendations; Local Survey: Service Areas, Book Stock, Location, Contiguous Land Uses, Building Size, Total Main Floor Space, Meeting Room Space, Parking Facilities; Financing.

1318. Glenn, Neva, "An Experiment in Reclamation," *Westways,* Vol. 58, No. 3, March 1966, pp. 40-42.

"A Southern California community of 12,000 recently . . . solved two of the most formidable problems of urbanization at once: sewage disposal and recreational needs."

1319. Hogerton, John F., "The Arrival of Nuclear Power," *Scientific American*, Vol. 218, No. 2, February 1968, pp. 21-31.

"Electric power obtained from nuclear fission has made a decisive market breakthrough and now accounts for nearly half of all the new power-generating capacity being ordered by U.S. utilities."

1320. Milliman, J. W., "Policy Horizons for Future Urban Water Supply," *Land Economics*, Vol. XXXIX, No. 2, May 1963, pp. 109-132.

Present Status of Urban Water Economics; Reasons for Lack of Interest; Some Future Problems—Change in Outlook; Price Policy for Urban Water Supply; Peak Loads and Urban Water Use; Determination of the General Rate Level; Overview.

1321. Revelle, Roger, "Water," *Scientific American*, Vol. 209, No. 3, September 1963, pp. 92-100, 102, 104, 106, 108.

"Men need water to drink and for many other purposes, but by far the largest amount of water they have available must go to agriculture. Again the basic need in the proper utilization of water is education."

1322. Teitz, Michael B., "Cost Effectiveness: A Systems Approach to Analysis of Urban Services," *Journal of the American Institute of Planners*, Vol. XXXIV, No. 5, September 1968, pp. 303-311.

Diffusion of PPB; PPB and Systems Analysis; Specifying Objectives; System Structure; Cost and Effectiveness; Cost Analysis; Effectiveness Analysis; Measurement; Conclusion.

1323. Warren, Robert, "A Municipal Services Market Model of Metropolitan Organization," *Journal of the American Institute of Planners*, Vol. XXX, No. 3, August 1964, pp. 193-204.

Traditional Model of Decentralized Metropolitan Organization; Municipal Services Market Model; Economics of Scale; Population Growth and Service Choices; Incentives for Decentralization; Pricing of Services; Reconsideration of Decentralized Government; Centralization and Regional Planning; Order and Change Within Metropolitan Government.

1324. Wolman, Abel, "The Metabolism of Cities," *Scientific American*, Vol. 213, No. 3, September 1965, pp. 179-188, 190; also in: Scientific American (Periodical), *Cities*, New York (Knopf), 1965, 211 pp. Paperback. [65-28177]

"In the U.S. today attention is focused on shortages of water and the pollution of water and air. There is plenty of water, but supplying it requires foresight. Pollution calls for public economic decisions."

Books, Reports, Pamphlets

1325. Advisory Commission on Intergovernmental Relations, *Intergovernmental Responsibilities for Water Supply and Sewage Disposal in*

Metropolitan Areas, Report A-13, Washington, D.C. (Superintendent of Documents), October 1962, 135 pp. Paper.

Scope of the Report; Setting of the Urban Water Problem; Patterns and Problems at the Local Level; Metropolitan Approaches to the Water Problem; The States and Urban Water; Role of the Federal Government; Recommendations.

1326.   Aerojet-General Corporation, (California) Department of Public Health, *California Integrated Solid Wastes Management Project—A Systems Study of Solid Wastes Management in the Fresno Area,* Berkeley, Cal. (Bureau of Vector Control), 1968, c. 280 pp. Paper.

Summary; Problems of Solid Wastes; Performance-Scoring Procedure; Ancillary Effects Scoring Procedure; Operating Conditions; Conceptual Design of Systems; Scoring and Costs; Selected System Concepts; Application to Other Regions.

1327.   American Public Works Association, *Municipal Refuse Disposal,* Chicago (Public Administration Service), 1966, 528 pp. [66-25574]

Introduction; Quantities and Composition of Refuse; Selecting Disposal Methods; Sanitary Landfills; Central Incineration; On-Site Incineration; Grinding Food Wastes; Feeding Food Wastes to Swine; Composting; Salvage and Reclamation; Refuse Disposal Management; Selected Bibliography, pp. 507-515.

1328.   American Public Works Association, *Refuse Collection Practice,* Chicago (Public Administration Service), Third Edition, 1966, 524 pp. [66-25573]

Refuse Collection Problem; Refuse Materials; Preparation of Refuse for Collection; Factors Affecting Refuse Collection Cost; Refuse Collection Methods, Equipment; Planning Refuse Collection Systems; Supplemental Transportation of Refuse; Special Refuse Collection Problems; Municipal, Contract, or Private Collection of Refuse; Financing Refuse Collection Operations; Organization; Personnel; Equipment Management; Reporting, Cost Accounting, and Budgeting; Selected Bibliography, pp. 505-512.

1329.   Atlanta-Fulton County Joint Planning Board, *Community Facilities, City of Atlanta,* Atlanta, Ga. (City Hall, 30303), 1968, 33 pp. Paper

Neighborhood; General; Service; Maps.

1330.   Belknap, Ivan; Heinle, John G., *The Community and Its Hospitals: A Comparative Analysis,* Syracuse, N.Y. (Syracuse University), 1963, 218 pp. [63-14406]

". . . the planner will learn of the intricacies of the influence of local government upon semi-public health and welfare activities, the complex nature of private and public voluntary non-profit hospitals, and . . . why programs for interns and medical education are essential in achieving and maintaining high quality medical care in community hospitals."

1331.   Beuscher, J. H., *Water Rights,* Madison Wis. (College Printing & Typing Co., 1967, c. 425 pp.

Water Regimes and Water Law; Law of Diffused Surface Water; Law of Percolating Ground Water; Private Rights in Watercourses; Limitation Imposed on Private Water Rights by Assertions of Public Interest; Water Pollution Abatement; Allocations of Water Between States and Interstate Water Compacts.

1332. Black & Veatch, *Solid Waste Disposal Study For the Washington Metropolitan Region,* Washington, D.C. (Northern Virginia Regional Planning Commission; Metropolitan Washington Council of Governments; The Maryland-National Capital Park and Planning Commission), 1967, c. 275 pp. Paper.

Summary of Findings and Recommendations; History of Disposal of Solid Wastes in the Metropolitan Washington Region; Current Status of Solid Waste Programs; Refuse Quantities; Alternative Disposal Methods; Land Requirements for Disposal; Inventory of Potential Disposal Sites; Transportation of Solid Wastes; Recommended Disposal Programs; Administrative and Financing Considerations; Sediment Disposal; Appendix.

1333. Bower, Blair T.; Larson, Gordon P.; Michaels, Abraham; Phillips, Walter M.; Anderson, Richard T. (Editor), *Waste Management—Generation and Disposal of Solid, Liquid and Gaseous Wastes in the New York Region,* Regional Plan Association, New York (230 West 41st St., 10036), March 1968, 107 pp. Paper. [A 38358]

Foreword; Preface and Some Conclusions; Introduction; Underlying Concepts and Principles; Procedures for Analysis of Waste Management; Waste Generation: Present and Future; Waste Management Methods and Residual Wastes Discharged: Present and Future; Implications for Governmental Organization; Summary and Suggestions; Appendices: Waste Management Costs, Assumptions Used in Developing Variants for 2000, Background Data—Generation and Handling of Wastes, Uncertainty in Environmental Quality Management.

1334. Butler, George D., *Introduction to Community Recreation,* New York (McGraw-Hill), Fourth Edition, 1967, 612 pp. [67-10874]

Recreation—Its Nature, Extent, and Significance: What Is Recreation?, Importance of Recreation, Agencies Providing Recreation, Recreation—Function of Local Government, History of Municipal Recreation in the United States; Leadership: Recreation Leadership, Recreation Department Personnel, Education for Recreation Leadership, Selecting and Maintaining the Leadership Staff, Volunteer Service in the Recreation Department; Areas and Facilities: City Planning for Recreation, Design and Equipment of Recreation Areas, Planning Special Areas and Structures; Activities and Program Planning: Recreation Activities, Program-Planning Principles and Methods, Organizing and Conducting Recreation Activities; Operation of Areas and Facilities: Playgrounds, Recreation Buildings and Indoor Centers, Recreation Facilities; Program Features and Services: Arts and Crafts, Athletics, Games, and Sports, Drama, Music, Nature, Gardening, and Outing Activities, Other Program Features, Service to Special Groups, Typical Community Recreation Programs; Organization and Administration Problems: Legal Aspects of Community Recreation, Municipal Organization for Recreation, Organization of the Recreation Department, Financing Recreation, Records, Reports, and Research, Public Relations, Cooperation and Relationships with Public and Private Groups; Bibliography, pp. 582-592.

1335. California, Department of Water Resources, Resources Agency, *Municipal and Industrial Water Use,* Bulletin No. 166-1, Sacramento, Cal. (401 Public Works Bldg., 1120 N. St., 95805), August 1968, 106 pp. Paper.

Introduction; Factors Affecting Urban Water Use: Climatic, Man-Made; Urban Per Capita Water Use: Considerations in Using Results, Results and Discussion; Trends in Per Capita Water Use: Variability and Trends in Monthly Values, Trends in Annual Values; Appendices.

1336. Campbell, H. C., *Metropolitan Public Library Planning Throughout the World,* International Series of Monographs in Library and Information Science, Vol. 5, London (Pergamon), 1967, 168 pp. [66-29370]

The Public Library in the Metropolitan Area; Municipal Government Responsibilities and the Public Library; Metropolitan Public Library Planning in Action; Library Planning in Four Large Metropolitan Areas, in Seven Metropolitan Areas with Populations of 2-4 Million, in Eight Metropolitan Areas with Populations of 1-2 Million; Metropolitan Library Planning in the Future.

1337. Conant, Ralph W. (Editor), *The Public Library and the City,* Cambridge, Mass. (M.I.T.), 1965, 216 pp. [65-27504]

Library Consumer (Philip H. Ennis); Social-Class Perspectives (Allison Davis); Noncollege Youth (Howard S. Becker); Public Library in Perspective (Herbert J. Gans); Library in the Community (Nathan Glazer); Library—Instrument for Metropolitan Communications (Richard L. Meier); Public Nature of Libraries (Charles M. Tiebout; Robert J. Willis); Needed—A Public Purpose (Edward C. Banfield); Dissemination of Print (Dan Lacy); Reflections of a Library Administrator (Jerome Cushman); Trends in Urban Politics and Government—Effect on Library Functions (Robert H. Salisbury); Trends in Urban Fiscal Policies— Effect on Library Functions (William F. Hellmuth); The Library Faces the Future (Emerson Greenaway); Some Research Questions (Ralph W. Conant; Ralph Blasingame); The Critic Speaks (John E. Burchard); Annotated Bibliography of Items Relating to Library Problems in Metropolitan Areas (Leonard Grundt), pp. 196-210.

1338. Deering, Francis R.; Jewell, Don; Lueddeke, Lindsley C., *Planning and Management of Public Assembly Facilities,* Chicago, Ill. (International Association of Auditorium Managers, Inc.), 1968(?), 102 pp. Paper.

Trends in Construction; Construction Financing; Problem of Location; Building Facilities; Ice Activities and Operations; Building Capacities; Income and Expenditures; Rental Rates; Special Equipment and Service Charges; Convention Practices; Licenses and Taxes; Insurance; Concessions Management; Auditorium Problems; Ticket Office Practices; Union Affiliations; University, College Centers; Memorial Buildings; The Manager.

1339. Downing, Paul B., *The Economics of Urban Sewage Disposal,* New York (Praeger), 1969, c. 225 pp. [69-15590]

". . . effects of grinding garbage to convert it into sewage, the cost of planning for peak loads, optimal size of utilities as related to population density and cost of transmission, effects of water quality standards, and the financing of urban sewage disposal utilities . . . basic guidelines for sanitation engineers and others in the field of urban planning."

1340. Eckel, Edwin B. (Editor), *Landslides and Engineering Practice,* Special Report 29, NAS-NRC Publication 544, Washington, D.C. (Highway Research Board), 1958, 232 pp. [58-60007]

Introduction (Edwin B. Eckel); Economic and Legal Aspects (Rockwell Smith); Landslide Types and Processes (David J. Varnes); Recognition and Identification of Landslides (Arthur M. Ritchie); Airphoto Interpretation (Ta Liang; Donald J. Belcher); Field and Laboratory Investigations (Shailer S. Philbrick; Arthur B. Cleaves); Prevention of Landslides (Arthur W. Root); Control and Correction (Robert F. Baker; Harry E. Marshall); Stability Analyses and Design of Control Methods (Robert F. Baker; E. J. Yoder); Trends (John D. McNeal).

1341. Gittell, Marilyn (Editor), *Educating an Urban Population,* Beverly Hills, Cal. (Sage), 1967, 320 pp. [67-18421]

Metropolitan Education Dilemma—Matching Resources to Needs (Alan K. Campbell; Philip Meranto); Chicago's Educational Needs—1966 (Robert J. Havighurst); Suburban Education—Fiscal Analysis (Seymour Sacks; David C. Ranney); Law and the Education of the Urban Negro (William G. Buss, Jr.), Fiscal Independence and Large City School Systems (T. Edward Hollander); Demographic Analysis As a Tool for Educational Planning (Charles A. Glatt; Arliss L. Roaden); School Desegregation and School Decision-Making (Robert L. Crain; David Street); Obstacles to School Desegregation in New York City—Benchmark Case (David Rogers); Pedagogues and Power—Descriptive Survey (Alan Rosenthal); Decision-Making in the Schools—New York City, Case Study (Marilyn Gittell); Patterns of White and Nonwhite School Referenda Participation and Support—Cleveland, 1960-1964; Planning Education Today for Tomorrow (Werner Z. Hirsch); Urban Educational Problems—Models and Strategies (Richard L. Derr); Preparation of School Personnel to Work in an Urban Setting (Gordon J. Klopf; Garda Bowman); Extending Educational Opportunities—School Desegregation (Thomas F. Pettigrew); Politics of Education in Large Cities (David W. Minar).

1342. Hirshleifer, Jack; De Haven, James C.; Milliman, Jerome W., *Water Supply—Economics, Technology, and Policy,* Chicago (University of Chicago), 1960, 378 pp. [60-14355]

Water Problem; Our Water Resources—Present Picture; Economics of Utilization of Existing Water Supplies; Criticisms of Market Allocations—Political Allocation Process; Municipal Water Rates; Investment in Additional Water Supplies; Practical Logic of Investment Efficiency Calculations; Technological Features and Costs of Alternative Supplies of Water; Water Law—Government Discretion or Property Rights?; New York's "Water Crisis"—Case Study of a Crucial Decision; Water for Southern California—Case Study of an Arid Region; Some Controversial Conclusions and Their Implications.

1343. Honolulu, Health and Hospital Planning Council, *Areawide Health Facilities and Service Plan for County of Honolulu, 1968-1985,* Honolulu, Hawaii (Suite 820, Ala Moana Bldg., 1441 Kapiolani Blvd., 96814), 1968, 104 pp. Paper.

Summary of Plan; Changing Health Care System; Population—County of Honolulu; Inpatient Care; Health Manpower; Economic Aspects of Patient Care; Education and Training; Outpatient Services; Progress Report—Council's Recommendations in 1965 Study; Future Need for Health Facilities and Health Services; Neighborhood Health Centers; Franchising and Statutory Control for Hospitals; Regional Areawide Comprehensive Health Planning.

1344. Kneese, Allen V.; Bower, Blair T., *Managing Water Quality: Economics, Technology, Institutions,* Baltimore, Md. (Johns Hopkins), 1968, 328 pp. [68-8290]

Nature of the Water Quality Problem; Economic Concepts and Policies for Controlling Individual Waste Discharges; Economic Concepts Relating to Regional Water Quality Management Systems With Collective Facilities; Institutional and Organizational Approaches to Regional Water Quality Management; Implementation and Management.

1345. Landsberg, Hans H.; Schurr, Sam H., *Energy in the United States—Sources, Uses, and Policy Issues,* New York (Random House), 1967, 242 pp. Paperback. [68-13467]

History and Prospects (Hans H. Landsberg): Energy in the Economy—Over-all View, Supply and Consumption—Changes Over Time; Dimensions of Future Demand; Nation's Conventional Fuel Resources; Energy from New Sources and Processes; Role of Electricity, Outlook for Adequacy; Public Policy Issues (Sam H. Schurr and Hans H. Landsberg): Some Issues of Public Policy.

1346. Leopold, Luna B., *Hydrology for Urban Land Planning—A Guidebook on the Hydrologic Effects of Urban Land Use,* Geological Survey Circular 554, Washington, D.C. (U.S. Geological Survey), 1968, 18 pp. Paper.

Planning Procedures and Hydrologic Variables; Availability of Data and Technique of Analysis; Effect of Urbanization on Increasing Frequency of Overbank Flow; Local Storage to Compensate for Peak Flow Increase; Sediment Production; Effect of Increased Peak Flows on Sediment Yield; Water Quality; Selected References, p. 18.

1347. Linsley, Ray K.; Franzini, Joseph B., *Water-Resources Engineering,* New York (McGraw-Hill), 1964, 654 pp. [63-13935]

Textbook. Introduction; Descriptive Hydrology; Groundwater Probability Concepts in Design; Water Law; Reservoirs; Dams; Spillways, Gates, and Outlet Works; Open Channels; Pressure Conduits; Hydraulic Machinery; Engineering Economy in Water-Resources Planning; Irrigation; Municipal and Industrial Water Supply; Hydroelectric Power; River Navigation; Drainage; Sewage Disposal and Water Quality Control; Flood Control; Planning for Water-Resources Development.

1348. Lund, William S.; McCarthy, James H.; McElyea, J. Richard, *Site Analysis and Selected Planning Factors for the Proposed Los Angeles*

World Zoo, South Pasadena, Cal. (Southern California Laboratories, Stanford Research Institute), June 1960, c. 164 pp.

Introduction; Summary and Conclusions; Analysis of Possible Sites; Public Use Analysis; Planning Factors; Administration and Organization; Financial Analysis; Appendix.

1349.   National Academy of Sciences—National Research Council, Committee on Pollution, *Waste Management and Control,* Publication No. 1400, Washington, D.C. (2101 Constitution Ave., 20418), 1966, 257 pp. [66-60050]

Introduction; Nature of the Problem; Legal, Legislative, and Institutional Problems; Areas of Inadequacy; Possible Improved Approaches; Recommendations; Appendices: Pollution Processes in Ecosystems; Criteria, Instrumentation, and Monitoring; Transport Systems; Residue Situation, Current and Future; Pollution, Abatement Technology; Legal and Public Administration Aspects; Public Policy and Institutional Arrangements; Brief Analysis of Pollution in the Delaware Survey; Bibliography, p. 252.

1350.   National Commission on Community Health Services, *Health Is a Community Affair,* Cambridge, Mass. (Harvard University), 1966, 252 pp. [66-27415]

Health and the Community; Comprehensive Personal Health Services; Comprehensive Environmental Health Services; Consumer; Health Manpower; Places for Personal Health Care; Organization and Management of Resources; Partners in Progress: Governments, Volunteers; Action-Planning; Future; Positions and Recommendations.

1351.   Nelson, W. Stewart, *Mid-Continent Area Power Planners: A New Approach to Planning in the Electric Power Industry,* East Lansing, Mich. (Graduate School of Business Administration, Michigan State University), 1968, 128 pp. [68-63563]

Introduction to MAPP; Economics of Interconnected Systems; Historical Development and Achievements of MAPP; Notes; Bibliography, pp. 125-128.

1352.   Netzer, Dick; Kaminsky, Ralph; Strauss, Katherine W., *Public Services in Older Cities,* New York (Regional Plan Association), May 1968, 56 pp. [A 38358]

Introduction and Summary; Conditions in the Older Cities; Local Government Expenditure in Old Cities; Poverty-Linked Services; Educational Needs in the Old Cities; Other Public Services; Financing the Improvements; Appendix: Comparative Quality of School Programs in Old Cities and High-Income Suburbs, Efforts to Improve Educational Performance in the New York City School System.

1353.   Oklahoma Health Sciences Foundation, *The Development Plan for the Oklahoma Health Center,* Oklahoma City, Okla. (Kermac Bldg.), Undated, 36 pp. Paper.

Health Center Concept; Background for Planning; Health Center Campus; Effecting the Plan.

1354. Salzman, Michael H., *New Water for a Thirsty World,* Los Angeles, Cal. (Science Foundation Press), 1960, 210 pp. Paperback. [60-9611]

Introduction; What Is Water?; New Water; Modern Science of Hydrology and Its Limitations; Dynamic Earth; Man's Challenge; References Cited, pp. 183-195.

1355. Scott, Mel., *Partnership in the Arts: Public and Private Support of Cultural Activities in the San Francisco Bay Area,* Berkeley, Cal. (Institute of Governmental Studies), 1963, 55 pp. Paper.

Summary and Recommendations; Community Responsibility; Municipal Contributions to Private Groups; Public Libraries; Cultural Programs of Municipal Recreation Departments; Museums; Scientific Institutions; Regional Council of the Arts; Townscape of the Public Environment.

1356. Sewell, W. R. Derrick, *Water Management and Floods in the Fraser River Basin,* Research Series No. 100, Chicago (Department of Geography, University of Chicago), 1965, 163 pp.

"The most serious flood problem in Canada is that in the Lower Fraser Valley of British Columbia. . . . There are now more than twice as many people living in the flood plain as there were in 1948. . . . The search for reasons for this persistent movement into the path of potential disaster provides a focus for this study."

1357. Smith, Stephen C.; Castle, Emery N. (Editors), *Economics and Public Policy in Water Resource Development,* Ames, Iowa (Iowa State University), 1964, 463 pp. [64-13370]

"Critical examination and analysis of water problems and their ramifications in law, economics, politics and other fields of public concern with contributions by 22 authors, each with special experience in certain phases of water resource development."

1358. Twin Cities Metropolitan Planning Commission, *Metropolitan Water Study, Part II,* Metropolitan Planning Report No. 6, St. Paul, Minn. (Griggs-Midway Bldg., University at Fairview), July 1960, 57 pp. Paper.

Summary of Findings and Recommendations; Water—A Basic Resource—Relationship of Water Uses; Organizational Aspects; Legal Aspects; Supply & Demand; Water Facilities; Action Programs at the Community Level; Recommended Comprehensive Water Program.

1359. U.S. Congress, Joint Economic Committee, Subcommittee on Economic Progress, *State and Local Public Facility Needs and Financing,* 89th Congress, 2nd Session, Vol. 1: *Public Facility Needs,* Vol. 2: *Public Facility Financing,* Washington, D.C. (Superintendent of Documents), 1966, Vol. 1: 693 pp., Vol. 2: 453 pp. Paperback.

Vol. 1. General Forces; Public Facility Categories—Facilities: Basic Community, Transportation, Educational, Health, Recreation and Cultural, Other Public Buildings.

Vol. 2. Trends in Public Facility Financing; Municipal Securities Market—Patterns,

Structure and Problems; Municipal Bond Interest Rates and Tax Exemption; Sources of Loan Funds.

1360. Weers, C. A.; Eby, R. I.; Myrick, M. J., *The Bel-Air–Brentwood Conflagration*, New York (National Board of Fire Underwriters), 1962, 20 pp.

Introduction; Bel-Air–Brentwood Area: General Description, Water Supply System, Other Sources of Water, Fire Fighting Forces; The Fire and Fire-Fighting Tactics: Weather Conditions, Origin of Fire, Southward Spread into Stone Canyon, Eastward Thrust Toward Beverly Glen, Sweep to Southwest, Final Stand at Mandeville Canyon, Fire-Fighting Tactics, Topanga-Santa Inez Fire, Benedict Canyon Fire; Water Supply and Operations; Operations by Other Departments; Analysis of Damage; Laws and Ordinances; Conclusions; Recommendations.

1361. White, Gilbert F., *Choice of Adjustment to Floods*, Research Paper No. 93, Chicago, Ill. (Department of Geography, University of Chicago), 1964, 159 pp.

"An examination of conditions in which managers of flood plain properties choose among possible adjustments to floods, based on an intensive study of LaFollette, Tennessee, and reconnaissance study of five other towns differing widely in physical characteristics of flooding."

Bibliographies

1362. American Public Works Association, *Municipal Refuse Disposal*, Selected Bibliography, pp. 507-515.

1363. American Public Works Association, *Refuse Collection Practice*, Selected Bibliography, pp. 505-512.

1364. Butler, George D., *Introduction to Community Recreation*, Bibliography, pp. 582-592.

1365. Lee, R. R.; Fleischer, G. A.; Roggeveen, V. J., *Engineering-Economic Planning of Transportation Facilities, Water Resources and Miscellaneous Subjects—A Selected Bibliography*, Stanford, Cal. (Stanford University), 1961. 3 Volumes.

1366. Salzman, Michael H., *New Water for a Thirsty World*, References Cited, pp. 183-195.

1367. Tennessee Valley Authority, *Flood Damage Prevention, An Indexed Bibliography*, Knoxville, Tenn. (Technical Library), Fifth Edition, May 1967, 39 pp. Paper.

"... majority of the titles ... emphasize flood damage prevention and flood plain regulation, with only selected items pertaining to flood control. Many ... can be found in public and university libraries ... most ... are available on inter-library loans from the TVA Technical Library."

1920–1939; 1940–1951; yearly through 1967. Subject Index: Building Codes; Channel Encroachment; Development Policies; Encroachment Status; Films; Flood Control–Development Policies, Prevention; Flood Information Reports; Flood Insurance; Flood Plain Regulations; Flood Proofing; Floods; Inter-Governmental Relations; Legal Aspects; Open-Space Recreation; Subdivision Regulations; Tax Adjustments; Urban Renewal; Use of Flood Plains; Water Resources Development; Zoning.

1368.   Tompkins, Dorothy Campbell, *Water Plans for California—A Bibliography*, Berkeley, Cal. (Bureau of Public Administration, University of California), 1961, 180 pp. Paperback.

State Plans for Water; Federal Interest in California Water; Bibliographies, pp. 178-180.

12. PARTICULAR FORMS OF URBAN PLANNING

Metropolitan Regional Planning, Urban Area Development

Articles, Periodicals

1369.   Burchard, John Ely, "Metropolitan Urbanity," *Technology Review*, Vol. 68, January 1966, pp. 15-21, 35-36, 38-39.

"The urban boats must be rocked if man is to match his machines and not creep dwarfed and debased below them."

1370.   Friedmann, John (Special Editor), Regional Development and Planning, *Journal of the American Institute of Planners*, Special Issue, Vol. XXX, No. 2, May 1964, 174 pp.

Regional Development in Post-Industrial Society (John Friedmann); Spatial Organization Theory as a Basis for Regional Planning (Edwin von Böventer); Establishing Goals for Regional Economic Development (Charles L. Leven); Use of Economic Projections in Planning (Sidney Sonenblum; Louis H. Stern); Planning for Depressed Areas: A Methodology Approach (Walter Stöhr); Transportation Investment and Regional Development (Arthur P. Hurter; Leon N. Moses); State Development Planning: The California Case (John W. Dyckman). Review Articles: French Regional Planning (Lawrence D. Mann; George J. Pillorgé); Natal Plans (D. Scott Brown).

1371.   Grove, William R., Jr., "Metropolitan Planning?," *University of Miami Law Review*, Vol. XXI, 1966, pp. 60-98.

Introduction; Some Myths in an Evolutionary Process; The Concept of Metropolitan Planning; Implications for the State Legislature; Summary.

1372. Hodge, Gerald, "Urban Structure and Regional Development," *The Regional Science Association PAPERS*, Vol. 21, 1968, pp. 101-123.

Regional Settings; Analytical Problem and Approach; Data Arrays; Comparative Urban Structures; Similarities in Urban Structure; Underlying Relationships in Urban Structure; Uniqueness in Regional Structure; Urban Structure and Urban Performance; Average Family Earnings, 1961; Population Change; Retail Change, 1951-61; Interrelations of Urban and Rural Structure; Some Propositions and Prospects; Prospects for Structural Analyses.

1373. Levin, Melvin R., "The Big Regions," *Journal of the American Institute of Planners*," Vol. XXXIV, No. 2, March 1968, pp. 66-80.

The Big Regions and America's Future; New Dimension; Historical Roots; Appalachian Prototype; Human Resource Programs; How To Run a Surplus—Case of New England; Criteria for Establishing Regions—Bad Fit?; Outmigration and the Regions; Conclusion—The New Regions as New Arenas.

1374. Mars, David, "Localism and Regionalism in Southern California," *Urban Affairs Quarterly*, Vol. 11, No. 4, June 1967, pp. 47-74.

"Extrapolation of existing data and predictions for the future indicate that lying ahead are more population, urbanization, mobility, automobiles, and almost everything that brings about municipal, county, metropolitan, and regional problems. Persons concerned about mobilizing efforts to solve some of these problems would do well to consider carefully the data reported in this study and try to understand the basic assumptions which these data in general bespeak."

1375. Perloff, Harvey S., "Key Features of Regional Planning," *Journal of the American Institute of Planners*, Vol. XXXIV, No. 3, May 1968, pp. 153-159.

Joint Defining of Regions and Planning Objectives; National Plans in Regional Terms; Larger Context of Regional Planning; Dealing with the Various Phases of The Planning Process; Hypothetical Example; Information-Analysis Phase; Planning-Programming Phase; Planning-Operational Phase.

1376. Zimmerman, Joseph F., "Metropolitan Ecumenism: The Road to the Promised Land?," *University of Detroit Journal of Urban Law*, Vol. 44, Spring 1967, pp. 433-457.

Metropolitan Reform Movement; Metropolitan Councils of Governments; Conclusions.

Books, Reports, Pamphlets

1377. Bogue, Donald J.; Beale, Calvin L., *Economic Areas of the United States*, New York (Free Press), 1961, 1162 pp. [61-9161]

". . . counties of the United States [grouped] into 506 State Economic Areas . . . in turn grouped into 121 Economic Subregions, the Subregions into 13 Economic Regions, and the regions into 4 Economic Provinces. The book comprises an outline of the methods of delimitation of these regions, a description of each economic subregion, a state-by-state description of each of the economic areas, and tables relating to the population, economy, agriculture, and occupational structure of the economic regions, subregions, and economic areas, plus very extensive cross-reference indexes."

1378. Bollens, John C.; Schmandt, Henry J., *The Metropolis: Its People Politics, and Economic Life*, New York (Harper & Row), 1965, 643 pp. [65-19489]

Metropolitan Backdrop; Nature and Dimensions of the Metropolitan Community; Metropolitan Models and Types; Social Anatomy of the Metropolis; Metropolitan Economy; Government in the Metropolis; Politics and Power in the Metropolis; Metropolitan Citizenry as Civic Participants; Social and Economic Maladies; Planning Challenge; Service Challenge; Financing Government in the Metropolis; Cooperative Approach: Agreements and Metropolitan Councils; One-Government Approach: Annexation and Consolidation; Two Level Approach: Districts, Comprehensive Urban County Plan, and Federation; Politics of Reform; States and the Nation; Metropolitan World; Shape of the Future; Commentary on Bibliography, pp. 599-608.

1379. Boudeville, J-R., *Problems of Regional Economic Planning*, Edinburgh, Scotland (Edinburgh University), 1966, 192 pp.

"The objective . . . is to evoke the interest of a wider public [in] French prospective regional planning, and of the French economic thought on this subject . . . not only addressed to economic specialists but to all responsible industrialists and civil servants."

Concepts and Definitions; Tools for Regional Economic Studies; Regional Economic Programmes; Regional Operational Models; Fourth French Plan; French Regional Planning.

1380. Doxiadis, Constantinos A. (Study Director), *Emergence and Growth of an Urban Region—The Developing Urban Detroit Area*, Detroit, Mich. (Detroit Edison Company), 1966, 335 pp. [66-29622]

Introduction; North American Setting; Great Lakes Megalopolis; Great Lakes Area; Urban Detroit Area; Summary and Conclusions; Appendix.

1381. French Embassy, *France, Town and Country Environment Planning*, New York (Press and Information Service), December 1965, 56 pp. Paper.

Dimensions of Town and Country Environment Planning: What Is Environment Planning?, Why Plan the Town and Country Environment?, Where is the Environment Being Planned?, When? Space-Time Continuum in Planning, How Did Environment Planning Originate?, Who Plans the Environment?, What Are the Problems in Environment Planning?; Guidelines for Environment Planning: Major Options for the France of 1985, Toward Better-Balanced Urbanization, Transportation Adapted to New Needs, Intellectual Decentralization, Industrialization Versus Decentralization and Conversion, Guidelines for Rural Planning, Conserving the Nation's Natural Resources; Large-Scale Regional Projects: National Company for the Rhône, Forerunner in Regional Development, Vaucluse, Example of Spontaneous

Development, Developing the Lower Rhône, Languedoc and Roussillon—Concerted Operation, Marseilles and the Gulf of Fos—Planning a Regional Metropolis, Toulouse Aerospace Complex—Example of Intellectual Decentralization, Bordeaux—Revitalizing the Southwest, Dunkirk—Catalyst in Reconverting the Industrial North, Further Examples of Regional Planning; Planning in the Region of Paris: Prospects for Paris Within Environment Planning, Master Plan for the Paris Region, Innovation and Renewal in the City of Paris.

1382. Friedmann, John; Alonso, William (Editors), *Regional Development and Planning: A Reader*, Cambridge, Mass. (M.I.T.), 1964, 704 pp. [64-25214]

"Intended primarily as a textbook for . . . urban and regional studies . . . . Brings into one volume many articles of high quality. . . ."
Space and Planning; Location and Spatial Organization; Theory of Regional Development: Resources and Migration, Role of the City, Problems of Rural Periphery; National Policy for Regional Development: Organization for Regional Planning, Objectives and Evaluation, Regional Development Strategies.

1383. Gulick, Luther Halsey, *The Metropolitan Problem and American Ideas*, New York (Knopf), 1962, 167 pp. [62-8675]

Significance of the Metropolitan Problem; American Ideals and Experience with Government; Further Ideas and Theories; Action Program.

1384. Kansky, K. J., *Structure of Transportation Networks: Relationships Between Network Geometry & Regional Characteristics*, Research Paper No. 84, Chicago (Department of Geography, University of Chicago), 1963, 155 pp.

"An analysis of the interaction existing between transportation network patterns and the economic characteristics of regions. Using graph theoretic measures for network geometry, and regression theory for statistical analysis, this study develops an exact, workable, predictive model of transportation systems for economic development and general planning purposes."

1385. Kaplan, Harold, *Urban Political Systems—A Functional Analysis of Metro Toronto*, New York (Columbia University), 1967, 320 pp. [67-29577]

Uses and Limits of Functional Theory; Metro System—Origins and Initial Development; System's Performance; Independent Policy-Making Subsystems; Social Context of Metro Politics; Roles in the Legislative Subsystem; Functional Assessment of the Metro System.

1386. League of Women Voters of California, *The Whole and Its Parts: State and Regional Planning in California*, Oakland, Calif. (355 Grand Ave.), 1962, 29 pp. Paper.

". . . introduction to the role of state government in regional and state planning, this booklet includes a review of some of the issues, a description of California's regional planning activities, and some of the changes that have been proposed to strengthen state and regional planning."

1387. Lowry, Ira S., *Portrait of a Region*, Pittsburgh, Penn. (University of Pittsburgh), 1963, 203 pp. [62-17227]

Regional Framework: Introduction, Geographic Setting, Patterns of Economic Opportunity, The Region's People; Place to Work: Manufacturing Industries, Wholesaling and Commodity Transportation, Administration and Auxiliary Services, Retail Trade and Consumer Services, Mining and Agriculture; Place to Live: Residential Location and Journey to Work, Residential Environment, Metropolis and Hinterland; Appendices.

1388. McKelvey, Blake, *The Emergence of Metropolitan America, 1915-1966*, New Brunswick, N.J. (Rutgers University), 1968, 308 pp. [68-18695]

Emergence of Metropolitan Dilemmas—1915-1920; Outburst of Metropolitan Initiative—1920-1929; Discovery of Metropolitan Inadequacy—1930-1939; Metropolis in War and Peace—1940-1949; Metropolis and the "Establishment"—1950-1959; Federal-Metropolitan Convergence in the 1960's; Notes, pp. 255-296.

1389. Martin, Roscoe C., *Metropolis in Transition: Local Government Adaptation to Changing Urban Needs*, Washington, D.C. (Office of Program Policy, Division of Urban Studies, U.S. Housing and Home Finance Agency), 1963, 159 pp.

"Local governments have developed many cooperative arrangements to meet the challenge of metropolitan growth. This report is a case history account of noteworthy programs illustrative of the leading approaches. . . . In a concluding section, the author presents a somber but probably accurate picture of the rough road leading to effective metropolitan action."

1390. Meier, Richard L., *Developmental Planning*, New York (McGraw-Hill), 1965, 420 pp. [64-25855]

Organization for Planning: Recent Experience with Economic Development, Political Preconditions and Development Strategy, Open Systems for Growth and Development, Constructing an Administration for Planning; Industrial Development: Directions and Strategies for Industrial Growth, Initiating Industrial Enterprises, Industrial Planning Procedures, Administration of an Incentive Program, Human Factor in Industrial Development, Industrial Location, International Specialization; Educational Development: Educational Function in Developmental Societies, Linkages for Educational Planning, Elementary Education, Vocational Education System, Planning for Higher Education, Community and Adult Education, When It Pays To Educate, Distribution of Facilities Over Space, Overview for Planning Strategy in Education; Further Integration of Plans: Rationalizing Urban Development.

1391. Mowitz, Robert J.; Wright, Deil S., *Profile of a Metropolis: A Case Book*, Detroit, Mich. (Wayne State University Press), 1962, 688 pp. [62-14069]

"Ten years in the making, this case book probes deeply into several development issues that came to prominence in the Detroit metropolitan area during the period 1945-1960 . . . . [Deals] with slum clearance, urban renewal, new sources of water, expressway construction, expansion of port facilities, and annexation of fringe areas. The authors find no master decision-makers, although they are impressed by the growing importance of the functional agencies in development decisions. They see a pluralistic, functionally-oriented system slowing down decision-making in the metropolis, but find nothing on which a single metropolitan governmental system can be built."

1392. Owen, Wilfred, *The Metropolitan Transportation Problem,* Washington, D.C. (Brookings Institution), Revised Edition, 1966, 266 pp. Paperback. [66-21151]

Metropolitan Transportation Dilemma; Adapting to the Motor Age; Crisis in Public Transportation; Role of Public and Private Transportation; Transport Financing Policy; Organizing Metropolitan Transportation; Transportation Demand and Community Planning; Facing the Transportation Problem.

1393. Pickard, Jerome P., *Dimensions of Metropolitanism,* Research Monograph 14, Washington, D.C. (Urban Land Institute), 1967, 93 pp. Paper. [67-31485]

Defining and Projecting the Spreading Dimensions of Metropolitanism; Historical Growth of Cities and Urbanized Areas; Growth of Urban Regions and Metropolitan Areas from 1920 to 1960 and Projected Growth to 2000; Regional Development and Growth of Urbanized Areas: 1920-1960, 1960-2000; Metropolitan Belt; Mid-West Region; Southern Regions; Western Regions; Implications of Growth and Future Development of Urbanization and Metropolitanization; Selected References, p. 93.

1394. Pickford, James H. (Study Director), *State Responsibility in Urban Regional Development,* Chicago, Ill. (Council of State Governments), 1962, 209 pp. Paperback. [62-63203]

Urban Development and the Role of the States; Providing the Framework for Action; Action in Two Critical Areas of Urban Development; Bibliography, pp. 193-209.

1395. Pittsburgh Regional Planning Association, *Region with a Future,* Pittsburgh, Penn. (University of Pittsburgh), 1963, 313 pp. [63-17227]

Introduction; As Goes the Nation; Past Extended; Assets and Liabilities; Prospects in Key Industries; Future Economic Structure of the Region; Geographical Shifts Within the Region; Challenge of the Future.

1396. Pittsburgh Regional Planning Association, *Region in Transition,* Pittsburgh, Penn. (University of Pittsburgh), 1963, 462 pp. [63-17227]

Region as a Region; Employment Patterns; Region as a Place to Work; Wages; Business Climate; Structure and Growth; Two Centuries of Industry; Transportation; Mineral Fuels; Primary Metals; Chemicals; Glass; Industry Structure and the Labor Market; Electrical Machinery; Administration and Research; Conclusions.

1397. Public Library Association, American Library Association, *Minimum Standards for Public Library Systems,* Chicago, Ill. (American Library Association), 1967, 69 pp. Paperback. [67-18362]

Role of the Public Library (Gerald W. Johnson); Services and Standards—An

Essential Partnership; Structure and Government of Library Service; Service; Materials—Selection, Organization, and Control; Personnel; Physical Facilities.

1398. San Francisco Bay Conservation and Development Commission, *San Francisco Bay Plan,* San Francisco, Cal. (507 Polk St., Rm. 320, 94102), January 1969, 41 pp. 19 plan maps. Paper. *Supplement,* 572 pp. Paper.

Summary; Objectives; The Bay as a Resource—Findings and Policies; Development of the Bay and Shoreline—Findings and Policies.
Tidal Movement; Sedimentation; Pollution; Fish and Wildlife; Marshes and Mud Flats; Flood Control; Smog and Weather; Salt, Sand, Shells and Water; Appearance and Design; Geology; Fill; Economic and Population Growth; Ports; Airports; Transportation; Recreation; Waterfront Industry; Waterfront Housing; Public Facilities and Utilities; Refuse Disposal; Ownership; Powers; Government; Review of Barrier Proposals; Oil and Gas Production.

1399. Senior, Derek (Editor), *The Regional City, An Anglo-American Discussion of Metropolitan Planning,* Chicago (Aldine), 1966, 192 pp. [66-25661]

Introduction; Context; Structure (Donald Foley; Derek Senior); Strategy (Peter Self); Machinery (Henry Cohen; J. D. Jones); Regional Studies (J. R. James; Frederick Gutheim); Implementation: Urban Renewal (William Slayton; Walter G. Bar); Implementation: New Towns (Robert E. Simon; Wyndham Thomas); Development Values and Controls (Louis Winnick; Nathaniel Litchfield; Daniel Mandelker; Richard May); Transport and Land Use (D. L. Munby; Britton Harris).

1400. Smallwood, Frank, *Metro Toronto: A Decade Later,* Toronto, Ontario (Bureau of Municipal Research), 1963, 41 pp. Paper.

Original Setting; Metro's Major Achievements; Metro's Major Difficulties; References, pp. 40-41.

1401. Sofen, Edward, *The Miami Metropolitan Experiment,* A Metropolitan Action Study, Garden City, N.Y. (Doubleday), Revised Edition, 1966, 375 pp. Paperback. [66-21016]

Background and Setting; Pre-Metro Consolidation; Creation of Metro; Metro Crisis; Progress of Metro; Metro-Problems and Prospects; Appendices; Bibliographical Data, pp. 305-320.

1402. Tobier, Emanuel; Pushkarev, Boris, *The Region's Growth,* New York (Regional Plan Association), May 1967, 143 pp. [A 38358]

Issue of Urban Size: Urbanizing World, Reasons for Urban Growth, Arguments Against Large Size, National Policy and Metropolitan Growth, Atlantic Urban Region, Conclusion, Atlantic Urban Region—Aerial Portrait; Study Area Forecasts—Employment and Population to the Year 2000: Employment, Population, Changes within the Region—Planning Implications of the Forecasts; Appendix.

1403. Towery, F. Carlisle, *Jamaica Center,* New York (Regional Plan Association), April 1968, 73 pp. [A 38358]

Conclusions; Why Centers?; Why Jamaica?; Proposals for Jamaica Center; Steps toward Jamaica Center; Appendix: Projections of Jobs in Office Buildings in the New York Metropolitan Region 1965-2000, Reasons for Locating York College in Jamaica Center, New York Medical College, Medical Care, Need for Rail Service to John F. Kennedy Airport, Jamaica Center—Credibility of Concept, Vehicle Circulation and Parking, Housing Density, Median Incomes and Residential Construction in the Jamaica Area, Daily Work Trips to Jamaica by Income and Origin, 20-Minute Zones of Influence by Mode.

1404. U.S. Department of Agriculture, *A Place to Live: The Yearbook of Agriculture, 1963,* Washington, D.C. (Superintendent of Documents), 1963, 578 pp.

"That [this is] an excellent book on regional planning is a testimonial to the forward-looking people in the Department of Agriculture as well as another indication of the broad scope of planning. Of the 79 articles . . . 70 are on topics directly related to regional planning."

1405. U.S. Housing and Home Finance Agency, *1964 National Survey of Metropolitan Planning,* Prepared for the Subcommittee on Intergovernmental Relations of the Committee on Governmental Operations of the U.S. Senate, Washington, D.C. (Superintendent of Documents), 8 March 1965, 121 pp. Paper.

1406. Vernon, Raymond, *Metropolis 1985—An Interpretation of the Findings of the New York Metropolitan Region Study,* Cambridge, Mass. (Harvard University), 1960, 252 pp. [60-15243]

Interpretation of the "analysis of the Region's probable development, assuming that economic and demographic forces in sight follow their indicated course and assuming that the role of government is largely limited to existing policies." Metropolis Today; Origins of a Metropolitan Region; Growth in the Region's Industries; Labor and Freight; External Economies; From Jobs to People to Jobs Again; Distribution of Jobs Within the Region; Jobs in Motion; From Tenement to Split Level; City Hall and Town Hall; Metropolis 1985.

1407. Williams, Oliver P.; Herman, Harold; Liebman, Charles S.; Dye, Thomas R., *Suburban Differences and Metropolitan Policies—A Philadelphia Story,* Philadelphia, Penn. (University of Pennsylvania), 1965, 363 pp. [64-24492]

Urban Differentiation and the Governing of Metropolitan Areas; Specialization and Differentiation in the Metropolitan Area; Fiscal Responses: Method of Analysis, Municipal Programs, School Revenues, Tax Policy; Land-Use Policy; Attitudes, Opinions and Local Policies; Inter-Local Cooperation; Functional Transfers— Two Cases; Urban Differentiation and the Future of Metropolitan Government.

Bibliographies

1408. Bollens, John C.; Schmandt, Henry J., *The Metropolis: Its People, Politics, and Economic Life,* Commentary on Bibliography, pp. 599-608.

1409. Foley, Donald L., *The Metropolitan Region: A Selective Bibliography*, Bibliography No. 38, Council of Planning Librarians, Monticello, Ill. (Exchange Bibliographies, P.O. Box 229, 61856), 1968, 9 pp.

1410. Golony, Gideon, *National and Regional Planning and Development in the Netherlands, An Annotated Bibliography*, Bibliography No. 47, Council of Planning Librarians, Monticello, Ill. (Exchange Bibliographies, P.O. Box 229, 61856), 1968, 38 pp.

1411. Hudson, Barbara; McDonald, Robert H., *Metropolitan Communities: A Bibliography*—With Special Emphasis Upon Government and Politics, Institute of Governmental Studies, University of California, Public Administration Service (1313 East Sixtieth St., Chicago, 60637), 1967. [67-21928]

Government and Politics: Functions and Problems, Governmental Organization, Politics in Metropolitan Communities; Socio-Economic Background: Social Structure and Process, Population, Metropolitan Community; Index.

1412. Jones, Victor; Hudson, Barbara, *Metropolitan Communities: A Bibliography*, With Special Emphasis upon Government and Politics, Supplement: 1955-1957, Chicago, Ill. (Public Administration Service), 1960, 229 pp. [56-13382]

Government and Politics in Metropolitan Areas: Functions and Problems, Governmental Organization in Metropolitan Areas, Politics in Metropolitan Communities; Socio-Economic Background: Social Structure and Process, Population, Metropolitan Economy.

1413. Pickford, James H. (Study Director), *State Responsibility in Urban Regional Development*, Bibliography, pp. 193-209.

1414. Stoots, Cynthia F., *Regional Planning, An Introductory Bibliography*, Bibliography No. 51, Council of Planning Librarians, Monticello, Ill. (Exchange Bibliographies, P.O. Box 229, 61856), 1968, 5 pp. Paper.

Regional Planning and Development; Regionalism; Regional Economic Development; Political and Administrative Considerations; Case Studies—U.S.; Case Studies—International.

Urban Renewal, Redevelopment, Rehabilitation

Articles, Periodicals

1415. Dixon, John Morris, "New Dimension in Urban Renewal," *Architectural Forum*, Vol. 129, No. 2, September 1968, pp. 44-56.

"The streets and shops of the city will be part of the vast structure proposed for the Tufts-New England Medical Center, which will eventually span several acres of Boston's South Cove area."

1416. *Economist* (England), "New Solutions for Old Cities," Vol. 214, 6 February 1965, pp. 541-544, 547-550, 553-557.

Rebirth for Cities?; Planning with Private Enterprise; The Dispossessed; Rehabilitation—No Cure-all; Renewal Coming of Age; The Quantifiers; Lessons for Britain. Boston Revisited; How Urban Renewal Works; Progress of a Community; Indianapolis Goes It Alone; Hail Columbia!; New Heart for Los Angeles; What Fiscal Zoning Looks Like; A Future for the Past.

1417. Glazer, Nathan, "The Renewal of Cities," *Scientific American*, Vol. 213, No. 3, September 1965, pp. 194-202, 204; also in: Scientific American (Periodical), *Cities*, New York (Knopf), 1965, 211 pp. Paperback. [65-28177]

"Many U.S. cities, with the aid of the Federal Government, are engaged in ambitious efforts to renew themselves. It is not certain, however, that the overall gains of these programs have outweighed the losses."

1418. Goldston, Eli; Hunter, Allen Oakley; Rothrauff, Guido A., Jr., "Urban Redevelopment—The Viewpoint of Counsel for a Private Developer," *Law and Contemporary Problems*, Vol. 26, No. 1, Winter 1961, pp. 119-177.

Introduction; Deciding Whether to Bid for a Project; Preparing to Bid for Sponsorship of a Project; Submission of the Bid; Unsuccessful and Successful Bidder; Preparing for Commencement of Construction; FHA and FNMA Aids to Urban Redevelopment; Occupancy and Operation.

1419. Mao, James C. T., "Efficiency in Public Urban Renewal Expenditures Through Benefit-Cost Analysis," *Journal of the American Institute of Planners*, Vol. XXXII, No. 2, March 1966, pp. 95-107.

Statement of the Problem; Theoretical Aspects: Benefits and Costs of Urban Renewal, Social Cost of Capital, Treatment of Intangible Benefits and Costs; An Application: Background and Location of the Project, Effect of Urban Renewal on Property Value, on Local Finance, on Housing Welfare, Total Cost of Renewal and Its Financing, Overall Evaluation; Summary and Conclusions.

1420. Montgomery, Roger, "Improving the Design Process in Urban Renewal," *Journal of the American Institute of Planners*, Vol. XXXI, No. 1, February 1965, pp. 7-20.

Early Experience in Detroit; Gaps in Design Process; "Open" Plans; Design Process Controls.

1421. Perloff, Harvey S., "New Towns Intown," *Journal of the American Institute of Planners*, Vol. XXXII, No. 3, May 1966, pp. 155-162.

"Suggesting a broad community plan involving staged rebuilding and rehabilitation, construction of a Lighted Center and a variety of public and private facilities, including a low-rent center for local artisans and merchants, and the creation of a development corporation to carry out the plan. Only the artisan-shopping center idea was actually implemented. . . ."

1422. Plager, Sheldon J.; Handler, Joel F., "The Politics of Planning for Urban Redevelopment: Strategies in the Manipulation of Public Law," *Wisconsin Law Review*, Vol. 1966, No. 3, Summer 1966, pp. 724-775.

Introduction; Center City; Strategies of Public Urban Redevelopment.

1423. Smith, Robert, "Renaissance of Cleveland, Ohio," *Management Services*, Vol. 3, No. 2, March-April 1966, pp. 44-51.

"One branch of management services leads inevitably to another and the scope of such services continues to expand . . . consider the leading role taken by an accounting firm in [Cleveland]."

Economic Approach; Erieview; New Plan Suggested; Central Group Stipulated; Plans Geared to 1980; Planning Areas; Eminent Domain Needed; Retail Core; Financing Redevelopment, Progress to Date.

1424. Wallace, David A. (Editor), Planning the City's Center, *Journal of the American Institute of Planners*, Special Issue, Vol. XXVII, No. 1, February 1961, 73 pp.

Downtown Decay and Revival (Charles Abrams); The Role of Theory in CBD Planning (Dennis Durden, Duane Marble); Approaches to Analysis (John Rannells); Transportation Planning for Central Areas (J. Douglas Carroll, Jr., Roger L. Creighton, John R. Hamburg); Space for CBD's Functions (Larry Smith); Downtown Housing (Morton Hoffman); The Form of the Core (Blanche Lemco Van Ginkel).

1425. Wilson, James Q., "Planning and Politics: Citizen Participation in Urban Renewal," *Journal of the American Institute of Planners*, Vol. XXIX, No. 4, November 1963, pp. 242-249.

Hyde Park–Kenwood Experience; Social Differences in Citizen Participation; Community Organization Strategies; Implications for Renewal Programs; Political Effects.

Books, Reports, Pamphlets

1426. Abrams, Charles, *The City Is the Frontier*, New York (Harper & Row), 1965, 394 pp. [64-25145]

Urban Problems in Our Urban Age: City at Bay, Slum, Housing Problem, Racial Upheaval in Cities; Prescription–Urban Renewal: Federal Role in Housing and Urban Renewal, Urban Renewal Is Renewed, Rosier Side of Renewal Investment, Dislocation and Relocation, Some Blessings of Urban Renewal, Rehabilitation and a Tale of Two Cities; Larger Perspectives: Implications of the New Federal System, Housing Problem–Need for Objectives, Blueprints for American Cities, The City Faces the Future.

1427. Anderson, Martin, *The Federal Bulldozer*, New York (McGraw-Hill), 1964, 272 pp. Paperback. [64-7546]

Introduction—Renewal by Government Decree; How the Program Works; The Program Mushrooms; Consequences; Land Lies Vacant; New Buildings; Private Developer; Where Does the Money Come From?; Rehabilitation; Tax Increase Myth; Effect on the National Economy; Urban Renewal and the Constitution; Changes in Housing Quality; Conclusion—Repeal the Urban Renewal Program; Bibliography, pp. 262-266.

1428.   Angle, Paula (Editor), *City in a Garden: Homes in the Lincoln Park Community,* Chicago, Ill. (Lincoln Park Conservation Association), 1963, 64 pp.

"The conservation of a community characterized in 1948 as 'predominantly in a state of deterioration' is shown in this picture story. . . . Private and institutional owners spent an estimated $6.7 million in the year 1962 alone for improvement, repairs, new additions and new dwellings. Small gardens and attention to architectural details have been prominent features in the restoration."

1429.   Davies, J. Clarence III, *Neighborhood Groups and Urban Renewal,* New York (Columbia University), 1966, 235 pp. [65-27764]

Introduction; Urban Renewal in New York City; Seaside-Hammels; West Village; West Side Urban Renewal Area; Formation of Neighborhood Attitudes; Neighborhood and Non-neighborhood Actors; Neighborhood Access to Government; Neighborhood Groups and the Public Interest.

1430.   Doxiadis C. A., *Urban Renewal and the Future of the American City,* Chicago, Ill. (Public Administration Service), 1966, 174 pp. [65-27611]

Introduction; Background of Urban Renewal; Urban Renewal Problems in the United States; Evolution of Human Settlements; Future of Human Settlements; Future Urban Renewal Policies, Programs, and Projects; Strategy of Urban Renewal; In Conclusion.

1431.   Duggar, George S., *Renewal of Town and Village I: A World-Wide Survey of Local Government Experience,* The Hague, Netherlands (Martinus Nijhoff), 1965, 243 pp.

". . . Survey of renewal in human settlements around the world which covers the whole range of towns from metropolis to village. . . . Based largely on questionnaire replies sent by 31 countries, including five in Africa, nine in Asia, ten in Europe, three in North America and four in Latin America."

1432.   Dyckman, John W.; Isaacs, Reginald R., *Capital Requirements for Urban Development and Renewal,* New York (McGraw-Hill), 1961, 334 pp. [60-14448]

Introduction—Problem Areas in an Economy of Abundance; The Nation Approaches Renewal; Urban Renewal in a Growing Economy; Estimating Renewal Needs from the Case Study of a Metropolitan Area; Case Study—Model for Planning; Case City Implications for Nationwide Capital Replacement; Limits of Renewal—National Economic Growth; Financing Urban Renewal: Taxing and Current Spending, Problems of Debt; Conclusions; Appendices: Note on Case Method, Methodology of

Case Analysis, Description of Case City Metropolitan Areas and Suburbs, Cost Estimates, Paying the Bill.

1433. Frieden, Bernard J., *The Future of Old Neighborhoods*, Rebuilding for a Changing Population, Cambridge, Mass. (M.I.T.), 1964, 209 pp. [64-17322]

". . . This book . . . balances careful analysis with well-thought out recommendations for public policy and action." Communities in Decline; Central City in the 1950's; Rebuilding the Declining Areas—Locational Preferences for New Housing; Economics of New Housing in Old Neighborhoods; Regional Framework; Policies for Rebuilding; Appendices.

1434. Grebler, Leo, *Urban Renewal in European Countries, Its Emergence and Potentials*, Philadelphia, Penn. (University of Pennsylvania), 1964, 136 pp.

"The problems that propel West European countries into urban renewal and the solutions they seek are contrasted with [the U.S.] program, which has been in existence since 1949. . . . Describes the status of urban renewal programs in . . . France, Denmark, Netherlands, Great Britain, Sweden, West Germany and Italy, and deals extensively with the legal and financial tools employed in land assembly and disposition, and with the relocation of displaced residents and business firms."

1435. Greer, Scott, *Urban Renewal and American Cities*, The Dilemma of Democratic Intervention, Indianapolis, Ohio (Bobbs-Merrill), 1965, 201 pp. [65-26544]

Overview of the Program; Urban Renewal as a Theory; Urban Renewal and the Local Community; The LPA and State and Local Government; Shape of the Program—Local Projects and Federal Control; Program in the Light of Social Trends; Urban Renewal and the Future—What Can Be Done?; Greatest Need.

1436. Gruen, Victor, *The Heart of Our Cities*—The Urban Crisis: Diagnosis and Cure, New York (Simon and Schuster), 1964, 368 pp. Paperback. [64-13607]

The City: What Makes a City?, Planning—Waste or Wisdom?, Anatomy of the City, Death or Transfiguration; The Anti-City: Spread, Sprawl and Scatterization, Flight and Blight, Tired Hearts of Our Cities, False Friends of the City, Full Speed Ahead on a Dead-End Road, Land Wasters, Sins of Omission and Commission, Chaotic State of Architecture, Appearance or Disappearance—That Is The Question; The Counterattack: Search, Separate and Meld, Taming of the Motorcar, Pedestrian and Other Future Modes of Transportation, Emerging New Urban Pattern, Rebirth of the Heart, Are We Equipped?

1437. Horwood, Edgar M.; Others, *Using Computer Graphics in Urban Renewal*, Community Renewal Program Guide No. 1, Urban Renewal Service, Urban Renewal Administration, Housing and Home Finance Agency, Washington, D.C. (U.S. Dept. of Housing and Urban Development, 451 7th St., S.W., 20410), January 1963, 281 pp. Paper.

Introduction and Glossary; Card Mapping Program; Array Program; BCD to Binary Tape Transfer Program; Distribution Program; Tape Mapping Program; Appendices: Data System Design for the Spokane CRP, Principles of Multiphasic Screening, Model Constructs in Urban Analysis, Block Diagrams of the Computer Program.

1438. Kaplan, Harold, *Urban Renewal Politics—Slum Clearance in Newark,* New York (Columbia University), 1963, 219 pp. [63-19076]

Introduction; NHA—Strategy of Slum Clearance; Politicos; Civic Leaders—Neighborhood Rehabilitation, Economic Development; Planners; Grass Roots; Urban Renewal System in Newark; Postscript—1963.

1439. Levitan, Sar A., *Federal Aid to Depressed Areas—An Evaluation of the Area Redevelopment Administration,* Baltimore, Md. (Johns Hopkins), 1964, 268 pp. [64-16310]

Politics and Economics of Depressed Area Legislation; ARA Sets Up Shop; Area Designation and Characteristics; Industrial and Commercial Loans; Loans and Grants for Public Facilities; Training the Unemployed for Jobs; Community Planning—Unlocking of Resources; Package of Tools; Concluding Observations—Policy Dilemma.

1440. Little, Arthur D., Inc., *Community Renewal Programming: A San Francisco Case Study,* New York (Praeger), 1966, 235 pp. [66-18906]

Need for Urban Renewal; Changing City; Meeting the Challenge of Change; Long-Range Policies for Housing and Renewal; Program of Action: 1966-72; Financing Renewal; Need for an On-Going Information Service; CRP Model; Renewal Attitudes Survey.

1441. Little, Arthur D., Inc., *CRP, San Francisco Community Renewal Program,* Cambridge, Mass. (Arthur D. Little, Inc.), October 1965, 173 pp. Paper.

Need for Urban Renewal: Quality of the Residential (Living) Environment, of the Employment (Working) Environment; Changing City: Population, Housing, Industry and Employment Composition, Industrial, Office and Commercial Construction, Personal Income; Meeting the Challenge of Change: Renewal Goals, Objectives, and Targets, Barriers to Improvement and Ways to Overcome Them; Long-Range Policies for Housing and Renewal: Overall Policy, Key Elements of Renewal Policy, Types of Residential Treatment Areas, Policies, and Approach, Industrial-Commercial Treatment Areas, Policies, and Approach; Program of Action, 1966-1972: Goals and Targets, Assessing Appropriate Mechanisms for All Neighborhoods, Guidelines for Selecting Renewal Program Areas, Designation of Program Areas, Supporting Public Actions; Financing Renewal: Past and Current Methods, Potential Sources of Non-Cash Credits, Cost of the Six-Year Program, Ability to Finance the Six-Year Program; Need for an On-Going Information Service: Up-to-Date Information, Key Symptomatic Indicators; Appendix A: The CRP Model.

1442. Little Tokyo Redevelopment Association, in cooperation with the Los Angeles City Planning Department, *General Plan for Little Tokyo,*

Los Angeles, Cal. (City Planning Dept., City Hall, 90012), November 1963, 21 pp.

Summary; General Plan (Map); Historical Perspective; General Location Map; Relative Location Map; Inventory; Air Photo; 1963 Land Use Inventory (Map); General Plan (Text); Thoroughfare Plan (Map); Proposed Street Improvements (Map); Thoroughfare Plan (Text); Implementation Program.

1443. McFarland, M. Carter; Vivrett, Walter K. (Editors), *Residential Rehabilitation,* Minneapolis, Minn. (School of Architecture, University of Minnesota), 1966, 331 pp.

Residential Rehabilitation—Overview (M. Carter McFarland); Elements of Urban Renewal Rehabilitation: Residential Rehabilitation Potential Through the Urban Renewal Program (Robert E. McCabe), Selection, Delineation and Planning of Residential Rehabilitation Areas (Thomas H. Jenkins; Walter L. Smart), Preparing and Organizing a Neighborhood for Rehabilitation (Grace C. Gates; Rose Jones), Rehabilitation Standards (M. Carter McFarland), Feasibility of Property Rehabilitation Financing Under FHA Section 220 (Alfred W. Jarchow), Economic Feasibility of Rehabilitation (Charles Shannon), Some Observations on Neighborhood Characteristics and How To Deal with Them (John R. Rothermel), Introduction to Mortgage Financing (Seymour Baskin), Techniques of Property Owner Servicing (Robert B. McGilvray), Assisting Property Owners with Financing—Baltimore Approach (Edgar M. Ewing), Special Approach to Absentee Owned Properties (Vernon W. Stull, Jr.), Urban Renewal Rehabilitation—Problems and Promise (William L. Slayton); Case Studies in Urban Renewal Rehabilitation: Harlem Park Project in Baltimore, Maryland (Richard L. Steiner)—Some Administration Considerations at Harlem Park (Edgar M. Ewing), Role of the Area Office in Harlem Park (James H. Gilliam, Jr.), Washington Park Project in Boston, Massachusetts (Thaddeus J. Tercyak), Wooster Square and Dixwell Projects in New Haven, Connecticut (Melvin J. Adams; Donald Kirk: Louis Onofrio); Novel Approaches to Rehabilitation: Physical and Human Rehabilitation in a Single Block in New York's Harlem (Neal J. Hardy), Revolutionary Approach to Rehabilitation Technology (Edward K. Rice), Big Industry Tackles Rehabilitation (Bowen Northrup).

1444. Messner, Stephen D., *A Benefit-Cost Analysis of Urban Redevelopment,* Indiana Business Report No. 43, Bloomington, Ind. (Bureau of Business Research, Graduate School of Business, Indiana University), 1967, 115 pp. Paperback.

Urban Renewal—Public Policy and Investment Criteria; Benefit-Cost Analysis Applied to Urban Renewal; Indianapolis Redevelopment; Benefits and Costs of Indianapolis Redevelopment; Conclusions and Recommendations; Appendices; Bibliography, pp. 109-115.

1445. Miller, J. Marshall (Editor), *New Life for Cities Around the World,* International Handbook on Urban Renewal, Alhambra, Cal. (Books International), 1959, 233 pp.

Background and Conclusions; Seminar Proceedings—Selected Papers and Reports; Country and City Reports on Urban Renewal Programs.

1446. Ministry of Housing and Local Government, *The Deeplash Study—Improvement Possibilities in a District of Rochdale, London,* England (Her Majesty's Stationery Office), 1966, 74 pp. Paper.

Introduction; Physical Survey; Social Survey; Improvement Possibilities; General Conclusions.

1447. Niebauck, Paul L., *Relocation in Urban Planning: From Obstacle to Opportunity,* Philadelphia, Penn. (University of Pennsylvania), 1968, 123 pp. [68-21556]

Growing Concern for Adequate Relocation; Condition of the Elderly in Urban America; Knowledge Gained Through Demonstration Studies; Toward a Program of Positive Relocation; Incomplete Understandings.

1448. Providence City Plan Commission, *College Hill, A Demonstration Study of Historic Area Renewal,* Providence, R.I. (City Hall, 02903), Second Edition, 1967, 230 pp. Paper. [65-28006]

Summary; Preservation in America; Survey Techniques; Renewal of College Hill; Progress Since 1959.

1449. Rossi, Peter H.; Dentler, Robert A., *The Politics of Urban Renewal: The Chicago Findings,* New York (Free Press), 1962, 318 pp. [59-13865]

". . . documents the interplay of citizen, university and local government in shaping an urban renewal program for the Hyde Park-Kenwood (Univ. of Chicago) area on Chicago's South Side. Of special interest to those who urge greater citizen participation in the renewal process."

1450. Rothenberg, Jerome, *Economic Evaluation of Urban Renewal—Conceptual Foundation of Benefit-Cost Analysis,* Washington, D.C. (Brookings), 1967, 277 pp. [67-19190]

Scope and Purpose of Urban Renewal: Introduction and Summary, Criteria and Method, Elimination of Blight and Slums, Other Goals of Urban Renewal, Governmental Profits and Intergovernmental Subsidies, Structure of Benefit-Cost Comparisons; Measurement of Benefits: Model of Redevelopment Impact, Internalization of Externalities, Effects of Changes in the Housing Stock, Reduction in the Social Costs of Slum Living; Numerical Example, Issues of Public Policy: Major Criticisms of the Redevelopment Program, Some Alternatives to Redevelopment; Bibliography, pp. 259-265.

1451. Van Huyck, Alfred P.; Hornung, Jack, *The Citizen's Guide To Urban Renewal,* West Trenton, N.J. (Chandler-Davis), October 1962, 160 pp. Paperback. [62-21693]

What Is Urban Renewal?; Why Undertake Urban Renewal?; Should Urban Renewal Be Undertaken?; How To Begin; Who Should Carry Out The Program?; Survey and Planning Application; Program for Community Improvement; Hiring The Executive Director; Hiring Technical Consultants; Citizen Participation; Urban Renewal Plan; Planning For Buying Properties; Planning for—Family Relocation, Business Relocation, Disposition of Project Land; Urban Renewal Financing; Conservation; From Planning to Action; Community Renewal Program; Appendices.

1452. Wilson, James Q. (Editor), *Urban Renewal: The Record and the Controversy*, Publication of the Joint Center for Urban Studies, Cambridge, Mass. (M.I.T.), 1966, 683 pp. [66-14344]

Economics of Cities and Renewal: Changing Economic Function of the Central City (Raymond Vernon), Housing Markets and Public Policy (William G. Grigsby), Economics of Urban Renewal (Otto A. Davis and Andrew B. Whinston); Urban Renewal Background and Goals: Federal Urban Renewal Legislation (Ashley A. Foard and Hilbert Fefferman), Legal and Governmental Issues in Urban Renewal (William S. Sogg and Warren Werthheimer), Operation and Achievements of the Urban Renewal Program (William L. Slayton); Urban Renewal in Practice, Three Cases: Urban Renewal in Newark (Harold Kaplan), in Boston (Walter McQuade), The Industrial Corporation in Urban Renewal (Hubert Kay); Relocation and Community Life: Housing of Relocated Families (Chester Hartman), Summary of a Census Bureau Survey (U.S. Housing and Home Finance Agency), Comment on the HHFA Study of Relocation (Chester Hartman), Grieving for a Lost Home, Psychological Costs of Relocation (Marc Fried), The Small Businessman and Relocation (Basil Zimmer); Government and Citizen Participation in Urban Renewal: Planning and Politics, Citizen Participation in Urban Renewal (James Q. Wilson), Local Government and Renewal Policies (Norton E. Long); Planning and Design: Cities, Planners, and Urban Renewal (William Alonzo), Improving the Design Process in Urban Renewal (Roger Montgomery); Challenges and Responses: Federal Bulldozer (Martin Anderson), Urban Renewal Realistically Reappraised (Robert P. Groberg), Federal Bulldozer, A Review (Wallace F. Smith), Failure of Urban Renewal (Herbert J. Gans), Some Blessings of Urban Renewal (Charles Abrams); Future of Urban Renewal: Policies for Rebuilding (Bernard Frieden), General Strategy for Urban Renewal (William G. Grigsby), New Directions in Urban Renewal (Robert C. Weaver).

1453. Younger, George D., *The Church and Urban Renewal*, Philadelphia, Penn. (Lippincott), 1965, 216 pp. [65-14899]

Fame of Urban Renewal; Search of Metropolis; Issues Raised by the Renewal of Metropolis; Politics of Urban Renewal; Churches in Urban Renewal; Church's Role; Theological Afterword; Selective Bibliography, pp. 208-212.

Bibliographies

1454. Anderson, Martin, *The Federal Bulldozer*, Bibliography, pp. 262-266.

1455. Kaufman, Jerome L.; Vance, Mary, *Community Renewal Programs: A Bibliography*, Bibliography No. 32, Council of Planning Librarians, Monticello, Ill (Exchange Bibliographies, P.O. Box 229, 61856), May 1965, 23 pp.

General References; Case Studies.

1456. Messner, Stephen D., *A Benefit-Cost Analysis of Urban Redevelopment*, Bibliography, pp. 109-115.

1457. Rothenberg, Jerome, *Economic Evaluation of Urban Renewal—Conceptual Foundation of Benefit-Cost Analysis*, Bibliography, pp. 259-265.

1458. Wheaton, William L. C.; Baer, William C.; Veeter, David M., *Housing Renewal and Development Bibliography*, Bibliography No. 46, Council of Planning Librarians, Monticello, Ill. (Exchange Bibliographies, P.O. Box 229, 61856), 1968, 38 pp.

Housing Problems and Policies; Housing Industry; Analysis of Housing Requirements, Housing Markets and the Environmental Effects of Housing.

1459. Younger, George D., *The Church and Urban Renewal*, Selective Bibliography, pp. 208-212.

New Towns–Communities, Demonstration–Model Cities

Articles, Periodicals

1460. Atkinson, J. R., "Washington New Town, England," *Washington University Law Quarterly*, Vol. 1965, No. 1, February 1965, pp. 56-70.

Introduction; County Durham; Selection of a Site at Washington; The Public Inquiry; Problems of Developing the New Town.

1461. Buder, Stanley, "The Model Town of Pullman: Town Planning and Social Control in the Gilded Age," *Journal of the American Institute of Planners*, Vol. XXXII, No. 1, January 1967, pp. 2-10.

Building a Model Town; Management of Pullman; Social Criticism of the Town; Demise of a Model Town; Vision Versus Reality.

1462. Bull, D. A., "New Town and Town Expansion Schemes, Part II: Urban Form and Structure," *The Town Planning Review*, Vol. 38, No. 3, October 1967, pp. 165-186.

Principles Influencing Urban Form; The Individual Reports; Summary and Conclusions.

1463. Gladstone, Robert, " 'New Towns' Role in Urban Growth Explored," *Journal of Housing*, Vol. 23, No. 1, January 1966, pp. 29-36.

"Public policy issues examined—form of local government, land use controls, relationship to central cities." Central City, New Communities; New Community Concepts; Present Challenge; New Community Values; Metropolitan Goals; Government and Social Issues; Possible Approach; Outlook; Variety; Industry; Low-Cost Housing.

1464. Gruen, Victor, Associates, "Valencia, A Planned New City," *Arts & Architecture*, Vol. 83, No. 10, November 1966, pp. 18-23.

1465. Hancock, Macklin L.; Harrison, Peter, Policies for New Towns, in: *Planning 1965*, Chicago, Ill. (American Society of Planning Officials), 1965, pp. 264-278. Paperback. [39-3313]

"Policies, Problems, and Prospects in Legislation, Design, and Administration" (Macklin L. Hancock); "Canberra—Case Notes on a New Town": Economic Base, Private Enterprise, Paying Off, Long-Term Planning, Satellite Towns or New Cities (Peter Harrison).

1466. *House and Home*, "El Dorado Hills: New Model for Tomorrow's Satellite Cities," Vol. XXIII, No. 3, March 1963, pp. 107-115.

Step-by-Step Analysis Assures Best Use of the Land; Community Plan Combines City, Village, and Resort Life; Village Plans Put All Houses on Quiet, Safe Streets; Giant Cluster Plan Allows a Full Range of Facilities; Model Houses Were Designed to Fit the Land—and the Demand; El Dorado's High Standards Are Based on Six Development Concepts—Plus the Owner's Willingness to Invest Heavily and Wait for Profits.

1467. *House and Home*, "New Towns: Are They Just Oversized Subdivisions —With Oversized Problems?," Vol. XXIX, No. 6, June 1966, pp. 92-103.

Problems: Location, Financing, Sales, Industry, Government; Who Builds?, Residents; Reston—Fine Concept and Handsome Start—Both Hamstrung by Management Problems; El Dorado Hills—Beautiful Land and a Good Plan Hurt by a Sickly Market; Clear Lake City—Unhappy Mixture of Instant Sprawl and Builder Problems; Columbia—Planned to the Nth Degree, It Has Yet to Face Its First Real Test; In the Face of All These Problems, Will New Towns Ever Make Economic Sense?

1468. Janis, Jay, "Model Cities, Their Role Is Vital in Developing an Overall Urban Strategy," *Nation's Cities*, Vol. 6, No. 9, September 1968, pp. 9-12.

Program Has Constraints; Community Links Vital; Focus on a Neighborhood; Can Concept Be Effective?; Means to Experiment; Rx for Urban Change.

1469. *Journal of the American Institute of Planners*, Creating New Communities: A Symposium on Process and Product, Vol. XXXIII, No. 5, November 1967, pp. 370-409.

Comparative New Community Design (David R. Godschalk); Social Research and New Communities (Peter Willmott); A Sketch of the Planning-Building Process for Columbia, Maryland (Morton Hoppenfeld).

1470. Ketchum, Morris, Jr. (Jury Chairman), *Stockholm, Tapiola, Cumbernauld*, Jury Report, R. S. Reynolds Memorial Award for Community Architecture, *AIA Journal*, Vol. XLVIII, No. 3, July 1967, pp. 36-58. (Also available as reprint from Reynolds Metals Co., P.O. Box 2346, Dept. AD, Richmond, Va. 23218.)

Stockholm: Site, City Planning Policy, Objectives, Urban Design, Implementation, Planning Process, Comprehensive Approach, Revitalized City Core, Rapid Transit System, Highway System, New Neighborhoods, Summary—Concept, Program, Regional Concept; Tapiola: Regional Concept, Organization, Urban Design, Neighborhoods, Town Center, Circulation, Financing, Summary—Concept, Program, Architecture; Cumbernauld: Regional Concept, Site, Objectives, Urban Design Factors, Solution, Density, Circulation, Housing, Industry, Center, Planned Future Development, Summary—Concept, Program, Architecture; Urban Characteristics Common to Stockholm, Tapiola and Cumbernauld.

1471. Penfold, Anthony H., "Ciudad Guayana, Planning a New City in Venezuela," *The Town Planning Review*, Vol. 36, No. 4, January 1966, 225-248.

Guayana Region; City: Site, Immediate Action Programme, Growth Targets, Physical Planning Goals, Major Steps in the Planning Process, Development Strategy; Future Perspective.

1472. Peterson, William, "The Ideological Origins of Britain's New Towns," *Journal of the American Institute of Planners*, Vol. XXXIV, No. 3, May 1968, pp. 160-169.

Garden City—The Blueprint; On the Myth of Victorian Smugness; Nature v. the City; Community; Influence of Kingsley and Ruskin; Liberalism, Conservation, and Social Conscience; Opposition to "Mass Society"; Conclusion.

1473. Rodwin, Lloyd, "Ciudad Guayana: A New City," *Scientific American*, Vol. 213, No. 3, September 1965, pp. 122-130, 132; also in: Scientific American (Periodical), *Cities*, New York (Knopf), 1965, 211 pp. Paperback. [65-28177]

"Venezuela is building a metropolis as a key part of a plan to advance its national economy. The effort is complicated by the fact that the impoverished residents of the countryside are 'imploding' to the site."

1474. *Town and Country Planning*, Special Issue, Vol. 36, No. 1-2, January-February 1968, 136 pp. Paperback. (28 King St., Covent Garden, London WC2, England)

Achievement and Objectives (Wyndham Lewis); Parliament Debates the Bill (Lewis Silkin); Legislative Framework (A. E. Telling); Ministerial Views (Lord Reith; Lord Silkin; Lord Brooke); Political Leaders and the Act (Anthony Greenwood; Geoffrey Ripon; Jeremy Thorpe); The Prime Minister at Stevenage (Harold Wilson); Administration and New Towns (William Hart); Land for Towns—New and Old (Robin H. Best); Value of the Private Garden (Arnold Whittick); Gold-Mines of the Future (Henry Wells); Whose Heritage? (Maurice A. Ash); The Shape of Cities to Come (D. E. C. Eversley); Landscape Policy and Achievement (Bodfan Gruffydd); Planting at Peterlee (Ian Watson); Milton Keynes—London's Latest New Town (Lord Campbell); New Town Statistics; New Towns Financial Results (F. J. Osborn); New Town Notebook; New Town Forum; Role of Private Enterprise; Building Societies and New Towns—Note on Home Ownership (C. John Dunham); Cold Comfort Plan (David Hall); Dawley—Preservation and

Planned Change (Arnold Whittick); Craigavon—the "rural city" (A. H. Bannerman); Early Days in a New Town (F. J. Osborn); A Theatre for Harlow (Derek Hawes); From Coal-Face to Razor-Blades (P. S. Johnson); Communicare at Killingworth (Bernice Broggio); Publications on Britain's New Towns, p. 136.

1475. *Washington University Law Quarterly*, "The Administration of the English New Towns Program," Vol. 1965, No. 1, February 1965, pp. 16-55.

Introduction; Summary of the Program; Major Aspects of the English Program: Site Selection and Designation, Town Development, Land Acquisition, Finance, Governmental Structure; Conclusion; Appendix.

Books, Reports, Pamphlets

1476. Creese, Walter L. (Editor), *The Legacy of Raymond Unwin: A Human Pattern for Planning*, Cambridge, Mass. (M.I.T.), 1967, 234 pp. [67-12087]

Introduction—Ideals and Assumptions of Sir Raymond Unwin (Walter L. Creese); Quotations Cherished by Unwin; "The Dawn of a Happier Day"; "Gladdening v. Shortening the Hours of Labour"; The Art of Building a Home; Cottage Plans and Common Sense; Town Planning in Practice—Introduction to the Art of Designing Cities and Suburbs; Nothing Gained by Overcrowding! How the Garden City Type of Development May Benefit Both Owner and Occupier; "Higher Building in Relation to Town Planning"; "Regional Planning With Special Reference to the Greater London Regional Plan"; "Housing and Town Planning Lectures at Columbia University"; "Land Values in Relation to Planning and Housing in the United States."

1477. Eichler, Edward P.; Kaplan, Marshall, *The Community Builders*, Berkeley, Cal. (University of California), 1967, 196 pp. [67-13601]

Planning and the Critique of Urban Development; From Frontier to Megalopolis; Breaking Ground for the New Communities; New Breed; Political Relations; Market; Community Building as a Market and Political Strategy; Taxation; Community Building as a Business; New Communities and Public Policy; Appendix: New Community Developments, 1964 . . .

1478. Evenson, Norma, *Chandigarh*, Berkeley, Cal. (University of California), 1966, 116 pp. & Plates. [66-11037]

Introduction To a Capital; Creation of the City; Albert Mayer—Master Plan; Matthew Nowicki—Final Effort; Second Team; Le Corbusier—Master Plan; Getting Under Way—The "Three Disciplines"; The Sector; Commercial Center; Educational and Cultural Areas; Capitol Complex; Chandigarh—End of the Beginning; Selected Bibliography, pp. 111-113.

1479. International City Managers' Association, *New Towns—New Dimension of Urbanism*, Chicago, Ill. (1313 East 60th St., 60637), November 1966, 54 pp. Paper.

New Towns in Great Britain (G.W.G.T. Kirk); Europe Offers New Town Builders Experience (Frederick Gutheim); New Towns Solve Problems of Urban Growth (Robert Gladstone; Harold F. Wise); Federal Proposals May Solve City Problems (Robert C. Weaver); Prototype City—Design for Tomorrow (Wayne E. Thompson); Urban Growth Challenges New Towns (Stanley Scott); What New Towns Ought to Be (David S. Arnold).

1480. London County Council, *The Planning of a New Town,* Data and Design Based on a Study for a New Town of 100,000 at Hook, Hampshire, London, England (Information Bureau, The County Hall, S.E. 1), 1961, 182 pp.

Selection of the New Town Site; Main Aims of the Town; People; Master Plan; Residential Areas; Central Area; Industry; Community Services; Recreation and Open Space; Communications; Engineering Services; Costs; Programming; Appendices: Population and Household Projections, Projection of Occupied Population, Distribution of Employment in Year 50, Projection of Housing Need, Use of the Town Garden, Phasing of Non-Retail Uses in Central Areas of New Towns, Method of Estimating Shopping Requirements, Standard Industrial Classification, Service Industry in Year 50—Land Needs and Employment, Assumptions Made in Estimating Employment in Manufacturing Industry in Residential Areas, Manufacturing Industry—Particulars of Industrial Estates in 26 Towns, Those Suitable for Location in Residential Areas, Programme of Land Requirements for Industry, Industrial Land in Years 15 and 50—Breakdown into Uses for Costing Purposes, Regional Road Design Factors, Town Roads—Journey to Work, Pedestrian Movement Analysis, Increased Costs per Person Resulting from Smaller Average Family Size, Cost of Road System Compared with that of Existing New Towns, "All-In" Development Costs at Different Densities, Percentage Reduction on London Housing Costs for Building at Hook, Summary of Major Road Costs, etc. To Be Apportioned, Highway Grants, Apportionment of Major Road Costs, etc., Housing Subsidies, Comparison of Development Costs at Different Densities; Bibliography, pp. 177-181.

1481. Ministry of Housing and Local Government, Welsh Office Sub-Committee of the Central Housing Advisory Committee, *The Needs of New Communities,* A Report on Social Provision in New and Expanding Communities, London, England (Her Majesty's Stationery Office), 1967, 123 pp. Paperback.

Types of New Communities; Moving to a New Community; Needs of New Communities; Providing for Community Needs; Finance of Social and Community Facilities; Conclusions and Recommendations; Appendices.

1482. Osborn, Frederick J.; Whittick, Arnold, *The New Towns, The Answer to Megalopolis,* New York (McGraw-Hill), 1963, 376 pp. [64-14633]

New Towns in Modern Times; Functions and Failings of Towns; Some Data on Town Growth; Experimental New Towns; Town Growth and Governmental Intervention; Evolution of the New Towns Policy; Legislation for New Towns; Finance of New Towns; Town and Country Pattern; Antagonisms to New Towns; Achievement, Emulation and Prognostic; Stevenage; Crawley; Hemel Hempstead; Harlow; Newton Aycliffe; East Kilbride; Peterlee; Welwyn Garden City; Hatfield; Glenrothes; Basildon; Bracknell; Cwmbran; Corby; Cumbernauld; Skelmersdale; Livingston; Dawley; Aesthetic Aspect of Urban Environment in New Towns; Selected Bibliography, pp. 359-361.

1483. Rodwin, Lloyd, *The British New Towns Policy—Problems and Implications,* Cambridge, Mass. (Harvard University), 1956, 252 pp. [56-6512]

New Towns—Promise and Reality; Garden City Idea—Historical Appraisal; New Towns and Town Development—Problems of Execution; Perspective on New Towns—Implications and Prospects; Appendices: Achilles Heel of British Town Planning, British Local Government Organization and Functions, Look of the New Towns.

1484. Stein, Clarence S., *Toward New Towns for America,* Cambridge, Mass. (M.I.T.), 1957, 263 pp. Paper. [57-6538]

Introduction (Lewis Mumford); Toward New Towns for America; Sunnyside Gardens; Radburn; Chatham Village; Phipps Garden Apartments, I and II; Hillside Homes; Valley Stream Project; Greenbelt: Greenhills, Greenbrook, Greendale; Baldwin Hills Village; Indications of the Form of the Future; Bibliography, pp. 249-254.

1485. United Nations, Department of Economic and Social Affairs, *Planning of Metropolitan Areas and New Towns,* United Nations Publication, Sales No.: 67.IV.5, New York (Room 1059, 10017), 1967, 255 pp. Paper.

Conclusions and Recommendations; Urbanization: Processes of World Urbanization (Barbara Ward), Urbanization in Latin America (Latin American Demographic Center, Santiago), Urbanization in South-East Asia (Demographic Training and Research Center, Bombay); Development Problem: Metropolitan Planning and Development, Planning For New Towns; Metropolitan Planning: Physical Factors of Metropolitan Planning With Special Reference to Stockholm (Bo Fredzell), Community Structure and Metropolitan Planning (C. Doxiadis; Athens Technological Institute), Economic Factors of Metropolitan Planning (M. J. Wise), Administrative Factors of Metropolitan Planning in Western Europe (A. Prothin), Administrative Considerations of Metropolitan Planning and Development (Luther Gulick), Control of Urbanization (Boleslaw Malisz), Some Policy Issues Regarding Urban Renewal (Housing Committee, Economic Commission for Europe); Reviews of Metropolitan Planning: Regional Plan For the Stockholm Area (C. F. Ahlberg), Metropolitan Planning Problems in the Netherlands (G. A. Wissink), Metropolitan Planning in: Poland (Julius Gorynski), Yugoslavia (Dusan Stefanovic), India (G. Mukharji), Scope and Purpose of Metropolitan Planning and Development in the United States (Robert C. Weaver), Provision of Welfare and Cultural Facilities and Public Utilities in the Cities of the Union of Soviet Socialist Republics (P. N. Blokhine), Outlines of Metropolitan Development Problems of Prague (Jiri Novotny; J. Stván), Urban Problems and Planning With Special Attention to Transport Problems in Tokyo (Susumu Kobe); New Towns Planning and Development: Economic Problems in Developing New Towns and Expanded Towns (Lloyd Rodwin), New Towns in Britain (William A. Robson), Argentina's Nineteenth-Century New Town (Jorge E. Hardoy), Contemporary New Towns

(Athens Technological Institute), Social Factors Involved in the Planning and Development of New Towns (Lesley E. White), Physical Planning for the Development of New Towns (Boleslaw Malisz), Building New Towns (N. V. Baranov), Economic Considerations in the Planning and Development of New Towns (P. B. Desai; Ashish Bose), Financing the Construction of New Towns (P. A. Stone), Planning of New Capital Cities (Jorge E. Hardoy).

Bibliographies

1486. Allen, Muriel I. (Editor), *New Communities: Challenge for Today*, Background Paper No. 2, Washington, D.C. (American Institute of Planners), October 1968, 39 pp. Paper. Suggested Readings on New Communities, pp. 37-39.

Readings Focused on New Communities; Readings on Policy Context for New Communities; Bibliographies on New Communities.

1487. American Institute of Architects, Urban Programs Department, "Selected References on Planned Communities," Washington, D.C. (1735 New York Ave., N.W., 20006), 1969, 5 pp.

Bibliography (2); Books (11); Selected Planned Cities—Books (7); Periodical Articles (22); Selected Periodical References on Selected U.S. New Towns (15).

1488. London County Council, *The Planning of a New Town*, Bibliography, pp. 177-181.

Mainly British references, many available only from England. Existing New Towns; Civic Design Generally; Regional Studies; Social Studies; Population and Housing Need; Housing; Employment; Industry; Shopping; Education; Recreation and Play; Landscape; Roads and Traffic; Public Transport; Economics; Miscellaneous.

1489. Menges, Gary L., *Model Cities*, Bibliography No. 48, Council of Planning Librarians, Monticello, Ill. (Exchange Bibliographies, P.O. Box 229, 61856), 1968, 13 pp.

Demonstration Cities and Metropolitan Development Act of 1966—Legislative History; General; Case Studies.

1490. Osborn, Frederick J.; Whittick, Arnold, *The New Towns, The Answer to Megalopolis*, Selected Bibliography, pp. 359-361.

1491. *Town and Country Planning*, Special Issue, Vol. 36, No. 1-2, January-February 1968, Publications on Britain's New Towns, p. 136.

1492. U.S. Housing and Home Finance Agency, Library, *New Communities: A Selected, Annotated Reading List*, Washington, D.C. (Office of the Administrator, 20410), January 1965, 24 pp. Paper.

Bibliography (292 References); Index of Secondary Authors; Geographic Index.

13. GENERAL URBAN PLANNING BIBLIOGRAPHIES

1493. American Society of Planning Officials, *Motion Picture Films on Planning, Housing, and Related Subjects,* Chicago, Ill. (1313 East 60th St., 60637), 1966, 21 pp. Paper.

". . . contains 131 entries—84 of them not heretofore included. Subjects covered, in addition to the films on general aspects of planning, are: commercial and industrial, conservation and recreation, governmental services, sanitation and health, transportation, and renewal and housing. Several films from Canada, England as well as from other countries are included. A brief description of each film is followed by information on availability from the distributor, including rental and purchase arrangements."

1494. Bestor, George C.; Jones, Holway R., *City Planning, A Basic Bibliography of Sources and Trends,* Sacramento, Calif. (California Council of Civil Engineers and Land Surveyors), 1962, 195 pp. [61-15577]

On the Nature and Form of Cities; History of Cities and City Planning; Contemporary City Planning—Nature, Function, Process: Nature of City Planning and Its Function in Local Government, General Plan—Theory and Example, City Planning Analysis, Elements of the General Plan—Analysis and Design, Other Design Considerations, Urban Planning in the Metropolitan Context, Effectuation—Translating the Plan into Action; Education for Planning; General Bibliographies.

1495. Dyckman, John W., *An Individual Review of Current Planning Literature,* Bibliography No. 36, Council of Planning Librarians, Monticello, Ill. (Exchange Bibliographies, P.O. Box 229, 61856), 1967, 18 pp.

1496. Goodman, William I.; Freund, Eric C. (Editors), *Principles and Practice of Urban Planning,* Municipal Management Series, Washington, D.C. (International City Managers' Association), Fourth Edition, 1968, Selected Bibliography, pp. 585-604.

Antecedents of Local Planning; Intergovernmental Context of Local Planning; Population Studies; Economic Studies; Land Use Studies: Land Use Theories, Models, and Trends, Bibliographies, Mapping, Land Use Survey—Classification, Recording, and Analysis, Forecasting Space Requirements, Industrial and Wholesale, Central Business District and Commercial, Civic, Institutional, Recreational, and Open Space, Transportation, Residential and Neighborhood Facilities, Welding the Land Use Plan; Transportation Planning: General, Highway Transportation Studies, Trucking, Railroads, Ports, Airports; Open Space, Recreation, and Conservation; Government and Community Facilities: General Higher Education, Health, Administrative Centers, Libraries, Fire Stations, Cemeteries and Crematoria, Water and Sewers, Water for Industry, Electric Power Systems, Overhead and Underground Utility Wires, Refuse and Maintenance Facilities; City Design and City Appearance; Quantitative Methods in Urban Planning; Social Welfare Planning; Defining Development Objectives; Comprehensive Plan: Conceptual, Examples; Programming Community Development: Capital Improvement, Extension of Services, Annexation, Official Mapping; Zoning; Land Subdivision: Regulation Manuals, Written Provisions, Design—General, Regulatory Law, Improvement Requirements, Design Innovations, Special Problems; Urban Renewal; Local Planning Agency—Internal Administration; Planning and the Public.

1497. Guttenberg, Albert Z., *Content Analysis for City Planning Literature,* Bibliography No. 37, Council of Planning Librarians, Monticello, Ill. (Exchange Bibliographies, P.O. Box 229, 61856), 1967, 13 pp.

Introduction; Thematic Content of Planning Literature; Intellectual Sources; Style.

1498. Shillaber, Caroline, *References on City and Regional Planning,* Cambridge, Mass. (M.I.T.), 1959, 41 pp. Paper.

"A guide for libraries in developing their city planning and architecture departments."

General References; History of Cities and City Planning; Theory; Administration; Survey and Analysis; Elements in Planning Practice; General Planning; Urban Design; Legal Factors; The Profession; Periodical and Serial Publications; Bibliographies; Agencies.

1499. Shillaber, Caroline, "Review Article: A Review of Planning Bibliographies," in: *Journal of the American Institute of Planners,* Vol. XXXI, No. 4, November 1965, pp. 352-360.

Bibliographies on City Planning: Metropolitan Areas, New Towns, Planning Theory, Transportation, Urban Renewal, Land Use, Recreation, Social Problems, Aesthetics; Research Projects; Index; Other Publications Lists.

1500. U.S. Housing and Home Finance Agency, *Sixty Books on Housing and Community Planning,* Washington, D.C. (U.S. Department of Housing and Urban Development, 20410), 1963, 21 pp. Paper.

RELATED MATERIALS

Municipal and Metropolitan Planning Agencies in the United States and Canada

As mentioned in the Introduction and Explanation on page 20, established municipal and metropolitan-regional planning agencies comprise a special "bibliographical" resource, in that they are a source of information and experience which is not made generally available in published form. From another point of view, these agencies *are* governmental city planning. Together with the individual planning practitioner, consulting and other private organizations, and universities with planning programs, they constitute the field and institutionalized endeavor of city and metropolitan-regional planning. The hundreds of municipalities and metropolitan areas served by governmental planning agencies comprise the end-product and purpose of the urban planning effort.

In 1968, more than 51 million city-dwellers were served directly by reporting municipal planning agencies. Many additional millions were served by agencies which have not reported their activities in one of the several national surveys, and by combined city-county planning bodies. Most recently, numerous Councils of (Local) Governments have been established throughout the United States to coordinate the planning activities of separate municipal jurisdictions, toward solving the many problems and needs which are metropolitan-regional in nature and do not conform to artificial legal-political boundaries. Unfortunately, there is no complete compilation and listing of all these official municipal, metropolitan-regional, and other forms of urban governmental planning.

The American Society of Planning Officials annually surveys the planning agencies which subscribe to its Planning Advisory Service [Faraci, Piero, *Expenditures, Staff, and Salaries of Planning Agencies*, Report No. 245, ASPO Planning Advisory Service, Chicago, Ill. (1313 East 60th St., 60637), April 1969, 55 pp. Paper]. Since this list is limited to those "city, county, and 'combined' planning agencies" that subscribe to its service, and only approximately three-quarters of these respond, this survey is not intended to be complete.

A second source of information on official city planning, also prepared by the staff of the American Society of Planning Officials, is included every few years in the yearbook of the International City Managers' Association. The 1965 issue is the most recent reporting the activities of city planning agencies [Nolting, Orin F.; Arnold, David S. (Editors); Powers, Stanley P. (Managing Editor); Webb, Walter L. (Assistant Editor), *The Municipal Yearbook, 1965*, International City Managers' Association, Washington, D.C. (1140 Connecticut Ave., N.W., 20036), 1965, pp. 311-343 (34-27121).]

Of these two national surveys, the yearbook data are more complete but three to four years older. The differences in the numbers of cities reporting in the two surveys are as follows:

Population Category			ASPO (1968)	ICMA (1964)
1,000,000	and	Over	6 ⎱	18
500,000	to	1,000,000	17 ⎰	
250,000	to	500,000	19	25
100,000	to	250,000	56	62
50,000	to	100,000	100	154
25,000	to	50,000	75	—
10,000	to	25,000	38	—

One or more planning agencies may combine with a principal city or several cities to form a county, regional, metropolitan, or area planning body responsible for a large urban expanse. Such organizations seek to coordinate, to the extent possible, the plans of separate political jurisdictions and different functional agencies interrelated areally or geographically. They will usually plan unincorporated county territory; they may also do city planning for smaller municipalities under contract, or advise them in various ways. There are 112 combined agencies of this type, and 118 county planning agencies which take the ASPO Planning Advisory Service, and are reported in the 1969 annual survey of the American Society of Planning Officials, referenced above.

Councils of governments are formal attempts toward greater collaboration between separate local governments with historical boundaries unrelated to the derivation and widespread areal impact of many modern municipal problems. They seek to achieve, by voluntary governmental cooperation, effective planning for the entire contiguous urbanized expanse, within which many separate municipal jurisdictions have become geographically anachronistic parts. Most have been formed in recent years. Specific information is contained in the Directory of Regional Councils [National Service to Regional Councils, 1968 Directory of Regional Councils, Washington, D.C. (1700 K. St., N.W., 20006), February 1969, 44 pp. Paper].

Urbanized county and metropolitan-regional planning agencies, together with councils of governments, comprise the official governmental effort to plan for contiguous urban areas rather than legally separate municipalities. They describe urban rather than city or municipal planning.

The reference sources discussed briefly above can be used to contact specific agencies for information and experience which is either unpublished or presented locally in various report forms, usually in very limited numbers. If a city or county which is not listed in one of the above surveys is of particular interest, its "planning commission" or "planning department" can be addressed in care of City Hall or County Offices. It may be that this city or county has no official planning agency, but the chances are that it will have a formal function of this sort. For, although city, county, and metropolitan-regional planning agencies vary widely in their budgets, staff, activities, and effectiveness, they have by now been established—at least nominally—far and wide throughout the United States.

Colleges and Universities in the United States and Canada Offering Graduate Programs in Urban and Regional Planning and Traditionally Allied Fields 1968

The list of planning programs beginning on page   is drawn from information collected annually about planning education by the American Society of Planning Officials (1313 East 60th St., Chicago, Ill. 60637).

An asterisk preceding the name of the college or university indicates that the graduate program is recognized by the American Institute of Planners (917 Fifteenth St., N.W., Washington, D.C. 20005) for membership credit. This may be some years after the planning program was inaugurated.

The date in parentheses, immediately after the name of the educational institution, indicates the year the graduate planning program was established.

The first number thereafter is student enrollment at the university to the nearest hundred, the second, after the hyphen, is the number of faculty, both as reported in *The World Almanac, 1968*. Although quantity is rarely a criterion of quality, there is some relationship between size of university and faculty resources, and what is likely to be represented in a planning program. Especially for prospective students and researchers abroad wishing to establish contact, size of enrollment and faculty may prove useful.

Colleges and universities with undergraduate planning programs only are not included in this list. However, when the educational institution with a graduate planning program also offers an undergraduate curriculum in planning, the latter is included as extra information.

Sixty United States and Canadian colleges and universities award degrees in planning. Of these programs, twenty-eight are recognized by the American Institute of Planners as providing the equivalent of experience for certain categories of membership—the closest approximation at present to accreditation in other professional fields. The purposes and content of these selected references relate most directly to this group of twenty-eight which constitute the higher quality degrees in planning as a distinct professional field. It should be borne in mind, however, that the field of city planning is still in a period of change and resolution, as discussed briefly in the Introduction and Explanation. Not only are the curricula of graduate planning programs and the titles of their degrees subject to further change, but degrees awarded in related fields emphasizing planning are likely to assume greater importance. Curricula focusing on comprehensive city planning as represented by this bibliography are still the exception rather than the rule, even among the accredited group of planning programs.

Unless otherwise noted, graduate master's degrees in planning normally require two years. When the time programmed for the degree is either less than or more than two years, this is shown in parentheses. As can be seen by examining Figure 1 below, there is considerable variation in the name of the planning degrees awarded by the sixty United States and Canadian educational institutions with graduate planning programs. In most instances, this reflects different conceptual views of planning as a field, established university tradition, or varying intents for the future—rather than descriptive differentiation between distinct curricula.

In Figure 1 can be seen a trend of professional planning education. To the original emphasis on physical land-use planning have been added planning degrees representing urban studies, ecology, and environment in a broader areal and substantive sense. Different interrelated programs—between city and regional planning and: transportation and civil engineering, regional and social science, operations research, public administration and politics—signify the participatory role of numerous related fields in comprehensive urban planning.

The kinds of educational units awarding graduate degrees in planning are listed in Figure 2.

These educational units conducting planning programs are within departments, schools, colleges, or divisions at the next higher organizational level as shown in Figure 3. Nineteen affiliations with architecture and the arts continue an historical association, but twelve affiliations with various other units represent the recent trend toward greater diversity. Possibly as many as twenty-seven planning programs stand separately within graduate schools or divisions, rather than particular schools or colleges.

Figure 1

TYPES OF PLANNING AND RELATED GRADUATE DEGREES – 1968

Planning Degrees

Designation of Degree	No.	Master's Master's	M.S.	M.A.	Ph.D. No.
City (Urban, or Community) and				
Regional Planning	24	12	8	4	6
in Transportation Planning	—	—	—	—	1
City (Urban, or Community)				
Planning	17	15	2	—	4
Urban Studies (or Affairs)	4	1	1	2	2
Urban (or City) Design	3	—	—	3	—
Planning	6	1	4	1	—
Urban and Regional				
Environment	1	1	—	—	—
Urban Ecology	1	1	—	—	—
and Urban Design	1	—	—	1	—
	57	31	15	11	13
Regional Planning	6	4	1	1	2
and Resources Development	1	—	—	1	1
Environmental Planning	1	—	1	—	—
	8	4	2	2	3
Totals	65	35	17	13	16

Joint or Combined Fields

Designation of Degree	No.	Master's Master's	M.S.	M.A.	Ph.D. No.
City Planning and—				
Landscape Architecture	1	1	—	—	—
Transportation Engineering	1	1	—	—	—
Regional Science	1	1	—	1	—
Operations Research	1	—	1	—	—
Regional Planning and—				
Civil Engineering	1	—	1	—	—
Social Science	—	—	—	—	1
Totals	5	3	2	1	1

Figure 1 *(Continued)*

Related Fields: Combinations and Options

Designation of Degree	No.	Master's	M.S.	M.A.	Ph.D. No.
Architecture—					
in (and) Urban Design	3	2	—	1	—
Environmental Systems	1	1	—	—	—
Geography—					
major in Planning	3	1	2	—	—
Landscape Architecture—					
and Environmental Planning	1	1	—	—	—
major in Town and Regional Planning	1	—	1	—	—
Engineering—					
Environmental	1	—	1	—	—
Transportation and City and Regional Planning	1	—	1	—	—
Public Administration—					
in Urban Renewal	1	1	—	—	—
major in Urban Planning	—	—	—	—	1
Urban Politics and Planning	—	—	—	—	1
Social Science, option in Urban Planning	—	—	—	—	1
Totals	12	6	5	1	3

Figure 2

ORGANIZATIONAL UNIT AWARDING PLANNING AND RELATED GRADUATE DEGREES

Planning Programs

Department (Program, Division, Graduate Curriculum, School, Center, or Institute):

City (Urban, or Community) and Regional Planning (or Studies) 26
    and Housing and Urban Renewal 1
City (Urban, or Community) Planning 13
    and Area Development 1
    and Landscape Architecture 1
City Design 1
Urban Affairs (or Studies) 4
Planning 1
    Regional 1
Graduate Studies 2
(Graduate) Planning Program 2

Figure 2 *(Continued)*

Planning Programs Within Related Fields

Department (School, or Program):

Architecture	1
Geography (or Earth Science)	1
and Area Development	1
and Planning	1
Government (Public) and Business Administration	1
Landscape Architecture	2
and Environmental Planning	1
Public Administration and Urban Studies	1

Figure 3

ORGANIZATIONAL PARENT OF PLANNING PROGRAM

Department, School, or College 31

Administration: Public	1
Business and Public	1
Architecture	9
and Allied Arts	1
Arts, and Planning	1
and Engineering	1
and Planning	1
and Urban Planning	1
Art and Architecture	1
Civil Engineering	1
Design	1
Architecture and Art	1
Environmental	1
Fine Arts	2
and Applied Arts	1
Geography	1
Humanities and Arts	1
International and Public Affairs	1
Letters and Science	2
Social Science	2

Graduate School, Division, or Faculty; College;
University at large; or none indicated 31

LIST OF SCHOOLS OFFERING GRADUATE DEGREES IN URBAN AND REGIONAL PLANNING

United States

1. UNIVERSITY OF ARIZONA (1961), 21,400-1,616: M.S., major in Urban Planning.

    Chairman, Committee on Urban Planning, Department of Geography and Area Development, College of Business and Public Administration, University of Arizona, Tucson, Ariz. 85721.

2. AUBURN UNIVERSITY (1968), 11,326-843: Master of City and Regional Planning.

    Dean, Graduate School, Mary Martin Hall, Auburn University, Auburn, Ala. 36830.

3. *UNIVERSITY OF CALIFORNIA (1948), 26,000-2,328: Master of City Planning; Ph.D. in City and Regional Planning.

    Chairman, Department of City and Regional Planning, College of Environmental Design, Berkeley, Cal. 94720.

4. THE CATHOLIC UNIVERSITY OF AMERICA (1964), 6,800-466: Master in City and Regional Planning.

    Chairman, Program in City and Regional Planning, School of Architecture and Engineering, Catholic University of America, Washington, D.C. 20017.

5. *UNIVERSITY OF CINCINNATI (1961), 21,100-2,139: Bachelor in Community Planning; (a) Master in Community Planning; (b) Master of Community Planning.

    (a) Head, Department of Community Planning, College of Design, Architecture and Art, (b) Director, Graduate Division, Community Planning; University of Cincinnati, Cincinnati, Ohio 45221.

6. *COLUMBIA UNIVERSITY (1935), 16,000-3,911: M.S. in Urban Planning; Ph.D. in Urban Planning.

    Chairman, Division of Urban Planning, School of Architecture, 409 Avery Hall, Columbia University, New York, N.Y. 10027.

7. *CORNELL UNIVERSITY (1935), 13,900-2,276: Master of Regional Planning; Ph.D. in City and Regional Planning.

    Chairman, Department of City and Regional Planning, College of Architecture, Art and Planning, Sibley Hall, Cornell University, Ithaca, N.Y. 14850.

8. FLORIDA STATE UNIVERSITY (1965), 13,400-848: M.S. in Planning.

    Chairman, Department of Urban and Regional Planning, Florida State University, Tallahassee, Fla. 32306.

9. FRESNO STATE COLLEGE (1968), 9,100-549: M.A. in Urban and Regional Planning.

    Director, City and Regional Planning Program, Department of Geography, Fresno State College, Fresno, Cal. 93726.

10. GEORGE WASHINGTON UNIVERSITY (1968), 12,000-1,667: Master of Urban and Regional Planning.

    Dean, School of Government and Business Administration, George Washington University, Washington, D.C. 20006.

11. *Georgia Institute of Technology (1952), 7,400-490: *Master of City Planning; joint degree program in City Planning and Landscape Architecture; joint degree program in City Planning and Transportation Engineering.

    Chairman, Graduate City Planning Program, School of Architecture, Georgia Institute of Technology, 225 North Ave. N.W., Atlanta, Ga. 30332.

12. *Harvard University (1923), 13,700-6,730: *Master in City Planning; *Master in Regional Planning; M.A. in Urban Design; *Ph.D. in Regional Planning.

    Chairman, Department of City and Regional Planning, Graduate School of Design, Harvard University, Cambridge, Mass. 02138.

13. Howard University (1967), 8,200-1,002: Master of City Planning.

    Head, Graduate Program of City and Regional Planning, Department of Architecture, Howard University, Washington, D.C. 20001.

14. Hunter College (1966), 26,100-1,321: M.S. in Urban Planning.

    Director, Urban Planning Program, Graduate Division, Hunter College, 695 Park Ave., New York, N.Y. 10021.

15. Illinois Institute of Technology (1939), 8,200-678: B.S. in Architecture with planning option; M.S. in City and Regional Planning; Ph.D. in City and Regional Planning; M.S. in Environmental Engineering.

    Chairman, Department of City and Regional Planning, School of Architecture and Planning, Illinois Institute of Technology, Chicago, Ill. 60616.

16. *University of Illinois (1945), 40,100-3,114: Bachelor of Urban Planning; *Master of Urban Planning.

    Chairman, Department of Urban Planning, College of Fine and Applied Arts, University of Illinois, 208 Mumford Hall, Urbana, Ill. 61801.

17. Iowa State University (1947: M.S.; 1967: B.S.), 15,200-1,194: B.S. in Urban Planning; M.S., major in Town and Regional Planning.

    Head, Department of Landscape Architecture, Iowa State University, Ames, Iowa 50010.

18. University of Iowa (1964), 16,800-1,102: M.A. or M.S. in Urban and Regional Planning.

    Chairman, Graduate Program in Urban and Regional Planning, University of Iowa, 209 University Hall, Iowa City, Iowa 52240.

19. *Kansas State University (1957), 10,400-983: Master in Regional and Community Planning.

    Director, Graduate Curriculum in Regional and Community Planning, Seaton Hall, Kansas State University, Manhattan, Kan. 66502.

20. Mankato State College (1967), 10,300-446: B.S. in Urban Studies; M.A. in Urban Studies.

    Director, Urban Studies Center, Mankato State College, Mankato, Minn. 56001.

21. University of Massachusetts (1969), 14,586-799: Master of Regional Planning.

    Head, Department of Landscape Architecture, Wilden Hall, University of Massachusetts, Amherst, Mass. 01002.

22. *MASSACHUSETTS INSTITUTE OF TECHNOLOGY (1932), 7,600-1,581: B.S. in Art and Design, planning option; *Master in City Planning; *Ph.D. in City Planning; Ph.D. in Urban Politics and Planning.

> Head, Department of City and Regional Planning, Massachusetts Institute of Technology, 77 Massachusetts Ave., Cambridge, Mass. 02139.

23. MIAMI UNIVERSITY (1954), 12,700-658: Master in City Design.

> Director, Graduate Program in City Design, Department of Architecture, School of Fine Arts, Miami University, Oxford, Ohio 45056.

24. *UNIVERSITY OF MICHIGAN (1945), 33,000-3,772: (a) Bachelor of Architecture, planning option; *Master of City Planning; (b) Ph.D. in Urban and Regional Planning.

> (a) Chairman, Planning Committee, Graduate Planning Program, Department of Architecture, (b) Horrace H. Rackham School of Graduate Studies; University of Michigan, Ann Arbor, Mich. 48104.

25. *MICHIGAN STATE UNIVERSITY (1946), 38,100-2,750: *B.S. in Urban Planning; *Master of Urban Planning; Ph.D. in Social Science, option in Urban Planning.

> Director, School of Urban Planning and Landscape Architecture, College of Social Science, Michigan State University, East Lansing, Mich. 48823.

26. UNIVERSITY OF MISSISSIPPI (1958), 6,800-1,106: Master of Urban and Regional Planning.

> Head, Department of City Planning, Graduate School, 207 Lyceum Building, Box 193, University of Mississippi, University, Miss. 38677.

27. *NEW YORK UNIVERSITY (1959), 42,000-5,149: *Master of Urban Planning; Ph.D. in Public Administration, major in Urban Planning.

> Dean, Program in Urban and Regional Planning and Housing and Urban Renewal, Graduate School of Public Administration, New York University, Four Washington Square, North, New York, N.Y. 10003.

28. *UNIVERSITY OF NORTH CAROLINA (1946), 29,800-2,361: *Master of Regional Planning; *Ph.D. in Regional Planning; joint Master of Regional Planning, and M.S. in Civil Engineering.

> Chairman, Department of City and Regional Planning, University of North Carolina, Chapel Hill, N.C. 27514.

29. NORTHWESTERN STATE COLLEGE OF LOUISIANA (1967), 4,900-300: M.A. in Urban and Regional Planning.

> Program in Urban and Regional Planning, Department of Social Sciences, Northwestern State College of Louisiana, Natchitoches, La. 71457.

30. NORTHWESTERN UNIVERSITY (1962), 17,200-2,347: M.S. in Transportation Engineering and Urban and Regional Planning (one year); Ph.D. in Urban and Regional Planning; Ph.D. in Urban and Regional Planning in Transportation Planning.

> Director, Urban and Regional Planning Program, Department of Civil Engineering, Technological Institute, Northwestern University, Evanston, Ill. 60201.

31. *OHIO STATE UNIVERSITY (1958), 36,100-4,303: Master of City Planning.

> Chairman Division of City and Regional Planning, School of Architecture, Ohio State University, 190 West 17th Ave., Columbus, Ohio 43210.

32. *UNIVERSITY OF OKLAHOMA (1948), 16,000-850: Master of Regional and City Planning.
    Chairman, Department of Regional and City Planning, University of Oklahoma, Norman, Okla. 73069.

33. UNIVERSITY OF OREGON (1960), 11,700-1,150: Master of Urban Planning.
    Head, Department of Urban Planning, School of Architecture and Allied Arts, University of Oregon, Eugene, Ore. 97403.

34. *UNIVERSITY OF PENNSYLVANIA (1951), 18,100-4,000: *Master in City Planning; Master of City Planning and M.A. in Regional Science (three years); Master of City Planning and M.S. in Operations Research (three years); *Ph.D. in City Planning.
    Chairman, Department of City and Regional Planning, Graduate School of Fine Arts, University of Pennsylvania, Philadelphia, Penn. 19104.

35. PENNSYLVANIA STATE UNIVERSITY (1965), 37,500-2,200: M.S. of Regional Planning.
    Director, Graduate Program in Regional Planning, Pennsylvania State University, 318 Sackett Building, University Park, Penn. 16802.

36. *UNIVERSITY OF PITTSBURGH (1962), 20,600-1,445: *Master of Urban and Regional Planning; Master in Public Administration (Urban Renewal); Ph.D. in Urban Affairs.
    Director of Programs, Department of Urban Affairs, Graduate School of Public and International Affairs, University of Pittsburgh, Pittsburgh, Penn. 15213.

37. *PRATT INSTITUTE (1957), 4,170-444: Master in Urban Design (one year); *M.S. in Planning.
    Chairman, Department of City and Regional Planning, School of Architecture, Pratt Institute, Brooklyn, N.Y. 11205.

38. *UNIVERSITY OF RHODE ISLAND (1963), 6,400-433: Master of Community Planning.
    Director, Curriculum in Community Planning and Area Development, Graduate School, University of Rhode Island, Kingston, R.I. 02881.

39. RUTGERS UNIVERSITY (1948), 24,800-2,684: B.A., option in City and Regional Planning; M.S. in City and Regional Planning, three options (one year); Master in City and Regional Planning.
    Chairman, Department of City and Regional Planning, Livingston College, Rutgers — The State University, New Brunswick, N.J. 08903.

40. SAN DIEGO STATE COLLEGE (1968), 18,519-1,187: Master of City Planning.
    School of Public Administration and Urban Studies, San Diego State College, San Diego, Cal. 92115

41. *UNIVERSITY OF SOUTHERN CALIFORNIA (1955), 17,500-1,900: *Master of Planning (Urban and Regional Environment); Master of Planning (Urban Ecology); *M.S. in City and Regional Planning; Ph.D. in Urban Studies.
    Chairman, Graduate Department of Urban and Regional Planning, von KleinSmid Center for International and Public Affairs—Room 351, University of Southern California, University Park, Los Angeles, Cal. 90007.

42. SOUTHERN ILLINOIS UNIVERSITY (1967), 24,600-2,341: M.A. and M.S. in Geography, major in Planning (one and two years).

    Chairman, Earth Science Department, Southern Illinois University, Carbondale, Ill. 62901.

43. *SYRACUSE UNIVERSITY (1960), 22,300-1,223: *Master of Regional Planning; joint Master of Regional Planning and Doctor of Social Science.

    Chairman, Graduate Planning Program, School of Architecture, 417 Slocum Hall, Syracuse University, Syracuse, N.Y. 13210.

44. UNIVERSITY OF TENNESSEE (1965), 26,600-2,200: M.S. in Planning.

    Director, Graduate School of Planning, University of Tennessee, 1515 Cumberland Ave., Knoxville, Tenn. 37916.

45. TEXAS A. & M. UNIVERSITY (1965), 14,900-1,097: Master in Urban and Regional Planning.

    Director, Urban and Regional Planning, School of Architecture, Texas A. & M. University, College Station, Tex. 77843.

46. UNIVERSITY OF TEXAS (1959), 28,000-1,177: M.S. in Community and Regional Planning.

    Chairman, Interdepartmental Graduate Program in Community and Regional Planning, University of Texas, Austin, Tex. 78712.

47. UTAH STATE UNIVERSITY (1967), 7,300-470: B.S. in Environmental Planning; M.S. in Environmental Planning; Master of Landscape Architecture, Environmental Planning (3 years).

    Department of Landscape Architecture and Environmental Planning, College of Humanities and Arts, Utah State University, Logan, Utah 84321.

48. *UNIVERSITY OF VIRGINIA (Bachelor: 1958; M.A.: 1966), 17,000-829: *Bachelor of City Planning (five years); M.A. in Planning and Urban Design.

    Chairman, Division of City Planning, School of Architecture, University of Virginia, Charlottesville, Va. 22904.

49. *VIRGINIA POLYTECHNIC INSTITUTE (1959), 8,510-843: *M.S. in Urban and Regional Planning; Master of Architecture in Urban Design; Master of Architecture in Environmental Systems.

    Director, Center for Urban and Regional Studies, Virginia Polytechnic Institute, Magill House, Blacksburg, Va. 24061.

50. *UNIVERSITY OF WASHINGTON (1941), 26,400-1,698: *Master of Urban Planning; M.A. in Architecture and Urban Design; Ph.D. in Urban Planning.

    Chairman, Department of Urban Planning, College of Architecture and Urban Planning, University of Washington, Seattle, Wash. 98105.

51. WASHINGTON UNIVERSITY (1962), 11,200-2,605: Master in Architecture and Urban Design (one year).

    Dean, Graduate School of Architecture, Washington University, St. Louis, Mo. 63130.

52. *WAYNE STATE UNIVERSITY (1959), 30,800-1,927: Master of Urban Planning.

    Chairman, Department of Urban Planning, Wayne State University, 5757 Cass Ave., Detroit, Mich. 48202.

53. *University of Wisconsin (1944), 31,100-4,000: *M.A. and M.S. in Urban and Regional Planning; Ph.D. in Urban and Regional Planning.
   Chairman, Department of Urban and Regional Planning, College of Letters and Science, University of Wisconsin, Madison, Wis. 53706.
54. University of Wisconsin — Milwaukee (1963), 14,200-1,310: M.A. and M.S. in Urban Affairs (one year).
   Chairman, Department of Urban Affairs, College of Letters and Science, University of Wisconsin, 3203 Downer Ave., Milwaukee, Wis. 53201.
55. *Yale University (1951), 8,700-1,220:'*Master of City Planning (one year); Master of Urban Studies (one year).
   Chairman, Department of City Planning, School of Art and Architecture, Yale University, 180 York St., New Haven, Conn. 06520.

Canada

56. *University of British Columbia (1953), 18,800-1,238: M.A. or M.Sc. in Planning.
   Director, School of Community and Regional Planning, Faculty of Graduate Studies, University of British Columbia, Vancouver 8, British Columbia, Canada.
57. University of Manitoba (1950), 13,000-585: Diploma in City Planning (one year); Master in City Planning.
   Department of City Planning, Faculty of Graduate Studies, University of Manitoba, Winnipeg 19, Manitoba, Canada.
58. University of Montreal (1961), 15,900-1,050: Master of Planning (two and one-half years).
   Director, L'Institut d'Urbanisme, L'Université de Montreal, Montreal, Quebec, Canada.
59. University of Toronto (1956: Diploma; 1963: M.Sc.), 25,300-3,477: M.Sc. in Urban and Regional Planning; Diploma in Urban and Regional Planning (one year).
   Chairman, Department of Urban and Regional Planning, University of Toronto, Toronto, Ontario, Canada.
60. University of Waterloo (1966), 7,700-450: B.A. in honors, in Urban and Regional Planning; M.A. in Regional Planning and Resource Development; Ph.D. in Planning and Resource Development.
   Director, Planning Program, Department of Geography and Planning, University of Waterloo, Waterloo, Ontario, Canada.

List of Publishers and Sources with Addresses

ACADEMIC PRESS, INC., 111 Fifth Ave., New York, N.Y. 10003.

ACADEMY OF POLITICAL SCIENCE, COLUMBIA UNIVERSITY, 116th and Broadway, New York 10027.

ADVANCE PLANNING DEPARTMENT, CALIFORNIA DIVISION OF HIGHWAYS, 120 S. Spring St., Los Angeles, Cal. 90012.

ADVISORY COMMISSION ON INTERGOVERNMENTAL RELATIONS, 726 Jackson Pl., N.W., Washington, D.C. 20575.

AEROJET-GENERAL CORPORATION, 9100 E. Flair Dr., El Monte, Cal. 91734.

AGENCY FOR INTERNATIONAL DEVELOPMENT, 2201 C St., N.W., Washington, D.C. 20523.

ALDINE PUBLISHING Co., 320 W. Adams St., Chicago, Ill. 60606.

AMERICAN ACADEMY OF POLITICAL AND SOCIAL SCIENCE, 3937 Chestnut St., Philadelphia, Penn. 19104.

AMERICAN ASSOCIATION FOR THE ADVANCEMENT OF SCIENCE, 1515 Massachusetts Ave., N.W., Washington, D.C. 20005.

AMERICAN CONSERVATION ASSOCIATION, 30 Rockefeller Plaza, New York, N.Y. 10020.

AMERICAN DATA PROCESSING, INC., 19830 Mack Ave., Detroit, Mich. 48236.

AMERICAN ELSEVIER PUBLISHING Co., INC., 52 Vanderbilt Ave., New York, N.Y. 10017.

AMERICAN HOSPITAL ASSOCIATION, 840 N. Lake Shore Dr., Chicago, Ill. 60611.

AMERICAN INSTITUTE OF PARK EXECUTIVES, now: NATIONAL RECREATION AND PARK ASSOCIATION, 1700 Pennsylvania Ave., N.W., Washington, D.C. 20006.

AMERICAN INSTITUTE OF PLANNERS, 917 Fifteenth St., N.W., Washington, D.C. 20005.

AMERICAN LIBRARY ASSOCIATION, 50 East Huron St., Chicago, Ill. 60611.

AMERICAN MANAGEMENT ASSOCIATION, INC., 135 W. 50th St., New York, N.Y. 10020.

AMERICAN MUNICIPAL ASSOCIATION, 1612 K St., N.W., Washington, D.C. 20006.

AMERICAN SOCIETY OF AGRONOMY, 677 South Segoe Rd., Madison, Wis. 53711.

AMERICAN SOCIETY OF PLANNING OFFICIALS, 1313 East 60th St., Chicago, Ill. 60637.

AMERICAN SOCIETY FOR PUBLIC ADMINISTRATION, 1329 18th St., N.W., Washington, D.C. 20036.

APPLETON-CENTURY-CROFTS, INC., EDUCATIONAL DIVISION OF MEREDITH PUBLISHING Co., 440 Park Ave. South, New York, N.Y. 10016.

ARCHITECTURAL PRESS, 9–13 Queen Anne's Gate, London, S.W. 1, England.

ARTHUR D. LITTLE, INC., Acorn Park, Cambridge, Mass. 02140.

ATHERTON PRESS, 70 Fifth Ave., New York, N.Y. 10011.

AUTOMOBILE CLUB OF SOUTHERN CALIFORNIA, 2601 S. Figueroa St., Los Angeles, Cal. 90007.

AUTOMOTIVE SAFETY FOUNDATION, 200 Ring Bldg., 1200 18th St., N.W., Washington, D.C., 20036.

BAKER, VOORHIS AND Co., INC., Mount Kisco, N.Y. 10549.

BANCROFT-WHITNEY Co., 301 Brannon St., San Francisco, Cal. 94107.

BANTAM BOOKS, INC., 271 Madison Ave., New York, N.Y. 10016.

BARNES & NOBLE, INC., 105 Fifth Ave., New York, N.Y. 10003.

BASIC BOOKS, INC., 404 Park Ave. South, New York, N.Y. 10016.

BENDER, MATTHEW & Co., 235 E. 45th St., New York, N.Y. 10017. (Orders to 1275 Broadway, Albany, N.Y. 12201.)

BOARDMAN, CLARK, Co., LTD., 435 Hudson St., New York, N.Y. 10014.

BOBBS-MERRILL, INC., HOWARD W. SAMS & Co., INC., 4300 W. 62nd. St., Indianapolis, Ind. 46268.

BOOKS INTERNATIONAL, 134 Stockbridge Ave., Alhambra, Cal. 91801.

BRAZILLER, GEORGE, INC., 1 Park Ave., New York, N.Y. 10016.

BRITISH INFORMATION SERVICE, 845 Third Ave., New York, N.Y. 10022.

BROOKINGS INSTITUTION, 1775 Massachusetts Ave., N.W., Washington, D.C. 20036.

BUILDING RESEARCH INSTITUTE, 1424 Sixteenth St., N.W., Washington, D.C. 20036.

BUREAU OF BUSINESS RESEARCH, GRADUATE SCHOOL OF BUSINESS, INDIANA UNIVERSITY, Bloomington, Ind. 47401.

BUREAU OF MUNICIPAL RESEARCH, 12 Richmond, E. Toronto, Ontario, Canada.

BUREAU OF PUBLIC ADMINISTRATION, UNIVERSITY OF CALIFORNIA, Berkeley, Cal. 94720.

BUREAU OF VECTOR CONTROL, DEPARTMENT OF PUBLIC HEALTH, STATE OF CALIFORNIA, 2151 Berkeley Way, Berkeley, Cal. 94704.

BUSINESS AND PUBLIC ADMINISTRATION RESEARCH CENTER, UNIVERSITY OF MISSOURI, 312 B and PA Bldg., Columbia, Mo. 65201.

BUTTENHEIM PUBLISHING CORP., 757 Third Ave., New York, N.Y. 10017.

BUTTERWORTH, INC., 7300 Pearl St., Washington, D.C. 20014.

CALIFORNIA COUNCIL OF CIVIL ENGINEERS AND LAND SURVEYORS, 1107–9th. St., Sacramento, Cal. 95814.

CALIFORNIA TOMORROW, 334 Forum Bldg., Sacramento, Cal. 95814.

CAMBRIDGE UNIVERSITY PRESS, 32 E. 57th St., New York, N.Y. 10022.

CENTER FOR PLANNING AND DEVELOPMENT RESEARCH, UNIVERSITY OF CALIFORNIA, 316 Wurster Hall, Berkeley, Cal. 94720.

CENTER FOR REAL ESTATE AND URBAN ECONOMICS, INSTITUTE OF URBAN AND REGIONAL DEVELOPMENT, UNIVERSITY OF CALIFORNIA, 260 Stephens Hall, Berkeley, Cal. 94720.

CHAMBER OF COMMERCE OF THE UNITED STATES, ASSOCIATION SERVICE DEPT., 1615 "H" St., N.W., Washington, D.C. 20006.

CHANDLER-DAVIS PUBLISHING CO., Box 36, West Trenton, N.J. 08628.

CHANDLER PUBLISHING CO., 124 Spear St., San Francisco, Cal. 94105.

CHICAGO AREA TRANSPORTATION STUDY, 130 North Franklin St., Chicago, Ill. 60606.

CHILTON BOOK CO., 401 Walnut St., Philadelphia, Penn. 19106.

CITY OF CHICAGO, City Hall, 121 N. La Salle St., Chicago, Ill. 60602.

CITY AND COUNTY OF SAN FRANCISCO, City Hall, 400 Van Ness Ave., San Francisco, Cal. 94102.

CITY PLAN COMMISSION, 104 S. Main St., Wichita, Kan. 67202.

CITY PLANNING COMMISSION, 306-A 2nd St., S.W., Roanoke, Va. 24011.

THE CITY UNIVERSITY OF NEW YORK, 33 West 42nd St., New York, N.Y. 10036.

CLEARINGHOUSE FOR FEDERAL SCIENTIFIC & TECHNICAL INFORMATION, Springfield, Va. 22151. (Also CLEARINGHOUSE, U.S. DEPARTMENT OF COMMERCE.)

CLEAVER-HUME PRESS, LTD., care of MACMILLAN & CO., 4 Little Essex St., London, W.C. 2, England.

COLLEGE PRINTING & TYPING CO. INC., 453 W. Gilman, Madison, Wis. 53703.

COLLIER-MACMILLAN LTD., 10 South Audley St., London W.1, England.

COLUMBIA UNIVERSITY PRESS, 136 South Broadway, Irvington-on-Hudson, New York 10533.

COMMITTEE FOR ECONOMIC DEVELOPMENT, DISTRIBUTION DIVISION, 711 Fifth Ave., New York, N.Y. 10022.

COMMUNITY AND FAMILY STUDY CENTER, UNIVERSITY OF CHICAGO, 5801 S. Ellis Ave., Chicago, Ill. 60637.

COMMUNITY PLANNING, UNIVERSITY OF MANITOBA, Winnipeg, Manitoba, Canada.

COMPREHENSIVE HEALTH PLANNING UNIT, SCHOOL OF PUBLIC HEALTH, Earl Warren Hall, UNIVERSITY OF CALIFORNIA, Berkeley, Cal. 94720.

CONNECTICUT FEDERATION OF PLANNING & ZONING AGENCIES, 49 Pearl St., Hartford, Conn. 06103.

CORNELL UNIVERSITY PRESS, 124 Roberts Pl., Ithaca, N.Y. 14850.

COUNCIL OF STATE GOVERNMENTS, 1313 East 60th St., Chicago, Ill. 60637.

COUNTY GOVERNMENT CENTER, 401 Marshall St., Redwood City, Cal. 94063.

CREST PUBLICATIONS, (address unknown), Los Angeles, Cal.

CROWELL, THOMAS Y., CO., 201 Park Ave. South, New York, N.Y. 10003.

CROWN PUBLISHERS, Inc., 419 Park Ave. South, New York, N.Y. 10016.

DAUGHTERS OF ST. PAUL, 50 St. Paul Ave., Boston, Mass. 02130.

DAVID MCKAY CO., INC., 750 Third Ave., New York, N.Y. 10017.

DELACORTE, see DELL PUBLISHING CO., INC.

DELL PUBLISHING CO., INC., 750 Third Ave., New York, N.Y. 10017.

DEPARTMENT OF CITY AND REGIONAL PLANNING, UNIVERSITY OF NORTH CAROLINA, Chapel Hill, N.C. 27514.

DEPARTMENT OF DEVELOPMENT AND PLANNING, City Hall, 121 N. La Salle St., Chicago, Ill. 60602.

DEPARTMENT OF ENGINEERING, Boelter Hall, UNIVERSITY OF CALIFORNIA, Los Angeles, Cal. 90024.

DEPARTMENT OF GEOGRAPHY, UNIVERSITY OF CHICAGO, 5801 S. Ellis Ave., Chicago, Ill. 60637.

DEPARTMENT OF GOVERNMENT, INDIANA UNIVERSITY, Bloomington, Ind. 47401.

DEPARTMENT OF HIGHWAYS, Toronto, Canada.

DEPARTMENT OF POLITICAL SCIENCE, CALIFORNIA STATE POLYTECHNIC COLLEGE, San Luis Obispo, Cal. 93401.

DEPARTMENT OF URBAN STUDIES, NATIONAL LEAGUE OF CITIES, The City Bldg., 1612 K St., N.W., Washington, D.C. 20006.

DETROIT EDISON COMPANY, 2000 Second, Detroit, Mich. 48226.

DIABLO PRESS, 440 Pacific Ave., San Francisco, Cal. 94133.

DIVISION OF COMMUNITY HEALTH SERVICES, U.S. DEPARTMENT OF HEALTH, EDUCATION, AND WELFARE, 330 Independence Ave., S.W., Washington, D.C. 20201.

DOCUMENTS SECTION, STATE LIBRARY, STATE OF CALIFORNIA, Sacramento, Cal. 95814.

DODGE, LAWRENCE G., 794 Main St., West Newbury, Mass. 01985.

DORSEY PRESS INC., DIVISION OF RICHARD D. IRWIN, INC.

IRWIN, 1818 Ridge Rd., Homewood, Ill. 60430.

DOUBLEDAY AND CO., INC., 501 Franklin Ave., Garden City, N.Y. 11530.

DOVER PUBLICATIONS, INC., 180 Varick St., New York, N.Y. 10014.

DUKE UNIVERSITY PRESS, Box 6697, College Station, Durham, N.C. 27708.

DUTTON, E. P. AND CO., INC., 201 Park Ave. South, New York, N.Y. 10003.

ECONOMICS AND REQUIREMENTS DIVISION, OFFICE OF RESEARCH AND DEVELOPMENT, BUREAU OF PUBLIC ROADS, FEDERAL HIGHWAY ADMINISTRATION, Sixth and D St., S.W., Washington, D.C. 20591.

EDINBURGH UNIVERSITY PRESS, 1 George Sq., Edinburgh 8, Scotland.

EDUCATIONAL, SCIENTIFIC AND CULTURAL ORGANIZATION, UNESCO PUBLICATIONS CENTER, U.S.A., 317 E. 34th St., New York, N.Y. 10014.

ELSEVIER, see AMERICAN ELSEVIER PUBLISHING CO., INC.

ENO FOUNDATION FOR HIGHWAY TRAFFIC CONTROL, Saugatuck, Conn. 06853.

ESTATES GAZETTE LIMITED, 28 Denmark St., London, W.C.2., England.

FABER AND FABER LTD., 24 Russell Square, London, England.

FAIRCHILD PUBLICATIONS, INC., BOOK DIVISION, 7 East 12th St., New York, N.Y. 10003.

FIDES PUBLISHERS, INC., Box F, Notre Dame, Ind. 46556.

FORD FOUNDATION, 320 East 43rd St., New York, N.Y. 10017.

FRANKLIN INSTITUTE, Benjamin Franklin Parkway at Twentieth St., Philadelphia, Penn. 19103.

FREE PRESS, DIVISION OF CROWELL-COLLIER & MACMILLAN, 866 Third Ave., New York, N.Y. 10022.

IRA J. FRIEDMAN, INC., P.O. Box 270, Port Washington, N.Y. 11050.

GALE RESEARCH CO., Book Tower, Detroit, Mich. 48226.

GENERAL RADIO COMPANY, 22 Baker Ave., West Concord, Mass. 01781.

GLEERUP BOKFÖRLAG AB C.W.K., Oresundsvägen 1, Lund, Sweden.

(U.S.) GOVERNMENT PRINTING OFFICE, Washington, D.C. 20402.

GRADUATE PROGRAM IN URBAN AND REGIONAL PLANNING, UNIVERSITY OF SOUTHERN CALIFORNIA, VKC 382, University Park, Los Angeles, Cal. 90007.

GRADUATE SCHOOL OF BUSINESS ADMINISTRATION, HARVARD UNIVERSITY, Cambridge, Mass. 02138.

GRADUATE SCHOOL OF BUSINESS ADMINISTRATION, MICHIGAN STATE UNIVERSITY, East Lansing Mich. 48823.

GRADUATE SCHOOL OF BUSINESS ADMINISTRATION, UNIVERSITY OF CALIFORNIA, Los Angeles, Cal. 90024.

GRADUATE SCHOOL OF BUSINESS AND PUBLIC ADMINISTRATION, CORNELL UNIVERSITY, Ithaca, N.Y. 14850.

GREATER BOSTON ECONOMIC STUDY COMMITTEE, ASSOCIATE CENTER OF THE COMMITTEE FOR ECONOMIC DEVELOPMENT, 200 Berkeley St., Boston 17, Mass.

HAMPSHIRE TECHNICAL RESEARCH INDUSTRIAL COMMERCIAL SERVICE, Southhampton, England.

HARCOURT, BRACE & WORLD, INC., 757 Third Ave., New York, N.Y. 10017.

HARPER & ROW PUBLISHERS, INC., 49 East 33rd St., New York, N.Y. 10016. (Orders to Keystone Industrial Park, Scranton, Penn. 18512.)

HARVARD BUSINESS SCHOOL, 16 N. Harvard, Allston, Mass. 02134.

HARVARD UNIVERSITY PRESS, Kittredge Hall, 79 Garden St., Cambridge, Mass. 02138.

HER MAJESTY'S STATIONERY OFFICE, British Information Services, 845 Third Ave., New York, N.Y. 10022.

HIGHWAY RESEARCH BOARD, 2101 Constitution Ave., N.W., Washington, D.C. 20418.

HOLT, RINEHART & WINSTON, INC., 383 Madison Ave., New York, N.Y. 10017.

HOUGHTON MIFFLIN CO., 2 Park St., Boston, Mass. 02107; 53 West 43rd St., New York, N.Y. 10036.

HUTCHINSON & CO., LTD., 178–202 Great Portland St., London W1, England.

INDIANA UNIVERSITY PRESS, 10th & Morton Sts., Bloomington, Ind. 47401.

INDUSTRIAL COLLEGE OF THE ARMED FORCES, Fort Leslie J. McNair, 4th & P. Sts., S.W., Washington, D.C. 20315.

INSTITUTE FOR COMMUNITY DEVELOPMENT AND SERVICES, MICHIGAN STATE UNIVERSITY, East Lansing, Mich. 48823.

INSTITUTE FOR ENVIRONMENTAL STUDIES, UNIVERSITY OF PENNSYLVANIA, Philadelphia, Penn. 19104.

INSTITUTE FOR RESEARCH IN SOCIAL SCIENCE, UNIVERSITY OF NORTH CAROLINA, Chapel Hill, N.C. 27514.

INSTITUTE OF BUSINESS AND ECONOMIC RESEARCH, UNIVERSITY OF CALIFORNIA, Berkeley, Cal. 94720.

THE INSTITUTE OF ECONOMIC AFFAIRS LTD., Eaton House, 66A Eaton Square, London SWI, England.

INSTITUTE OF GOVERNMENT, UNIVERSITY OF NORTH CAROLINA, Box 990, Chapel Hill, N.C. 27514.

INSTITUTE OF GOVERNMENTAL STUDIES, UNIVERSITY OF CALIFORNIA, 109 Moses Hall, Berkeley, Cal. 94720.

INSTITUTE OF TRAFFIC ENGINEERS, 1725 De Sales St., N.W., Washington, D.C. 20036.

INSTITUTE OF TRANSPORTATION AND

TRAFFIC ENGINEERING, UNIVERSITY OF CALIFORNIA, Berkeley, Cal. 94720.

INTERNATIONAL ASSOCIATION OF AUDITORIUM MANAGERS, INC., Sherman House, Room 261, Chicago, Ill. 60601.

INTERNATIONAL CITY MANAGERS' ASSOCIATION, 1140 Connecticut Ave., N.W., Washington, D.C. 20036.

IOWA STATE UNIVERSITY PRESS, Press Bldg., Ames, Iowa 50010.

IRWIN, RICHARD D., INC., 1818 Ridge Rd., Homewood, Ill. 60430.

IRWIN AND DORSEY, see IRWIN, RICHARD D., INC.

JOHNS HOPKINS PRESS, Homewood Campus, Baltimore, Md. 21218.

JOINT CENTER FOR URBAN STUDIES OF THE MASSACHUSETTS INSTITUTE OF TECHNOLOGY AND HARVARD UNIVERSITY, 66 Church St., Cambridge, Mass. 02138.

KANSAS STATE UNIVERSITY PRESS, Manhattan, Kan. 66502.

KNOPF, ALFRED A., INC., 501 Madison Ave., New York, N.Y. 10022.

LAWYERS CO-OPERATIVE PUBLISHING COMPANY, Aqueduct Bldg., Rochester, N.Y. 14603.

LEAGUE OF WOMEN VOTERS, 1200 17th St., N.W., Washington, D.C. 20036.

LEICESTER UNIVERSITY, THE SECRETARY, Leicester, England.

LINCOLN PARK CONSERVATION ASSOCIATION, 741 W. Fullerton, Chicago, Ill. 60614.

LIPPINCOTT, J. B., CO., E. Washington Sq., Philadelphia, Penn. 19105.

LITTLE, BROWN & CO., 34 Beacon St., Boston, Mass. 02106.

LONDON SCHOOL OF ECONOMICS AND POLITICAL SCIENCE, Houghton St., Aldwych, London, W.C.2, England.

LOS ANGELES REGIONAL TRANSPORTATION STUDY, TRANSPORTATION ASSOCIATION OF SOUTHERN CALIFORNIA, P.O. Box 2304, Terminal Annex, Los Angeles, Cal. 90054

LOUISIANA STATE UNIVERSITY PRESS, Baton Rouge, La. 70803.

LUND HUMPHRIES & CO., LTD., 12 Bedford Sq., London, N.W.3, England.

McKAY, DAVID, COMPANY, INC., 750 Third Ave., New York, N.Y. 10017.

McGRAW-HILL BOOK CO., 330 W. 42nd St., New York, N.Y. 10036.

MACMILLAN CO., 866 Third Ave., New York, N.Y. 10022.

MARKHAM PUBLISHING CO., 863 N. Dearborn St., Chicago, Ill. 60610.

MARTINUS NIJHOFF, Lange Voorhout 9, Box 269, The Hague, Netherlands. (U.S. Agent: W. S. HEINMAN, 400 E. 72nd St., New York, N.Y. 10021.)

THE MARYLAND-NATIONAL CAPITAL PARK AND PLANNING COMMISSION, 8787 Georgia Ave., Silver Spring, Md. 20907.

MARYLAND STATE ROADS COMMISSION, 301 W. Preston St., Baltimore, Md. 21201.

MAYOR's OFFICE, City Hall, Dallas, Tex. 75201.

MENTAL HEALTH RESEARCH INSTITUTE, UNIVERSITY OF MICHIGAN, Ann Arbor, Mich. 48104.

MERRILL, CHARLES E., BOOKS, INC., 1300 Alum Creek Drive, Columbus, Ohio 43216.

METROPOLITAN HOUSING AND PLANNING COUNCIL OF CHICAGO, 53 W. Jackson Blvd., Chicago, Ill. 60604.

METROPOLITAN WASHINGTON COUNCIL OF GOVERNMENTS, 1701 Pennsylvania Ave., N.W., Washington, D.C. 20006.

MICHIE COMPANY, P.O. Box 57, Charlottesville, Va. 22902.

(M.I.T.) MASSACHUSETTS INSTITUTE OF TECHNOLOGY, PUBLICATIONS OFFICE, 77 Massachusetts Ave., Cambridge, Mass. 02139.

MOBILE HOMES MANUFACTURER's ASSOCIATION, 20 N. Wacker Drive, Chicago, Ill. 60606.

MUNICIPAL FINANCE OFFICERS ASSOCIATION OF U.S. AND CANADA, 1313 E. 60th St., Chicago, Ill. 60637.

MUNICIPAL REFERENCE LIBRARY, Room 1005, City Hall, Chicago, Ill. 60602.

MUSEUM OF NATURAL HISTORY, Central Park, W. and 79th St., New York, N.Y. 10024.

NATIONAL ACADEMY OF SCIENCES, 2101 Constitution Ave., Washington, D.C. 20418.

NATIONAL ASSOCIATION OF SOCIAL WORKERS, 2 Park Ave., New York, N.Y. 10016.

(NATIONAL BOARD OF FIRE UNDERWRITERS) AMERICAN INSURANCE ASSOCIATION, 110 William St., New York, N.Y. 10038.

(NATIONAL COMMISSION ON URBAN PROBLEMS) Superintendent of Documents, U.S. Government Printing Office, Washington, D.C. 20402.

NATIONAL LEAGUE OF CITIES, 1612 K St., N.W., Washington, D.C. 20006.

NATIONAL PLANNING ASSOCIATION, 1606 New Hampshire Ave., N.W., Washington, D.C. 20009.

(NATIONAL RECREATION ASSOCIATION) NATIONAL RECREATION AND PARK ASSOCIATION, 1700 Pennsylvania Ave., N.W., Washington, D.C. 20006.

NATIONAL RESEARCH COUNCIL, 2101 Constitution Ave., Washington, D.C. 20418.

NATIONAL SAND & GRAVEL ASSOCIATION, 900 Spring St., Silver Spring, Md. 20910.

NEW AMERICAN LIBRARY, INC., 1301 Ave. of the Americas, New York, N.Y. 10019.

NEW YORK GRAPHIC SOCIETY, 140 Greenwich Ave., Greenwich, Conn. 06830.

NEW YORK UNIVERSITY PRESS, Washington Sq., New York, N.Y. 10003.

NORTH CASTLE BOOKS, 212 Bedford Rd., Greenwich, Conn. 06830.

NORTHEASTERN ILLINOIS METROPOLITAN AREA PLANNING COMMISSION, 400 W. Madison St., Rm. 2500, Chicago, Ill. 60606.

NORTHERN VIRGINIA REGIONAL PLANNING COMMISSION, 6316 Castle Pl., Falls Church, Va. 22044.

NORTHWESTERN UNIVERSITY PRESS, 1735 Benson Ave., Evanston, Ill. 60201.

NORTON, W. W. & CO., INC., 55 Fifth Ave., New York, N.Y. 10003.

NTL INSTITUTE FOR APPLIED BEHAVIORAL SCIENCE, NATIONAL EDUCATION ASSOCIATION, 1201 16th St., N.W., Washington, D.C. 20036.

OAKLAND COUNTY PLANNING COMMISSION, One Lafayette St., Pontiac, Mich. 48053.

OCEANA PUBLICATIONS, INC., Dobbs Ferry, N.Y. 10522.

OFFICE OF CITY MANAGER, P.O. Box 5077, Tucson, Ariz. 85703.

OFFICE OF INDUSTRIAL ASSOCIATES, CALIFORNIA INSTITUTE OF TECHNOLOGY, 1201 E. California Blvd., Pasadena, Cal. 91109.

OFFICE OF METROPOLITAN DEVELOPMENT, URBAN TRANSPORTATION ADMINISTRATION, U.S. DEPARTMENT OF HOUSING AND URBAN DEVELOPMENT, 451 Seventh St., S.W., Washington, D.C. 20410.

(OFFICE OF PROGRAM POLICY, DIVISION OF URBAN STUDIES, HOUSING AND HOME FINANCE AGENCY) U.S. DEPARTMENT OF HOUSING AND URBAN DEVELOPMENT, 451 Seventh St., S.W., Washington, D.C. 20410.

OFFICE OF RESEARCH AND DEVELOPMENT, BUREAU OF PUBLIC ROADS, FEDERAL HIGHWAY ADMINISTRATION, Sixth and D Sts., S.W., Washington, D.C. 20591.

OPERATIONS RESEARCH SOCIETY OF AMERICA, Mt. Royal and Guilford Aves., Baltimore, Md. 21202.

OXFORD UNIVERSITY PRESS, INC., 200 Madison Ave., New York, N.Y. 10016. (Orders to 1600 Pollitt Dr., Fair Lawn, N.J. 07410.)

PANTHEON BOOKS, INC., 437 Madison Ave., New York, N.Y. 10022. (Orders to RANDOM HOUSE, INC., Westminster, Md. 21157.)

PEACOCK BOOKS, see PENGUIN BOOKS, INC.

PENGUIN BOOKS, INC., 7110 Ambassador Rd., Baltimore, Md. 21207.

PERGAMON PRESS, INC., 44–01 21st St., Long Island City, N.Y. 11101.

PITMAN AND SONS, LTD., Pitman House, Parker St., Kingsway, London, W.C.2, England.

PLANNING BOOKSHOP, (address unknown), London, England.

PLANNING DEPARTMENT (PROGRAM), PRATT INSTITUTE, Brooklyn, N.Y. 11205.

PLENUM PUBLISHING CORP., 227 W. 17th St., New York, N.Y. 10011.

POCKET BOOKS, INC., DIVISION OF SIMON AND SCHUSTER, 1 W. 39th St., New York, N.Y. 10018.

PORT OF OAKLAND, 66 Jack London Sq., Oakland, Cal. 94607.

PRAEGER, FREDERICK A., 111 Fourth Ave., New York, N.Y. 10013.

PRENTICE-HALL, INC., 70 5th Ave., New York, N.Y. 10011. (Orders to Englewood Cliffs, N.J. 07632.)

PRESS AND INFORMATION SERVICE, FRENCH EMBASSY, 972 Fifth Ave., New York, N.Y. 10021.

PRINCETON UNIVERSITY PRESS, Princeton, N.J. 08540.

PUBLIC ADMINISTRATION SERVICE, 1313 East 60th St., Chicago, Ill. 60637.

QUADRANGLE BOOKS, INC., 12 E. Delaware Pl., Chicago, Ill. 60611.

RAND CORPORATION, 1700 S. Main St., Santa Monica, Cal. 90406.

RAND MCNALLY & CO., P.O. Box 7600, Chicago, Ill. 60680; 405 Park Ave., New York, N.Y. 10022; 423 Market St., San Francisco, Cal. 94105.

RANDOM HOUSE, INC., 457 Madison Ave., New York, N.Y. 10022.

REAL ESTATE RESEARCH PROGRAM, INSTITUTE OF BUSINESS AND ECONOMIC RESEARCH, UNIVERSITY OF CALIFORNIA, Berkeley, Cal. 94720.

REGIONAL PLAN ASSOCIATION, 230 W. 41st St., New York, N.Y. 10036.

REGIONAL PLANNING COMMISSION, 1015 E. Main St., Richmond, Va. 23219.

REGIONAL SCIENCE RESEARCH INSTITUTE, G.P.O. Box 8776, Philadelphia, Penn. 19101.

REINHOLD PUBLISHING CORP., 430 Park Ave., New York, N.Y. 10022.

RESEARCH, TRAINING AND INFORMATION SECTION, CENTRE FOR HOUSING, BUILDING AND PLANNING, Publications, Sales Section, UNITED NATIONS, Rm. 1059, New York, N.Y. 10017.

RESOURCES FOR THE FUTURE, INC., 1755 Massachusetts Ave., N.W., Washington, D.C. 20036.

RIVERSIDE PRESS, 840 Memorial Dr., Cambridge, Mass. 02138.

RONALD PRESS CO., 79 Madison Ave., New York, N.Y. 10016.

ROUTLEDGE & KEGAN PAUL, LTD., Broadway House, 68–74 Carter Lane, London, E.C.4, England.

ROYAL VANGORCUM, Assen, Netherlands.

RUSSELL SAGE FOUNDATION, 230 Park Ave., New York, N.Y. 10017.

RUTGERS UNIVERSITY PRESS, 30 College Ave., New Brunswick, N.J. 08903.

SAGE PUBLICATIONS, INC., 275 South Beverly Drive, Beverly Hills, Cal. 90212.

SCARECROW PRESS, INC., 52 Liberty St., Box 656, Metuchen, N.J. 08840.

SCHOOL OF ARCHITECTURE, UNIVERSITY OF MINNESOTA, Minneapolis, Minn. 55455.

SCIENCE FOUNDATION PRESS, P.O. Box 945, Main Office, Los Angeles, Cal. 90053.

SCRIBNERS, CHARLES SONS, 597 Fifth Ave., New York, N.Y. 10017.

SECTION OF LOCAL GOVERNMENT LAW, AMERICAN BAR ASSOCIATION, 1155 East 60th St., Chicago, Ill. 60637.

SIMON & SCHUSTER, 630 Fifth Ave., New York, N.Y. 10020. (Orders to 1 W. 39th St., New York, N.Y. 10018.)

SMALL HOMES COUNCIL — BUILDING RESEARCH COUNCIL, UNIVERSITY OF ILLINOIS, Urbana, Ill. 61803.

SOCIETY OF INDUSTRIAL REALTORS, NATIONAL ASSOCIATION OF REAL ESTATE BOARDS, 36 S. Wabash Ave., Chicago, Ill. 60603.

SOIL SCIENCE SOCIETY OF AMERICA, 677 South Segoe Rd., Madison, Wis. 53711.

STANFORD RESEARCH INSTITUTE, 820 Mission St., South Pasadena, Cal. 91030.

STATE-LOCAL FINANCES PROJECT, GEORGE WASHINGTON UNIVERSITY, 1145 19th St., N.W., Washington, D.C. 20036.

ST. LOUIS UNIVERSITY, 221 N. Grand Blvd., St. Louis, Mo. 63103.

ST. MARTIN'S PRESS, INC., 175 Fifth Ave., New York, N.Y. 10010.

SUPERINTENDENT OF DOCUMENTS, U.S. GOVERNMENT PRINTING OFFICE, Washington, D.C. 20402.

TECHNICAL LIBRARY, TENNESSEE VALLEY AUTHORITY, New Sprankle Building, Knoxville, Tenn. 37902.

TRAILER COACH ASSOCIATION, 1340 W. Third St., Los Angeles, Cal. 90017.

TRAIL-R-CLUB OF AMERICA, 3211 Pico Blvd., Santa Monica, Cal. 90405.

TRANSPORTATION ASSOCIATION OF SOUTHERN CALIFORNIA, P.O. Box 2304, Terminal Annex, Los Angeles, Cal. 90054.

TRANSPORTATION CENTER, NORTHWESTERN UNIVERSITY, Evanston, Ill. 60204.

TRANSPORTATION RESEARCH INSTITUTE, CARNEGIE-MELLON UNIVERSITY, Pittsburgh, Penn. 15213.

UNITED RESEARCH CORPORATION, Cambridge, Mass.

UNIVERSITY OF ALABAMA PRESS, Drawer 2877, University, Ala. 35488.

UNIVERSITY OF CALIFORNIA PRESS, Berkeley, Cal. 94720. (Orders to 2223 Fulton St., Berkeley, Cal. 94720, or 25 W. 45th St., New York, N.Y. 10036.)

UNIVERSITY OF CHICAGO PRESS, 11030 South Langley Ave., Chicago, Ill. 60628.

UNIVERSITY OF CINCINNATI PRESS, Cincinnati, Ohio 45221.

UNIVERSITY OF FLORIDA PRESS, 15 N.W. 15th St., Gainesville, Fla. 32601.

UNIVERSITY OF GEORGIA PRESS, Athens, Ga. 30601.

UNIVERSITY OF ILLINOIS PRESS, Urbana, Ill. 61801.

UNIVERSITY OF MASSACHUSETTS PRESS, Munson Hall, Amherst, Mass. 01002.

UNIVERSITY OF NEBRASKA, Lincoln, Nebr. 68508.

UNIVERSITY OF NORTH CAROLINA, Chapel Hill, N.C. 27514.

UNIVERSITY OF PENNSYLVANIA, Philadelphia, Penn. 19104.

UNIVERSITY OF PITTSBURGH, Fifth Ave. and Bigelow Blvd., Pittsburgh, Penn. 15213.

UNIVERSITY OF PITTSBURGH PRESS, 3309 Cathedral of Learning, Pittsburgh, Penn. 15213.

UNIVERSITY OF TORONTO PRESS, Front Campus, University of Toronto, Toronto 5, Ontario, Canada, or 1061 Kensington Ave., Buffalo, N.Y. 14215.

UNIVERSITY OF VIRGINIA, Charlottesville, Va. 22903.

UNIVERSITY OF WISCONSIN PRESS, Box 1379, Madison, Wis. 53701.

URBAN LAND INSTITUTE, 1200 18th St., N.W., Washington, D.C. 20036.

URBAN STUDIES CENTER, RUTGERS UNIVERSITY, New Brunswick, N.J. 08903.

U.S. DEPARTMENT OF HOUSING AND URBAN DEVELOPMENT, 451 Seventh St., S.W., Washington, D.C. 20410.

U.S. GEOLOGICAL SURVEY, General Services Bldg., 18th and F Sts., N.W., Washington, D.C. 20242.

VAN NOSTRAND, D. CO., INC., Princeton, N.J. 08540.

VANDERBILT UNIVERSITY PRESS, Nashville, Tenn. 37203.

VIKING PRESS, INC., 625 Madison Ave., New York, N.Y. 10022.

VINTAGE BOOK, see RANDOM HOUSE, INC.

VIRGINIA COUNCIL OF HIGHWAY INVESTIGATION AND RESEARCH, VIRGINIA DEPARTMENT OF HIGHWAYS, 1221 E. Broad St., Richmond, Va. 23219.

VOCATIONAL GUIDANCE MANUALS, 235 E. 45th St., New York, N.Y. 10017.

WALCK, HENRY Z., INC., 19 Union Sq. W., New York, N.Y. 10003.

WAYNE STATE UNIVERSITY PRESS, 5980 Cass Ave., Detroit, Mich. 48202.

WHITEFRIARS PRESS, LTD., 26 Bloomsbury Way, London, W.C.1, England.

WILEY, JOHN & SONS, INC., 605 Third Ave., New York, N.Y. 10016.

WILLIAM L. PEREIRA & ASSOCIATES, 5657 Wilshire Blvd., Los Angeles, Cal. 90036.

WORLD PUBLISHING CO., 119 W. 57th St., New York, N.Y. 10019.

YALE UNIVERSITY PRESS, 92a Yale Station, New Haven, Conn. 06520. (Orders to 149 York St., New Haven, Conn. 06511.)

YEAR, INC., 20 W. 45th St., New York, N.Y. 10036.

YOUNG MEN'S CHRISTIAN ASSOCIATIONS, 422 9th Ave., New York, N.Y. 10001.

Periodicals Referenced

(Number of times cited indicated by numbers in parentheses)

ADMINISTRATIVE SCIENCE QUARTERLY (1)
AIA JOURNAL (1)
AIP NEWSLETTER (1)
AMERICAN BAR ASSOCIATION JOURNAL (2)
THE AMERICAN BEHAVIORAL SCIENTIST (1)
AMERICAN BUILDER (1)
AMERICAN HOSPITAL (1)
AMERICAN JOURNAL OF SOCIOLOGY (1)
AMERICAN POLITICAL SCIENCE REVIEW (1)
THE ANNALS OF THE AMERICAN ACADEMY OF POLITICAL AND SOCIAL SCIENCE (4)
APPRAISAL JOURNAL (1)
ARCHITECTURAL FORUM (4)
ARTS & ARCHITECTURE (1)
ASPO NEWSLETTER (2)
BUSINESS MANAGEMENT (1)
BUSINESS WEEK (1)
CALIFORNIA ENGINEER (1)
CALIFORNIA HIGHWAYS AND PUBLIC WORKS (1)
CALIFORNIA MANAGEMENT REVIEW (2)
C-E NEWSLETTER (1)
COLUMBIA LAW REVIEW (1)
DAEDALUS (11)
DUKE LAW JOURNAL (1)
THE ECONOMIC JOURNAL (1)
ECONOMIST (1)
EKISTICS (2)
FORTUNE (10)
FRONTIER (1)
HARPERS (1)
HARVARD BUSINESS REVIEW (21)
HARVARD TODAY (1)
HIGHWAY RESEARCH RECORD (1)
HOUSE & HOME (4)
HUMAN RELATIONS (1)
IEEE TRANSACTIONS ON ENGINEERING MANAGEMENT (1)
INTERNATIONAL REVIEW OF ADMINISTRATIVE SCIENCES (1)
INTERNATIONAL SCIENCE AND TECHNOLOGY (1)
JOURNAL OF THE ACADEMY OF MANAGEMENT (1)
JOURNAL OF THE AMERICAN INSTITUTE OF PLANNERS (117)

JOURNAL OF THE AMERICAN MEDICAL ASSOCIATION (1)
JOURNAL OF THE AMERICAN SOCIETY OF ARCHITECTURAL HISTORIANS (1)
JOURNAL OF THE FRANKLIN INSTITUTE (1)
JOURNAL OF HOUSING (1)
JOURNAL OF THE INSTITUTE OF TRANSPORT (1)
JOURNAL OF THE STANFORD RESEARCH INSTITUTE (1)
JOURNAL OF THE URBAN PLANNING AND DEVELOPMENT DIVISION, PROCEEDINGS OF THE AMERICAN SOCIETY OF CIVIL ENGINEERS (2)
KENTUCKY LAW JOURNAL (1)
LAND ECONOMICS (2)
LAW AND CONTEMPORARY PROBLEMS (4)
LAND USE CONTROLS — A QUARTERLY REVIEW (3)
LIFE (1)
LOOK (1)
MANAGEMENT SERVICES (1)
MANAGEMENT SCIENCE (28)
MANAGEMENT TECHNOLOGY (2)
MUNICIPAL REFERENCE LIBRARY NOTES (1)
NATION'S CITIES (1)
NEW YORK TIMES, International Edition (1)
THE NEW YORK TIMES MAGAZINE (2)
NOISE CONTROL (2)
NORTHWESTERN UNIVERSITY LAW REVIEW (1)
OPERATIONS RESEARCH (6)
PHOTOGRAMMETRIC ENGINEERING (1)
PLANNING AND CIVIC COMMENT (1)
POLITICAL SCIENCE QUARTERLY (1)
PRINCETON ALUMNI WEEKLY (1)
PSYCHOLOGICAL REVIEW (1)
PUBLIC ADMINISTRATION REVIEW (6)
THE PUBLIC INTEREST (1)
PUBLIC MANAGEMENT (1)
REGIONAL SCIENCE ASSOCIATION PAPERS (2)
SATURDAY REVIEW (4)
SCIENCE (15)
SCIENTIFIC AMERICAN (47)

SDC Magazine (6)
Socio-Economic Planning Sciences (2)
Space/Aeronautics (3)
Stanford Research Institute Journal (1)
Steelways (1)
Technology Review (1)
Texas Law Review (1)
Town and Country Planning (1)
The Town Planning Review (4)
Traffic Engineering and Control (1)
Traffic Quarterly (3)
Trends (1)
UCLA Law Review (2)
UNESCO International Social Science Journal (1)

University (2)
University of Detroit Journal of Urban Law (5)
University of Illinois Law Forum (1)
University of Miami Law Review (1)
University of Pennsylvania Law Review (3)
Urban Affairs Quarterly (3)
The Wall Street Journal, Pacific Coast Edition (8)
Washington University Law Quarterly (6)
Western City (1)
Westways (1)
Wisconsin Law Review (2)
The Yale Law Journal (1)

INDEXES

COMPILER'S NOTE

All numbers in the following indexes refer to a principal entry in the Selected References. They do not refer to page numbers. *Italicized* numbers indicate that a work by that author, under that title, or on that subject constitutes *a chapter or part* of the principal entry bearing that Selected Reference number.

Subject Index

Author Index

Aberbach, Joel D.: *791*

Abrams, Charles: *28, 81,* 480, *517,* 1102, 1149, *1424,* 1428

Abruzzi, Adam: 527

ABT Associates Inc.: 278

Ackerman, Edward A.: *747, 784*

Ackoff, Russell L.: *511,* 602

Adams, Bert N.: 760, 803

Adams, Frederick J.: 528

Adams, Melvin J.: *1443*

Adams, Robert McC.: *1049*

Addams, Jane: *549*

Adelson, Marvin: 279

Ad Hoc Committee of Librarians: 1297

Adrian, Charles R.: 17, *549, 927, 930,* 945

Advisory Commission on Intergovernmental Relations: 471, 545, 546, 547, *843,* 1325

Aerojet-General Corporation (California), Dept. of Public Health: 1326

Ahlberg, C. F.: *1485*

AIP Newsletter: 529

Air Conservation Commission: 1010

Aird, John S.: *747*

Alesch, Daniel J.: 552

Alexander, Christopher: 184, *368, 1027, 1109*

Alexander, John W.: *408*

Alexander, Robert E.: 472

Alexander, Tom: 1315

Alexis, Marcus: *614*

Alford, Robert R.: *413, 512*

Alinsky, Saul: *701*

Allee, David: *989*

Allen, J. Knight: *8*

Allen, K. J.: *361*

Allen, Muriel I.: 1486

Almon, Clopper, Jr.: 346

Alonso, William: *28,* 185, 280, *379,* 1382

Altshuler, Alan: 448, *480, 549,* 812, 922, 977

Alyea, Paul E.: 845

American Bar Association, National Institute: 899, 914

American Builder Special Report: 1087

American Institute of Architects: 528, 1103, 1487

American Institute of Consulting Engineers: 529

American Institute of Planners: 529

American Medical Association: 1011

American Public Works Association: 1327, 1328, 1362, 1363

American Society of Civil Engineers: 529

American Society of Landscape Architects: 529

American Society of Planning Officials: 241, 664, 665, 876, 877, 900, 1012, 1070, 1493

Anderson, David L.: 350

Anderson, Desmond L.: 666

Anderson, Martin: 1427, 1454

Anderson, Odin W.: *480*

Anderson, Richard T.: 1333

Anderson, Robert B.: 1252

Anderson, Robert M.: 667

Anderson, Robert T.: *747*

Anderson, Stanford: 214, 347, 423, 1071

Andrews, Richard B.: *408*

Andrzejewski, Adam: *127*

Angle, Paula (Editor): 1428

Annual Report of the Council of Economic Advisers, 1966: 1029

Anshen, Melvin: *495, 503, 607*

Ansoff, H. Igor: 281, *607*

Anthony, Harry Antoniades: 1

Appleton, J. H.: *18*

Appleyard, Donald: 1013

Archibald, Russell D.: 473, 520

Architectural Forum: 1202

Aregger, Hans: 1104

Argyris, Chris: 548

Arnold, David S.: 253

Aron, Raymond: *906*

Aronoff, Leah: 268

Arpaia, Anthony F.: *1271*

Arrow, Kenneth J.: *187*

Artle, Roland: *1194*

Ash, Maurice A.: *1474*

Ashby, Lowell D.: *190*

Title Index

COMPARATIVE URBAN RESEARCH: The Administration and Politics of Cities

Edited by ROBERT T. DALAND, *University of North Carolina*

This volume is sponsored by the Committee on Urban Administration and Politics of the Comparative Administration Group (ASPA)

CONTENTS

L.C. 69-18751   362 pp.   SBN 8039-0012-0   June, 1969

SAGE PUBLICATIONS, INC. / 275 South Beverly Drive / Beverly Hills, Calif. 90212

The Metropolitan Community

ITS PEOPLE AND GOVERNMENT

AMOS H. HAWLEY
University of North Carolina

BASIL G. ZIMMER
Brown University

Focussing on attitudes toward consolidation of suburban with central city governmental units and services, THE METROPOLITAN COMMUNITY, is a thorough and rigorous analysis of the economic, political and social characteristics of metropolitan residents. More than 50 tables summarize such factors as: education, age, occupation, income of household heads; church attendance; voter registration; attitudes toward present government, schools, public services, and toward proposed changes; and attitudes toward officials and their competence.

CONTENTS

(Available Clothbound and Paper)

L.C. 77-92358   160 pp.   February, 1970

SAGE PUBLICATIONS, INC. / 275 South Beverly Drive / Beverly Hills, Calif. 90212

CHECKLIST OF SAGE PROFESSIONAL JOURNALS

[ ] AMERICAN BEHAVIORAL SCIENTIST

> Published bi-monthly. Each issue devoted to a significant topic or area of interdisciplinary research.

[ ] COMPARATIVE POLITICAL STUDIES

> Published quarterly. Theoretical and empirical research articles in cross-national comparative studies.

[ ] URBAN AFFAIRS QUARTERLY

> Published quarterly. International and interdisciplinary focus on all areas of urban research.

[ ] SIMULATION AND GAMES

> Published quarterly. International journal of theory, design and research.

[ ] URBAN RESEARCH NEWS

> Published bi-weekly. Reports current developments in all areas of urban research.

[ ] DESIGN METHODS GROUP NEWSLETTER

> Published monthly. A compilation of research news, abstracts, etc., from architecture, civil engineering, city and regional planning and regional science.

[ ] EDUCATION AND URBAN SOCIETY

> Published quarterly. Social research with implications for public policy.

[ ] ENVIRONMENT AND BEHAVIOR

> Published quarterly. Concerned with study, design and control of the physical environment and its interaction with human behavior.

[ ] COMPARATIVE GROUP STUDIES

> Published quarterly. Research and theory in all fields of small group study, including therapy groups.

[ ] JOURNAL OF COMPARATIVE ADMINISTRATION

> Published quarterly. Cross-national, interdisciplinary research on public organizations.

[ ] LAW AND SOCIETY

> Published quarterly. Studies of law as a social and political phenomenon and as an instrument of public policy.

[ ] URBAN EDUCATION

> Published quarterly. Empirical and theoretical papers aimed at improving education in the city.

SAGE PUBLICATIONS / 275 South Beverly Drive / Beverly Hills, California 90212

Lewis and Clark College - Watzek Library
Z5942 .B7   wmain
Branch, Melville Ca/Comprehensive urban

3 5209 00408 6076